Advances in Intelligent Systems and Computing

Volume 1047

The series "Advances in Intelligent Systems and Computing" contains publications on theory, applications, and design methods of Intelligent Systems and Intelligent Computing. Virtually all disciplines such as engineering, natural sciences, computer and information science, ICT, economics, business, e-commerce, environment, healthcare, life science are covered. The list of topics spans all the areas of modern intelligent systems and computing such as: computational intelligence, soft computing including neural networks, fuzzy systems, evolutionary computing and the fusion of these paradigms, social intelligence, ambient intelligence, computational neuroscience, artificial life, virtual worlds and society, cognitive science and systems, Perception and Vision, DNA and immune based systems, self-organizing and adaptive systems, e-Learning and teaching, human-centered and human-centric computing, recommender systems, intelligent control, robotics and mechatronics including human-machine teaming, knowledge-based paradigms, learning paradigms, machine ethics, intelligent data analysis, knowledge management, intelligent agents, intelligent decision making and support, intelligent network security, trust management, interactive entertainment, Web intelligence and multimedia.

The publications within "Advances in Intelligent Systems and Computing" are primarily proceedings of important conferences, symposia and congresses. They cover significant recent developments in the field, both of a foundational and applicable character. An important characteristic feature of the series is the short publication time and world-wide distribution. This permits a rapid and broad dissemination of research results.

** **Indexing: The books of this series are submitted to ISI Proceedings, EI-Compendex, DBLP, SCOPUS, Google Scholar and Springerlink** **

More information about this series at http://www.springer.com/series/11156

Radek Silhavy · Petr Silhavy ·
Zdenka Prokopova
Editors

Computational Statistics and Mathematical Modeling Methods in Intelligent Systems

Proceedings of 3rd Computational Methods in Systems and Software 2019, Vol. 2

 Springer

Editors
Radek Silhavy
Faculty of Applied Informatics
Tomas Bata University in Zlin
Zlin, Czech Republic

Petr Silhavy
Faculty of Applied Informatics
Tomas Bata University in Zlin
Zlin, Czech Republic

Zdenka Prokopova
Faculty of Applied Informatics
Tomas Bata University in Zlin
Zlin, Czech Republic

ISSN 2194-5357 ISSN 2194-5365 (electronic)
Advances in Intelligent Systems and Computing
ISBN 978-3-030-31361-6 ISBN 978-3-030-31362-3 (eBook)
https://doi.org/10.1007/978-3-030-31362-3

This Springer imprint is published by the registered company Springer Nature Switzerland AG
The registered company address is: Gewerbestrasse 11, 6330 Cham, Switzerland

Preface

This book constitutes the refereed proceedings of the Computational Methods in Systems and Software 2019 (CoMeSySo 2019), held on October 2019.

CoMeSySo 2019 conference intends to provide an international forum for the discussion of the latest high-quality research results in all areas related to cybernetics and intelligent systems. The addressed topics are the theoretical aspects and applications of Computational Statistics and Mathematical Modelling Methods in Intelligent Systems.

The papers address topics as software engineering cybernetics and automation control theory, econometrics, mathematical statistics or artificial intelligence.

CoMeSySo 2019 has received (all sections) 140 submissions, 84 of them were accepted for publication.

The volume Computational Statistics and Mathematical Modelling Methods in Intelligent Systems discusses a novel statistical, mathematical or intelligent approaches and methods to real-world problems.

The editors believe that the readers will find the following proceedings interesting and useful for their research work.

July 2019

Radek Silhavy
Petr Silhavy
Zdenka Prokopova

Organization

Program Committee

Program Committee Chairs

Petr Silhavy — Department of Computers and Communication Systems, Faculty of Applied Informatics, Tomas Bata University in Zlin, Czech Republic

Radek Silhavy — Department of Computers and Communication Systems, Faculty of Applied Informatics, Tomas Bata University in Zlin, Czech Republic

Zdenka Prokopova — Department of Computers and Communication Systems, Tomas Bata University in Zlin, Czech Republic

Krzysztof Okarma — Faculty of Electrical Engineering, West Pomeranian University of Technology, Szczecin, Poland

Roman Prokop — Department of Mathematics, Tomas Bata University in Zlin, Czech Republic

Viacheslav Zelentsov — Doctor of Engineering Sciences, Chief Researcher of St. Petersburg Institute for Informatics and Automation of Russian Academy of Sciences (SPIIRAS), Russian Federation

Lipo Wang — School of Electrical and Electronic Engineering, Nanyang Technological University, Singapore

Silvie Belaskova — Head of Biostatistics, St. Anne's University Hospital Brno, International Clinical Research Center, Czech Republic

International Program Committee Members

Pasi Luukka	North European Society for Adaptive and Intelligent Systems & School of Business and School of Engineering Sciences Lappeenranta University of Technology, Finland
Ondrej Blaha	Louisiana State University Health Sciences Center New Orleans, New Orleans, USA
Izabela Jonek-Kowalska	Faculty of Organization and Management, The Silesian University of Technology, Poland
Maciej Majewski	Department of Engineering of Technical and Information Systems, Koszalin University of Technology, Koszalin, Poland
Alena Vagaska	Department of Mathematics, Informatics and Cybernetics, Faculty of Manufacturing Technologies, Technical University of Kosice, Slovak Republic
Boguslaw Cyganek	DSc, Department of Computer Science, University of Science and Technology, Krakow, Poland
Piotr Lech	Faculty of Electrical Engineering, West Pomeranian University of Technology, Szczecin, Poland
Monika Bakosova	Institute of Information Engineering, Automation and Mathematics, Slovak University of Technology, Bratislava, Slovak Republic.
Pavel Vaclavek	Faculty of Electrical Engineering and Communication, Brno University of Technology, Brno, Czech Republic
Miroslaw Ochodek	Faculty of Computing, Poznan University of Technology, Poznan, Poland.
Olga Brovkina	Global Change Research Centre, Academy of Science of the Czech Republic, Brno, Czech Republic
Elarbi Badidi	College of Information Technology, United Arab Emirates University, Al Ain, United Arab Emirates
Gopal Sakarkar	Shri. Ramdeobaba College of Engineering and Management, Republic of India
V. V. Krishna Maddinala	GD Rungta College of Engineering and Technology, Republic of India

| Anand N. Khobragade (Scientist) | Maharashtra Remote Sensing Applications Centre, Republic of India |
| Abdallah Handoura | Computer and Communication Laboratory, Telecom Bretagne, France |

Organizing Committee Chair

| Radek Silhavy | Tomas Bata University in Zlin, Faculty of Applied Informatics |
| | email: comesyso@openpublish.eu |

Conference Organizer (Production)

OpenPublish.eu s.r.o.
Web: http://comesyso.openpublish.eu
Email: comesyso@openpublish.eu

Conference Web site, Call for Papers:

http://comesyso.openpublish.eu

Contents

Static Compensator for Decentralized Control of Nonsquare Systems

František Dušek, Daniel Honc$^{(\boxtimes)}$, and Jan Merta

University of Pardubice, 532 10 Pardubice, Czech Republic
{frantisek.dusek,daniel.honc}@upce.cz

Abstract. The paper deals with the decentralized control problem for linear multivariable systems with the number of manipulated variables equal or greater than the number of controlled variables. Proposed static compensator ensures automatic creation of input/output pairs for individual control loops. The compensator provides steady state autonomy and unit gain for the control loops. Steady state gain matrix of the controlled system and vector of offsets are sufficient information for the compensator design. Estimation of the gain matrix and offsets from the measured data is also proposed in the paper.

Keywords: Multivariable system · Decentralized control · Static compensator

1 Introduction

A complication in the design of multivariable control of Multi-Input Multi-Output System (MIMO) over single-input and single-output (SISO) case is the fact that changing in one manipulated variable causes changes to more/all controlled variables. Two possible approaches for controlling of MIMO systems are shown in Fig. 1.

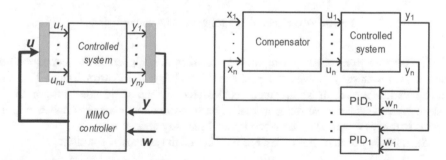

Fig. 1. Multivariable control structures – multivariable controller, decentralized control.

Multivariable controller is used in the first approach - it calculates the values of all manipulated variables simultaneously from all controlled variables and set-points. This approach is used by state-space "pole placement" controllers, quadratic optimal controllers or predictive controllers.

© Springer Nature Switzerland AG 2019
R. Silhavy et al. (Eds.): CoMeSySo 2019, AISC 1047, pp. 1–6, 2019.
https://doi.org/10.1007/978-3-030-31362-3_1

Autonomous control is the second approach - a special block of the dynamic compensator is placed in front of the controlled system to ensure the autonomy of the arising system. Autonomy means that changing a single input u_i^* will only cause a change to the corresponding output y_i. Then it is possible to use control loops with standard design methods.

In both cases, it is necessary to know the linear dynamical model of the MIMO system. Dynamic compensator ensuring autonomy in transient states can be complex and not physically feasible [1, 2]. Because of this only limited autonomy compensators are proposed.

Third approach is used for practical applications - decentralized control – second case without compensator – see e.g. [3]. The input u_j that affects output y_i at most is searched to create the control loops. The advantage is that it is not necessary to know the mathematical model of the controlled system and the control can be designed by trial/error methods. The problem with this approach is the need to find suitable pairs of manipulated/controlled variables. This may not be a trivial matter even with the same number of variables.

The paper deals with specific type of autonomous control of systems with equal number or with more inputs than outputs. Steady state autonomy is realized with simple static compensator - see Fig. 2.

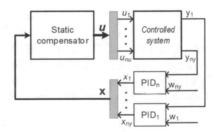

Fig. 2. Decentralized control with static compensator.

The introduction of new manipulated variables x_i assigned to the corresponding controlled variables y_i solves the problem of creating the control loops. The new control variables (vector **x**) are then recalculated by the static compensator to the system inputs (vector **u**) so the gain matrix between the new control variables and the system outputs (vector **y**) is identity matrix. In steady state **y** = **x**.

The authors consider the following benefits of the proposed solution:

- uniformity and relative simplicity of the design and implementation for MIMO systems
- it is sufficient to know only the gain matrix, which can be easily estimated from the experimental data
- automatic creation of input (y_i)/output (x_i) pairs for individual control loops

 – in the case of the system with more inputs than outputs, the approach allows to prefer the selected "optimal" input value combination (vector $\mathbf{u}_w$) independently to the controller tuning

 – process gain matrix together with the compensator is identity matrix - individual controllers are easier to tune.

The disadvantage is that the use of a static compensator may impair the frequency properties of the closed loop control circuit [4, 5].

The static compensator design can also be seen as a general procedure for the application of the multivariable Split Range method - see, e.g., [6]. However the Split Range does not generally provide guidance for a particular case. The proposal to divide the controller output into multiple manipulated variables is usually done ad hoc based e.g. on the properties of the controlled process. An example is the thermostat bath temperature control if separate cooling and heating is available. If the controller output is positive, then the cooling is zero and the heating power is proportional to the controller output and vice versa.

2 Gain Matrix Estimation

To determine the gain matrix of the controlled system we need experimental data - steady-state output values for different input combinations. The number of the samples must be greater than the number of inputs. The combination of input values should cover the entire working area together with the readings of the steady state outputs.

Let us suppose N samples of the steady state inputs and outputs of the linear MIMO system. We evaluate the data for a more general case, that for zero input there are non-zero outputs – there is so called offset $\mathbf{z}_0$ between the input and output. The relation between the input vector $\mathbf{u}$ ($\mathbf{u}$: $n_u \times 1$) and the output vector $\mathbf{y}$ ($\mathbf{y}$: $n_y \times 1$) in steady state is described by

$$\mathbf{y} = \mathbf{Z}.\mathbf{u} + \mathbf{z}_0 \tag{1}$$

where

    $\mathbf{Z}$ is the gain matrix ($\mathbf{Z}$: $n_y \times n_u$) and
    $\mathbf{z}_0$ is an offset vector ($\mathbf{z}_0$: $n_y \times 1$)

We arrange the data into a matrix of output values $\mathbf{Y}$ ($\mathbf{Y}$: $N \times n_y$) and extended matrix of input values $\mathbf{U}_r$ ($\mathbf{U}_r$: $N \times n_u + 1$) as follows

$$\mathbf{Y} = \begin{bmatrix} \mathbf{y}^T(1) \\ \vdots \\ \mathbf{y}^T(N) \end{bmatrix} = \begin{bmatrix} y_1(1) & \cdots & y_{ny}(1) \\ \vdots & \ddots & \vdots \\ y_1(N) & \cdots & y_{ny}(N) \end{bmatrix}, \quad \mathbf{U}_r = \begin{bmatrix} \mathbf{u}^T(1) & 1 \\ \vdots & \vdots \\ \mathbf{u}^T(N) & 1 \end{bmatrix} = \begin{bmatrix} u_1(1) & \cdots & u_{nu}(1) & 1 \\ \vdots & \ddots & \vdots & \vdots \\ u_1(N) & \cdots & u_{nu}(N) & 1 \end{bmatrix}$$

If the gain matrix $\mathbf{Z}$ and the offset vector $\mathbf{z}_0$ is constant then following matrix equation holds

$$\mathbf{Y}^T = [\mathbf{Z} \quad \mathbf{z}_0].\mathbf{U}_r = \mathbf{Z}_r.\mathbf{U}_r \tag{2}$$

Solution - the extended $\mathbf{Z}_r$ gain matrix is obtained as

$$\mathbf{Z}_r = \mathbf{Y}^T.\mathbf{U}_r.\left(\mathbf{U}_r^T.\mathbf{U}_r\right)^{-1} \tag{3}$$

The gain matrix $\mathbf{Z}$ is the first n_u columns and the vector of the offsets $\mathbf{z}_0$ is the last column of the extended matrix $\mathbf{Z}_r$.

3   Static Compensator Design

The design of the static compensator for the MIMO system with the inputs $\mathbf{u}$ ($\mathbf{u}$: $n_u \times 1$) and the outputs $\mathbf{y}$ ($\mathbf{y}$: $n_y \times 1$) with static behavior described by the matrix Eq. (1) is based on the idea of using the compensator $\mathbf{K}$ ($\mathbf{K}$: $n_u \times n_y$). Compensator together with the original system create new system with the same number of new inputs $\mathbf{x}$ ($\mathbf{x}$: $n_y \times n_y$) as the original outputs $\mathbf{y}$. The gain matrix of the new system is identity matrix $\mathbf{I}$ ($\mathbf{I}$: $n_y \times n_y$)

$$\mathbf{y}_\infty = \mathbf{Z}.\mathbf{u} + \mathbf{z}_0, \ \mathbf{u} = \mathbf{K}.\mathbf{x}, \mathbf{y}_\infty = \mathbf{x} \tag{4}$$

The solution for the same number of inputs and outputs - if $n_u = n_y$ is straightforward $\mathbf{K} = \mathbf{Z}^{-1}$. The solution for the case that the number of inputs is less than the number of outputs - if $n_u < n_y$ was published in [7] - solutions for all possible variants is studied in detail. We will present a solution for the most interesting case, where the number of inputs is greater than the number of outputs - if $n_u > n_y$. In this case, there are infinity combinations of inputs leading to the desired outputs. It is possible to introduce an additional requirement (preferred, optimum combination of inputs $\mathbf{u}_w$). Compensator calculation is done by minimizing the deviation of the input vector $\mathbf{u}$ from the preferred input vector $\mathbf{u}_w$, with the constraint respecting the dependence between inputs and outputs $\mathbf{y} = \mathbf{Z}.\mathbf{u} + \mathbf{z}_0$ and the requirement of the unit gain $\mathbf{y}_\infty = \mathbf{I}.\mathbf{x}$

$$\min_{\mathbf{u}} J(\mathbf{u}), \quad J(\mathbf{u}) = (\mathbf{u} - \mathbf{u}_w)^T.\mathbf{M}.(\mathbf{u} - \mathbf{u}_w)$$
$$\text{subject to}: \ \mathbf{y}_\infty = \mathbf{Z}.\mathbf{u} + \mathbf{z}_0 = \mathbf{I}.\mathbf{x} \tag{5}$$

where the optional matrix $\mathbf{M}$ ($\mathbf{M}$: $n_u \times n_u$) includes possible weighting requirements for the individual inputs or their combinations (can be chosen as an identity matrix). The matrix $\mathbf{M}$ and the vector of the preferred inputs $\mathbf{u}_w$ allows to include additional requirements to the control - in addition to the set-points following. Let's rewrite cost function in Eq. (5) - we introduce vector of Lagrange's multipliers λ and we write the cost function as

$$J(\mathbf{u}, \lambda) = (\mathbf{u} - \mathbf{u}_w)^T . \mathbf{M} . (\mathbf{u} - \mathbf{u}_w) + (\mathbf{Z} . \mathbf{u} + \mathbf{z}_0 - \mathbf{y}_\infty)^T . \lambda \qquad (6)$$

The Eq. (6) can be solved in a standard way, i.e. by placing partial derivatives according to the search parameters equal to zero and by solving the set of resulting linear equations. The solution is described by the matrix equation

$$\begin{bmatrix} \mathbf{u} \\ \lambda \end{bmatrix} = \begin{bmatrix} \mathbf{M}^T & \frac{1}{2}\mathbf{Z}^T \\ \mathbf{Z} & 0 \end{bmatrix}^{-1} . \begin{bmatrix} \mathbf{M}^T . \mathbf{u}_w \\ \mathbf{y}_\infty - \mathbf{z}_0 \end{bmatrix} \qquad (7)$$

Let us use the equality $\mathbf{y}_\infty = \mathbf{x}$, introduce the notation $\mathbf{S}$ ($\mathbf{S}$: $n_u + n_y \times n_u + n_y$) for the inverse matrix and denote two submatrices $\mathbf{R}$ ($\mathbf{R}$: $n_u \times n_u$) and $\mathbf{K}$ ($\mathbf{K}$: $n_u \times n_y$) as indicated in the equation

$$\begin{bmatrix} \mathbf{u} \\ \lambda \end{bmatrix} = \mathbf{S} . \begin{bmatrix} \mathbf{M}^T . \mathbf{u}_w \\ \mathbf{x} - \mathbf{z}_0 \end{bmatrix}, \qquad \mathbf{S} = \begin{bmatrix} \mathbf{R} & \mathbf{K} \\ \mathbf{X}_1 & \mathbf{X}_2 \end{bmatrix} \qquad (8)$$

Since we are interested in the solution with respect to the vector $\mathbf{u}$, we will only use the submatrices $\mathbf{R}$ and $\mathbf{K}$. The static compensator can be written in the matrix form as

$$\mathbf{u} = \mathbf{K} . (\mathbf{x} - \mathbf{z}_0) + \mathbf{R} . \mathbf{M}^T . \mathbf{u}_w \qquad (9)$$

and its bloc scheme is in Fig. 3.

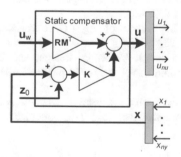

Fig. 3. Static compensator.

4   Conclusions

The paper describes design of a simple and practically usable decentralized MIMO control system for processes with different number of manipulated and controlled variables. The method does not require knowledge of the dynamical mathematical model of the controlled system. All necessary information for the design can be obtained by simple evaluation of the experimental data. It is shown how to estimate the gain matrix and how to calculate the matrices of the static compensator.

The use of the static compensator does not eliminate the dynamic influence of the individual control loops, nor does it ensure control stability. The main benefit is the general solution for the pairing of the manipulated and controlled variables for individual control loops in the case of MIMO systems with different number of inputs and outputs. In addition, the use of the static compensator will provide unity gain of each pair, their invariance in steady state and suppression of the offsets. These features make tuning of PID controllers easier.

The article does not deal with the issue of obtaining a static model of MIMO system from experimental data in the case of unstable systems or the issue of PID controllers tuning.

Acknowledgment. This research was supported by Institutional support of The Ministry of Education, Youth and Sports of the Czech Republic at University of Pardubice and SGS grant at Faculty of Electrical Engineering and Informatics.

References

1. Nordfeldt, P., Hägglung, T.: Decoupler and PID controller design of TITO systems. J. Process Control **10**, 923–936 (2006)
2. Waller, M., Waller, J.B., Waller, K.V.: Decoupling revisited. Ind. Eng. Chem. Res. **42**, 4575–4577 (2003)
3. Skogestad, S., Postlethwaite, I.: Multivariable feedback control: analysis and design, 2nd repr. edn, vol. xiv, 574 p. Wiley, Chichester (2008)
4. Lee, J., Kim, D.H., Edgar, T.F.: Static decouplers for control of multivariable processes. AIChE J. **51**(10), 2712–2720 (2005)
5. Skogestad, S., Morari, M.: Implications of large RGA elements on control elements. Ind. Eng. Chem. Fundam. **26**, 2323–2330 (1987)
6. Bequette, B.W.: Process Control: Modeling, Design, and Simulation (chap. 12. Ratio, selective, and split-range control), vol. xxix, 769 p. Prentice Hall PTR, Upper Saddle River (2003)
7. Dušek, F., Honc, D.: Transformace soustav s různým počtem vstupů a výstupů pro decentralizované řízení. Automatizace (51) 7–8, Automatizace s.r.o. Praha, pp. 458–462 (2008)

The Distributed Ledger-Based Technique of the Neuronet Training Set Forming

E. V. Melnik[1], A. B. Klimenko[2(✉)], and D. Y. Ivanov[2]

[1] Federal Research Centre the Southern Scientific Centre of the Russian Academy of Sciences, 41, Chehova st, 344006 Rostov-on-Don, Russia
[2] Scientific Research Institute of Multiprocessor Computer Systems of Southern Federal University, 2, Chehova st, 347928 Taganrog, Russia
anna_klimenko@mail.ru

Abstract. The generating of training datasets for machine learning projects is a topical problem. The cost of dataset formation can be considerably high, yet there is no guarantee of an acceptable quality of prepared data. The important issue of dataset generation is the labeling noise. The main causes of this phenomena are: expert errors, information insufficiency, subjective factors and so on. Labeling noise affects the learning stage of a neuronet and so increases the number of errors during the one's functioning. In the current paper the technique to decrease the labeling noise level is proposed. It is based on the principals of the distributed ledger technology. While there is a possibility to decrease the labeling errors number, the services integration on the basis of distributed ledger allows to improve the efficiency of dataset forming.

Keywords: Machine learning · Training dataset · Labeling errors · Data labeling

1 Introduction

Nowadays the field of machine learning (ML) is highly topical. Considering the numerous ML applications, i.e. image recognition, sound recognition, image segmentation, sentinel analysis and many others, the data sets forming seems to be an integral part of the ML-based business process.

As is mentioned in [1], a dataset, used for machine learning, is usually splitted into three parts: training set, test set and validation set. Training set is used for the training of model (in case of superwised learning), test set, which is not a subset of the training one, is used to evaluate the model in terms of its capacity of generalization. Validation set is used to tweak a model's hyperparameters, that can't be directly learned from data. The proportions of training and test sets are suggested as 80 to 20% respectively, while the size of dataset is considered individually for each project. The latter presents an issue for an ML-project because of a need of a quite large dataset, and the most of the data must be labeled.

Actually, the labeling of data is a stage of every ML project. In the current paper the neuronet projects are considered, where the supervised learning is used. The question of the amount of samples in a training set is quite sophisticated, yet, the rule of 10

© Springer Nature Switzerland AG 2019
R. Silhavy et al. (Eds.): CoMeSySo 2019, AISC 1047, pp. 7–15, 2019.
https://doi.org/10.1007/978-3-030-31362-3_2

exists, in which the amount of training data you need for a well performing model is 10x the number of parameters in the model [2].

Neural networks pose a different set of problems than linear models like logistic regression. To get the number of parameters in a neural network it is needed to:

- count the number of parameters used in the embedding layer;
- count the number of edges in your network.

The problem is the relationship between the parameters in a neural network is no longer linear, so the rule of 10 is a lower bound to the amount of training data needed.

There are some projects of labeled data storages. For example, Imagenet (more than 14 million images), [3] MS-COCO (more than 200000 labeled images), Visual QA (more than 250000 labeled images), The street View House Numbers (630420 images), CIFAR-10 (60000 images in 10 classes) and others.

Yet, taking into account the existing storages of labeled images, lots of particular projects need their own datasets, and this need has chances to be a kind of a cornerstone in the project's budget.

In this paper we consider the approaches to the labeled data forming. Some approaches in usage are considered with their pros and cons, and the distributed-ledger-based approach is presented and described. The approach presented is modeled and estimated in terms of the time needed to form the dataset.

The sections of this paper contain:

- an overview of the data labeling approaches;
- a brief summary of the distributed ledger technologies general components;
- the analysis of prospectiveness of the DL in terms of labeling efficiency enhancement;
- discussion and conclusion.

2   Data Labeling Approaches and the Labeling Noise

A comprehensive review on data labeling approaches is presented in [4]. The summary of the approaches is given in the Table 1.

The first four approaches listed above (in-house labeling, outsourcing, crowd-sourcing and the outsourcing companies involvement) are considered as a human-based ones and are in the scope of this paper. It is expedient to emphasize the crowdsourcing approach due to its speed and low costs. The crowdsourcing approach is presented by such platforms as Amazon Mechanical Turk and Clickworker, which guarantee the high affordability and low time consumption of the process, yet giving the risk of the moderate quality of the labeled data.

The key issue of approaches under consideration is the problem of different labeling errors, or as it defined in [5], the problem of labeling noise. The labeling noise is anything that obscures the relationship between the features of an instance and its class [6]. Possible sources of noise are considered in [4] and include the following situations:

Table 1. The summary of data labeling approaches.

Approach	Description	Pros	Cons
In-house labeling	Assignment of tasks to an in-house data science team	Predictable results of a good quality and the ability to track the process of labeling	The duration of the process and possibly high cost
Outsourcing	Temporary employees of freelance platforms, posting vacancies on social media and job search sites	The ability to evaluate applicant's skills	The need to organize the workflow
Crowdsourcing	Cooperation with the freelancers from crowdsourcing platforms	Cost savings and fast results	Quality of work can suffer
Specialized outsourcing companies	Hiring and external team for a specific project	Assured quality	Higher price compared to crowdsourcing
Synthetic labeling	Generating data with the same attributes of real data	Training data without mismatches and gaps; Cost- and time-effectiveness	High computational power required
Data programming	Using the software that programmatically label data to avoid manual work	Automation; fast results	Dataset of a relatively low quality

- the information which is provided to the expert may be insufficient to provide reliable labeling;
- the errors can occur in the expert's labeling process itself;
- the labeling task is subjective, as, for example, in medical applications and image data analysis;
- the presence of data encoding errors and communication problems.

As to the approaches developed to avoid the consequences of labeling noise, there is a considerable amount of them:

- label noise-robust methods; [7–9]
- probabilistic label noise-tolerant methods [10, 11];
- data cleansing methods [12–14];
- model-based label noise-tolerant methods [15, 16].

Despite the methods developed, the problem of labeling noise is topical and definitely has not been solved yet.

In this paper we address two general problems of training dataset forming:

- the large volume of data needed for neural net learning and the speed and cost of dataset forming;
- the labeling noise minimization.

Nowadays the new so-called distributed ledger technology emerged. We will analyze the applicability of the principles of the DLT to the problems described above.

3   Distributed Ledger and Its Applicability to the Data Labeling

A distributed ledger (or distributed ledger technology, DLT) is a consensus of replicated, shared, and synchronized digital data geographically spread across multiple sites, countries, or institutions [17]. A peer-to-peer network is required as well as consensus algorithms to ensure replication across nodes to be undertaken.

A considerable number of DLT-based systems are designed and developed till now, e.g., Bitcoin, Etherium, Nano, etc. Some systems use the blockchain data structure to store the transactions, and some use the relatively new blocklattice data structure, which can be considered as an extention of the blockchain.

So, the key elements of the DLT are a way of data storage and a consensus method. While the blocklattice [18] is quite new, the blockchain [19, 20] structure is used frequently with numerous consensus methods.

A comprehensive overview of the consensus methods used with blockchain is given in [19]. The consensus methods are subdivided into classes:

- Proof-based consensus algorithms (PoW, Prime number finding – based PoW, Generalized PoW, PoS, Hybrid forms of PoW and PoS, Proof of activity, Proof of burn, proof of space, proof of elapced time, proof of luck, multichain);
- Vote-based consensus algorithms (PBFT, Symbiont, Iroha with Sumeragi, Ripple, Stellar);
- Crach fault-tolerance – based consensus (Quorum with Raft, Chain).

So, the diversity of data structures and consensus methods used in DLT field makes it possible to consider the DLT application to the dataset forming process with DLT particular elements usage.

Consider the generic scheme of data labeling process: a software entity sends the content to be labeled to the expert, expert labels the content and saves it in the data storage.

The procedure of one-piece content labeling is an analogue of the transaction in cryptocurrency system in the context of this paper. So, as it is done in cryptocurrencies, having the piece of content labelled, the next step is to verify its correctness sending it to all participants and reaching the consensus about the correctness of the transaction. Obviously this stage decreases the labeling noise doing this before the labeled data are put into the storage. The well-known methods of data cleansing [12–14] propose some filters for data which are in the storage.

The situation will be as follows: besides his own content for labeling the expert receives the a considerable number of labeled content to be verified. As the system is asynchronous, the labeled content for verification can be received in a random order with delays, yet the consensus procedure must be conducted after the expert has received the labeled content from other experts.

Another way to conduct a verification is to give one piece of content to all experts, reach the consensus and to suppose the content to be verified after that.

Another issue we address in this paper is the rate of data storage filling up. A distributed ledger application allows to integrate the particular services and storages of labeled data and so to enhance the source coverage and the speed of data collecting. The DLT makes the storages universal, enhancing their applicability. So, the DL is quite a promising technology to improve the dataset growth rate.

The next section contains the simplified models to estimate the efficiency of the DLT application to the training dataset forming.

4 The Analysis and Estimation of DLT Application to the Dataset Forming

Define the "community" term as follows: community (of experts) is a set of employees who label data.

Consider the first case, when the content verification is conducted in the following way:

1. An expert receives the content to be labeled.
2. The expert labels the content and waits for the content to be verified from other community participants.
3. After all pieces of content are received and verified the consensus takes place.
4. A package of verified content is added to the block.

This procedure can be described by the following expression:

$$t_p = t + \xi N + (N - 1)t + N t_{cons},$$

where

t is the time needed for the content labeling;
ξN is the waiting time for content to be verified;
$(N - 1)t$ is the time of verification of content received from N − 1 experts;
$N t_{cons}$ is the time needed to reach a consensus for all label = led data

Considering $t_{cons} = kN$, where k is the ratio, which describes the dependency between consensus procedure speed and the number of participants, the time of package verification will be as follows:

$$t_p = t + \xi N + (N - 1)t + kN^2.$$

Consider the second case, when the community receives the same data for verification.

To verify 1 piece of content, the time needed is as follows:

$$t_p = t + kN.$$

To verify N pieces of content consequently, the time needed is:

$$t_N = Nt + kN^2.$$

Then, if we want to label M pieces of content, the time needed for this will be as follows:

$T = M(t + \xi + kN)$ for the first case, and $T = M(t + kN)$ for the second case.

In the pictures below the time diagrams are shown (Figs. 1 and 2).

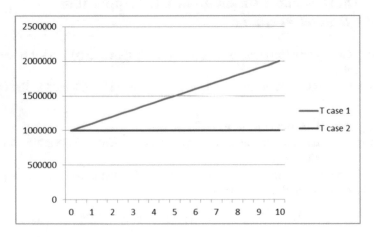

Fig. 1. The time needed to label M content pieces with ξ growth.

One can see that the case which presupposes the verification of the same piece of content by the community of experts is more prospective in terms of time consumption. It is because in case of package-like verification experts have to wait for the content from other participants to verify it and to reach the consensus from the labeling correctness point of view, while one-piece content verification presupposes that it is enough to receive the majority of the votes to make a decision about the label correctness.

Then, consider the situation, when one community fills a stand-alone dataset. For example, as is shown above, to verify M content pieces, the community needs time $T = M(t + kN)$. Transforming the data storage to the distributed ledger, and considering that every ledger replica is filled by N experts, the time needed is reduced in the following way:

$$T_{dl} = M(t + kN)/L,$$

where L is the number of services, which label data.

The diagrams of the time decrease with the services growth are shown in Fig. 3.

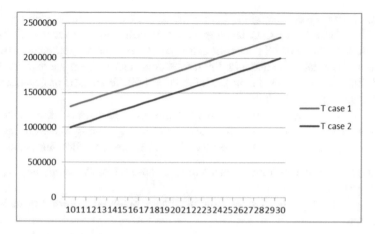

Fig. 2. The time needed to label M content pieces with N growth.

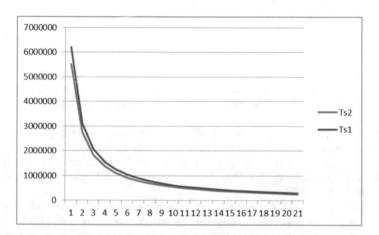

Fig. 3. The decrease of time for the dataset accomplishment in case of ledger node growth.

One can see that if a project, which collects the training dataset into the distributed ledger, shares the ledger with the other similar services, the storage filling rate increases considerably in all cases of labeled data verification. Yet due to the lack of delay the second case with the consequent manner of data labeling and verification shows better results.

5   Discussion and Conclusion

As machine learning-based projects are a considerable part of the contemporary software services, the question of training dataset forming is relevant. The key issues of dataset forming are: the need to gather, label and store really considerable amounts of information, and, besides, to label the data instances correctly.

In the current paper we address both of these issues, and examine the possibility to improve the efficiency of data labeling and data collecting by means of distributed ledger technologies. Labeled data verification can be conducted in a manner of parallel verification of the data instance with the farther consensus among the experts. The data storage filling rate increases due to the integration of different ML-projects, when each project has its copy of the ledger.

With simple models we have estimated the prospectiveness of the labeled data verification techniques and tested the prospects of storage filling rate with the usage of distributed ledger. The simulation results allow to make the following conclusions:

- the usage of consensus allow to clean the data automatically before they are put into the data storage;
- the usage of distributed ledger structure has a potential in terms of data storage filling rate.

Acknowledgements. The paper has been prepared within the RFBR project 18-29-22086 and 18-29-22046.

References

1. Machine Learning Project Structure: Stages, Roles, and Tools. https://www.altexsoft.com/blog/datascience/machine-learning-project-structure-stages-roles-and-tools/. Accessed 21 May 2019
2. How much training data do you need? https://medium.com/@malay.haldar/how-much-training-data-do-you-need-da8ec091e956. Accessed 21 May 2019
3. Open Datasets for Deep Learning Every Data Scientist Must Work With. https://www.analyticsvidhya.com/blog/2018/03/comprehensive-collection-deep-learning-datasets/. Accessed 21 May 2019
4. How to Organize Data Labeling for Machine Learning: Approaches and Tools. https://www.kdnuggets.com/2018/05/data-labeling-machine-learning.html. Accessed 21 May 2019
5. Frénay, B., Verleysen, M.: Classification in the presence of label noise: a survey. IEEE Trans. Neural Netw. Learn. Syst. **25**(5), 845–869 (2014)
6. Hickey, R.J.: Noise modeling and evaluating learning from examples. Artif. Intell. **82**(1–2), 157–179 (1996)
7. Beigman, E., Klebanov, B.B.: Learning with annotation noise. In: Su, K.-Y., Su, J., Wiebe, J., Li, H. (eds.) Proceedings of the Joint Conference of the 47th Annual Meeting ACL and 4th International Joint Conference on Natural Language Processing AFNLP, Suntec, Singapore, August 2009, vol. 1, pp. 280–287. World Scientific Publishing Co Pte Ltd, Singapore (2009)
8. Manwani, N., Sastry, P.S.: Noise tolerance under risk minimization. IEEE Trans. Cybern. **43**(3), 1–6 (2013)
9. Abellán, J., Masegosa, A.R.: Bagging schemes on the presence of class noise in classification. Expert Syst. Appl. **39**(8), 6827–6837 (2012)
10. Bouveyron, C., Girard, S.: Robust supervised classification with mixture models: learning from data with uncertain labels. Pattern Recogn. **42**(11), 2649–2658 (2009)

11. Evans, M., Guttman, I., Haitovsky, Y., Swartz, T.: Bayesian analysis of binary data subject to misclassification. In: Berry, D., Chaloner, K., Geweke, J. (eds.) Bayesian Analysis in Statistics and Econometrics: Essays in Honor of Arnold Zellner, IEEE Transactions on Neural Networks and Learning Systems, vol. 24, pp. 67–77. Wiley, New York (1996)
12. Gamberger, D., Lavrac, N., Dzeroski, S.: Noise detection and elimination in data preprocessing: experiments in medical domains. Appl. Artif. Intell. **14**, 205–223 (2000)
13. Segata, N., Blanzieri, E., Delany, S., Cunningham, P.: Noise reduction for instance-based learning with a local maximal margin approach. J. Intell. Inf. Syst. **35**(2), 301–331 (2010)
14. Karmaker, A., Kwek, S.: A boosting approach to remove class label noise. In: Nedjah, N., Mourelle, L.M., Vellasco, M.M.B.R., Abraham, A., Köppen, M. (eds.) Proceedings of the Fifth International Conference on Hybrid Intelligent Systems (HIS'05), Rio de Janeiro, Brazil. IEEE, Los Alamitos (2006)
15. Zhang, M.-L., Zhou, Z.-H.: CoTrade: confident co-training with data editing. IEEE Trans. Syst. Man Cybern. **41**, 1612–1626 (2011)
16. An, W., Liang, M.: Fuzzy support vector machine based on within-class scatter for classification problems with outliers or noises. Neurocomputing **110**(13), 101–110 (2013)
17. Distributed ledger technology: beyond block chain. https://www.gov.uk/government/news/distributed-ledger-technology-beyond-block-chain. Accessed 20 May 2019
18. Block lattice. https://github.com/nanocurrency/nano-node/wiki/Block-lattice. Accessed 21 May 2019
19. Nguyen, G., Kim, K.: A survey about consensus algorithms used in blockchain. J. Inf. Process. Syst. **14**(1), 101–128 (2018)
20. BlockChain Technology: Beyond Bitcoin. http://scet.berkeley.edu/wp-content/uploads/AIR-2016-Blockchain.pdf. Accessed 21 May 2019

Decision Support Systems for the Oil Fields with Cloud Multiagent Service

Donat Ivanov(✉) ⓘ, Sergey Kapustyan ⓘ, Anatoly Kalyaev ⓘ,
and Iakov Korovin ⓘ

Southern Federal University, 2 Chehova St., 3479328 Taganrog, Russia
donat.ivanov@gmail.com

Abstract. The paper is dedicated to the organization of agent's communities for cloud service of decision support system for the oil producing enterprise. There are the architecture of cloud service. An approach to the formation of communities of cloud service agents is proposed. An algorithm for creating communities of cloud service agents is considered. The results of experimental studies are presented. The results confirm the operability and effectiveness of the proposed approach.

Keywords: Cloud service · Decision support system · Agents societies · Bulleting boards · Oil-production

1 Introduction

Each oil-production enterprise faces an urgent problem of production increasing without purchase of additional equipment. The problem can be solved by means of proper and timely activities aimed at oil-production increasing [1–5] and improvement of oil-production efficiency, decreasing of equipment stoppage during an operating crew. Such activities require continuous processing of information about oil-production equipment functioning.

Moreover, current information about oil-production equipment functioning must be available to all those who is interested in it: from engineers to analysts and senior managers, who reside in offices far away from oil wells. This is a problem, based on cloud service principles [6, 7], which consists in development of a service for online processing of various data and presenting processing results in a digital form.

2 The Cloud Service Architecture

We suggest the cloud service architecture, based on a multi-agent dispatcher [8], which contains the following principal components (see Fig. 1):

- a user interface (development of tasks and user interaction with the cloud service), which interacts with a bulletin board;
- a bulletin board (BB), which provides interaction of agents, placed in computational nodes;

R. Silhavy et al. (Eds.): CoMeSySo 2019, AISC 1047, pp. 16–23, 2019.
https://doi.org/10.1007/978-3-030-31362-3_3

an agents society (monitoring of computational node resources, data exchange between subtasks, storing of intermediate data for correct resume of computational process in case of a subtask failure), which interacts with the bulletin board.

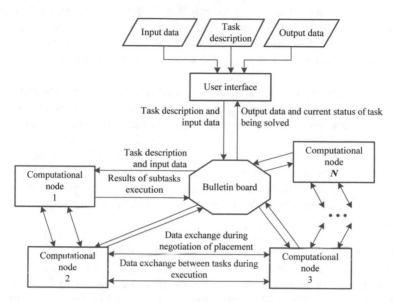

Fig. 1. The cloud service architecture

The user interface is intended for development of tasks for further execution within the cloud service based on a multi-agent dispatcher, and for interaction of the user and the owner of tasks with the cloud service during execution of the tasks. The user interface interacts directly only with the bulletin board. Any interaction with the society of agents, placed in the computational nodes, is performed only by means the bulletin board.

Owing to the graphic task editor, which is contained in the user interface, it is possible to make a complete task description in the form of an oriented graph and to define its execution parameters.

The task graph is a set of vertices (subtasks, which are separate applications), interconnected by edges (data transfer channels). A set of parameters is given for each subtask: its required resource, input/output identifications, start options (command-line arguments), files of input and output data, application files, which will be copied to the bulletin board (after placement of the task) and to a computational node (when an agent starts the subtask). The application files are both executable files, started by agents, and accompanying library files. Data files are any files, which are necessary for execution of the subtask.

When the task description is formed, the user places it on the bulletin board. Along with the task description, the user interface transfers all files (files of input data, executable program files, library files), necessary for execution of the subtasks, to the bulletin board.

The society of agents is placed in computational nodes – one agent in one node. The agent performs resource monitoring of the computational node (loading of each core of each processor, incoming and outgoing traffic for each network interface), interaction with the bulletin board during selection of the task and negotiation of subtasks placement, data exchange between the subtasks being executed, storing of intermediate data for correct resume of computational process in case of a subtask failure.

For solution of any task the agents self-assembly into a group. One and the same agent can be a member of several groups, organized for solution of different tasks. Exchange of data, concerning some certain task, is performed only within the corresponding group.

Each agent, placed into some computational node, asks regularly all known bulletin boards in order to find new unplaced tasks. If such task exists, the agents start to organize the group. The agents estimate the task and check, if it is possible to execute it during the required time on the total resource available in the computational nodes of all agents.

When the group is formed, each agent informs the bulletin board about the list of subtasks, selected for execution on its computational node. Then, each agent downloads all required files from the bulletin board and executes the subtasks.

During execution each subtask transfers its calculation results to the next subtask according to the task graph. The final subtask transfers (using the agent of the corresponding computational node) the results to the bulletin board, which transfers them further, to the user interface. When all subtasks are executed, the agents inform the bulletin board that the task is executed, and the task is marked as done.

For implementation of the cloud service of the decision support system, we have developed algorithms as follows:

- an algorithm which forms an agents society to perform a user task;
- an algorithm which distributes computational load in an agents society;
- an algorithm which distributes subtasks of a user task in an agents society;
- an algorithm of multi-agent resource scheduling in a heterogeneous cloud environment;
- an algorithm of an agent functioning.

3   The Suggested Approach to Creation of Agents Societies of the Cloud Service

For solution of all user tasks, placed on the bulletin boards, it is necessary to form agents societies in the system (executive computational units - ECU). The main aim of each society is to solve the specific user task during the specified time. Besides, each agent is aimed at getting the maximum possible score for solving the task within the society. Let us consider, that each agent A_j (and its corresponding ECU-"freelancer") has the specified computational performance S_j (which is defined as the number of computing operations performed by the ECU during the specified time), which can dynamically vary during its functioning period. Therefore, the society composition, which provides solution of the user task during the required time, can also dynamically vary with time.

Before we represent the algorithm of forming a society for solving a user task, let us give some basic principles.

To become a member of some society or to form it, the agents need information from bulleting boards about user tasks, which are in the system. Besides, the number of agents within the system can be, in general, much bigger than the number of bulletin boards. Therefore, if each agent is continuously asking the bulletin board, it can lead to congestion of the bulletin board link and, hence, access to the bulletin board will be blocked. To avoid blocking of the bulletin board, it is reasonable to let the agents ask the bulletin board with a specified pause T_{wait}.

On the other hand, if the agents ask the bulletin board with a big time delay, then the time of the user task solution within the system will considerably increase and exceed the required time limit. To avoid this shortcoming, we suggest use a principle of invitation: members of the society, which is solving the user task, can invite other agents in the society. It is supposed, that initially, when agents register in the system, they know no other agents, but during solution of the user tasks they save addresses of all other agents, whom they contacted with, and keep these addresses for a while. In addition, the agent, which finds an unsolved user task, can invite "known" agents for its solution without waiting that they will find this task on the bulletin board themselves.

When the agent finds an unsolved user task on the bulletin board or receives an invitation from some "known" agent to join the society, it has to evaluate reasonability of its participation in solution of the task. The agent can make such evaluation on basis of two parameters of the task: its total labor intensity Y (i.e. total computational complexity of all vertices on the task information graph $Y = \sum_{i=1}^{K} Y_i$, where K is the number of vertices of the task graph $\mathbf{G(Q, X)}$), and its price P (i.e. the score, which the user is ready to pay for its solution). On basis of the ratio P/Y (price to labor intensity), the agent can evaluate relative profit from solution of the task. If it exceeds a certain value, specified by its owner, the agent makes decision on joining the society for solving the task.

It should be noted, that the main aim of GRID, in general, is solution of the maximum possible number of incoming user tasks during limited time. For this, the society, which solves the tasks, must contain the minimum number of ECUs, which can solve the specified task during the specified time. In other words, it is required that the society, organized by agents for the task solution, has the minimum possible number of members. So, we suggest estimation of the time T_{sol}, which is required for the task solution by the agents society. If this estimation is less than the time $T_{res} = T_{Max} - T^*$ (where T^* is the time which has passed since the task appeared on the bulletin board, and T_{Max} is the total time, given by the user for the task solution), which is left for the task solution, then it is possible to say that the number of agents in the society is sufficient for solution of the user task during the required time. It is possible to suggest several versions of such estimation, but in conditions of a decentralized system, and dynamically varied contents and parameters of the ECUs of the society, it is reasonable to obtain such estimation as quick and often as possible, but, probably, to trade its precision. Therefore, we suggest the simplest method for estimation of the time T_{sol}, as a result of division of the total labor intensity (computational complexity) of currently unsolved subtasks of the task $Y_{res} = \sum_{i=1}^{L} Y_i$ being solved, where L is the number of unsolved subtasks, by the total performance of all ECUs of the society $S = \sum_{i=1}^{D} S_i$, where D is the number of ECUs in the society.

If the solution time T_{sol} of the task exceeds the time T_{res}, which is left for the task solution, and the number of agents in the society is less than M (the granularity of task parallelism), then the agent transfers invitations, which contain current information about the task, to all "known" agents to join the society. If these agents are free, i.e. they are not involved in solution of any other task, they evaluate the received information and, if the task is profitable, make decision concerning participation in the society for the task solution.

If the society is capable of solving the task during the specified time, then the task is marked on the bulletin board as "full" and other agents does not try to join the society for the task solution.

If the maximum number M (where M is the granularity of task parallelism) of the society members for the user task solution is gathered, but the task solution time exceeds the required value, then the agent with the lowest performance is excluded from the society after solution of its subtask.

If the agent is excluded from the society, or the performance of its ECU is changed so, that the task solution time exceeds the time limits specified by the user, then the task on the bulletin board is marked as "particle", and it is necessary to find additional agents for the society for the task solution.

Taking into account all things considered, we suggest an extended algorithm of the society creation for the user task solution in the cloud service.

4   The Algorithm of the Agents Society Creation

1. The user places the task descriptor on the bulletin board;
2. The bulletin board marks the received task as particle;
3. On the bulletin board the agent A_j searches the tasks marked as particle. For each task of such kind it gets data from the task descriptor as follows:
   (a) the task graph;
   (b) the price P and the task complexity Y, which is defined as the summary of complexity of the subtasks $\sum_{i=1}^{K} Y_i$, where K is a number of subtasks of the task;
   (c) the statuses q_i ($i = 1,2,...,K$) ("done"/"is being executed"/"free") of all the subtasks;

(d) the time, which is left to the task solution time, specified by the user, is $T_{res} = T_{Max} - T^*$, where T^* is the time passed since the moment when the task was announced on the bulletin board;

(e) the granularity of parallelism M of the task;

4. The agent A_j analyses possibility and reasonability of its participation in the society for solution of a certain task. For this:

(a) the agent A_j sorts the found tasks according to the ratio "price/complexity" P/Y (profit of the task);

(b) the agent A_j selects the task with the highest profit from the list of found tasks. If the list is empty, then the agent goes to item 5.

(c) if the task profit corresponds to the value, specified by the owner, then the A_j gets from the bulletin board the number of agents N, involved into the task solution, and their performance parameters.

(d) if $N = M$, where N is a number of agents in the society, M is the task granularity of parallelism, then the agent deletes the task of its list of found tasks and goes to item 4.b);

(e) if $T_{res} < T_{sol}$, where $T_{sol} = \dfrac{\sum_{i=1}^{L} Y_i}{\sum_{m=1}^{D} S_m}$, Y_i is a computational complexity of an unsolved subtask q_i, L is a number of unsolved subtasks, S_j is a current performance of an ECU j, D is a number of ECUs in the society, then the agent deletes the task from its list of found tasks and goes to item 4.b);

(f) the agent A_j joins the society and transfers parameters of its ECU to all members of the society and to the bulletin board;

(g) if the performance parameters of the members of the updated society is sufficient for solution of the task (i.e. $T_{res} \geq T_{sol}$), then it is marked on the bulletin board as full;

5. if the agent A_j joins the society, then go to item 7, else – it enters waiting mode for T_{wait};

5. if during waiting period the agent A_j receives an invitation to join the society from other agents, then it goes to item 4.c), else – to item 3;

7. if the task is particle and $N < M$, then the agent A_j sends invitations to take part in its solution to the known agents. Go to item 9;

8. if the task is particle and $N = M$, then the agent, which has the minimum performance, is excluded from the society. Go to item 7;

9. the agents start distribution of the subtasks of the user task among the members of the society and their solution (the algorithm of distribution of computational load in the agents society).

Owing to the abovementioned algorithm, it is possible to form agents societies in GRID for solution of incoming tasks. Composition of the society, which will solve the user task, will depend on the current performance of the ECU of its members. In other words, composition of the society will be continuously adapting to the dynamically changing parameters of the ECU-freelancers of its members.

5 The Results of Experimental Research

To analyse working ability and efficiency of the suggested multi-agent resource scheduling of heterogeneous cloud computational environment (CCE) for solution of the task flow **Z**, we have developed a software prototype. The software prototype simulates CCE functioning for various values of such initial parameters as:

- the number of resources in CCE (up to 1000);
- the number of various subtasks, solved by a separate resource of the CCE (up to 20);
- the performance of resources during solution of various subtasks, and the bandwidth of the links between them and cloud infrastructure;
- the graph of the user task (the number of vertices; the computational complexity of the subtasks, assigned to the vertices; the size of data, transferred between the subtasks and assigned to the edges of the graph);
- the required time moments, when the user tasks must be solved, and the reward score for their solution.

As criteria of efficiency of the CCE functioning during solution of the tasks flow we have accepted:

- the coefficient of efficiency (CE) – the ratio of the time, spent by the CCE resources for solution of the user tasks, to the total time of their functioning within the system;
- the coefficient of guarantee execution of the user task – the ratio of the number of the user tasks, solved to the required moment of time, to the total number of tasks, sent by users to the bulletin board.

The results of a set of experiments, performed for various initial parameters of the software prototype, prove that the value of the relative load coefficient always exceeds 70%, and the value of the guarantee execution coefficient always exceeds 90%.

Acknowledgement. The reported study was performed within the federal targeted program of the Ministry of Education and Science; the agreement's ID RFMEFI57517X0152.

References

1. Shah, D.O.: Improved Oil Recovery by Surfactant and Polymer Flooding. Elsevier, New York (2012)
2. Morrow, N., Buckley, J., et al.: Improved oil recovery by low-salinity waterflooding. J. Pet. Technol. **63**, 106–112 (2011)
3. Sorbie, K.S.: Polymer-Improved Oil Recovery. Springer, Heidelberg (2013)
4. Lake, L.W., Johns, R.T., Rossen, W.R., Pope, G.A.: Fundamentals of Enhanced Oil Recovery (2014)
5. Sheng, J.J.: Enhanced oil recovery in shale reservoirs by gas injection. J. Nat. Gas Sci. Eng. **22**, 252–259 (2015)

6. Tao, F., Zhang, L., Venkatesh, V.C., Luo, Y., Cheng, Y.: Cloud manufacturing: a computing and service-oriented manufacturing model. Proc. Inst. Mech. Eng. Part B J. Eng. Manuf. **225**, 1969–1976 (2011)
7. Bhardwaj, S., Jain, L., Jain, S.: Cloud computing: a study of infrastructure as a service (IAAS). Int. J. Eng. Inf. Technol. **2**, 60–63 (2010)
8. Kalyaev, A.I., Kalyaev, I.A.: Method of multiagent scheduling of resources in cloud computing environments. J. Comput. Syst. Sci. Int. **55**, 211–221 (2016)

Floating Data Window Movement Influence to Genetic Programming Algorithm Efficiency

Tomas Brandejsky[(✉)] [ID]

University of Pardubice, Studentska 95, 532 10 Pardubice, Czech Republic
tomas.brandejsky@upce.cz

Abstract. Presented paper deals with problem of large data series modeling by genetic programming algorithm. The need of repeated evaluation constraints size of training data set in standard Genetic Programming Algorithms (GPAs) because it causes unacceptable number of fitness function evaluations. Thus, the paper discusses possibility of floating data window use and brings results of tests on large training data vector containing 1 million rows. Used floating window is small and for each cycle of GPA it changes its position. This movement allows to incorporate information contained in large number of samples without the need to evaluate all data points contained in training data in each GPA cycle. Behaviors of this evaluation concept are demonstrated on symbolic regression of Lorenz attractor system equations from precomputed training data set calculated from original difference equations. As expected, presented results points that the algorithm is more efficient than evaluating of whole data set in each cycle of GPA.

Keywords: Genetic Programming Algorithm · Floating data window ·
Fitness function evaluation scheme · Efficiency · Floating data window movement

1 Introduction

This paper solves needed step of GPA (Genetic Programming Algorithm) [1] development to work with large and big data – the question if it is possible to replace repeated evaluation of typically relatively small training data vector by data window floating across long stream of data. While large data are organized and stored by standard database systems, so called big data systems are characterized by originally 3vs (volume, variety and velocity), formerly 4vs or 5vs (volume, variety, velocity, veracity and value) [2]. Especially data variety represents difference to really large data stored in warehouses – big data are not well organized and this fact complicates their application. Variety means that data are stored in different data types, in different representations, most of them are unstructured (text, images, video, voice, etc.) and thus them cannot be processed directly without prepossessing.

Significant difference between data warehousing and big data lies in allocation of data and processing resources. While data warehousing maximizes centralized solutions, big data are oriented to decentralized ones.

Comparing to statistical methods or other soft computing techniques like fuzzy sets and artificial neural networks, genetic algorithms are applied in large and big data

© Springer Nature Switzerland AG 2019
R. Silhavy et al. (Eds.): CoMeSySo 2019, AISC 1047, pp. 24–30, 2019.
https://doi.org/10.1007/978-3-030-31362-3_4

analytic rather sporadically. This situation is not caused by the lack of possible application. It is possible to mention such application domains as optimization, prediction, simulation, pattern recognition [3]. Significantly less frequent there are applications of Genetic Programming Algorithms (GPAs). It is possible to mention work [4] describing time dependent model development from big data information.

GPAs are applicable not only to program development but especially to symbolic regression – development of analytic model in the form of difference or differential equations describing data applicable in Big Data analytic. They can be also used for creating of complicated data mappings (transformations) and reductions. Evolutionary algorithms (genetic or genetic programming ones) are also applicable to development of alternative models like e.g. qualitative or fuzzy ones. In this paper the application of floating window in GPA will be discussed. Floating window is applicable both to centralized (data warehouse) and distributed (big data) approaches. Presented example will be taken from area of large data sets (30 million of samples). There are know many parallel implementations of GPAs for systems with shared memory, but these implementations are not relevant for our work because they solve especially problems of large sets of individuals and complicated fitness function evaluations. Large and big data require parallel GPA implementation in the sense of [5]. Nowadays there are thousands of works about parallel and distributed genetic programming algorithms, large collection is cited in this work. The book [5] also explains difference between parallel and distributed approach. Non-looking to this terminological problem (many so-called parallel GPAs are by [5] rather distributed), parallel algorithms use parallel computing of fitness function from the viewpoint of dividing of population into smaller groups.

From existing projects applying GPA onto big data the following ones can be referred. It is the project "Automatic Programming for Optimization Problems with Big Data" at School of Computing, University of Portsmouth listed suitable Big data problems for GPA application as especially Financial Forecasting, Traffic Optimization, Cloud Optimization and Scheduling [6]. It is possible to mention also any other potential application domains of genetic programming in the field of Big data as symbolic regression (search of model describing stored data), discovering game-playing strategies (and not only game-playing but also business ones), forecasting, induction of decision trees, etc. The work [7] describes interesting application of GPA in mining of big data obtained from the Large Hadron Collider. This work describes large set of sampling methods as random sampling, weighted sampling, incremental data selection, topology based subset selection, balanced sampling and many multilevel sampling and hierarchical sampling methods. Some authors applied genetic programming in big data domain indirectly in automatic synthesis of automatic classifiers of data, as [8].

There are well known distributed (parallel) implementations of GPA, one of the first was developed by pioneer of GPA research, prof. Koza [9]. The earlier work in this area was [10]. Nowadays there are thousands of works about parallel and distributed genetic programming algorithms, large collection is cited e.g. in the work [11]. This work also explains the difference between parallel and distributed approach. Non-looking to this terminological problem (many so-called parallel GPAs are by [9] rather distributed), these algorithms parallels computing of fitness function from the viewpoint of dividing of population into smaller groups.

The biggest problem of GPA application lies in its computational complexity given by repeated evaluation of fitness function in that case when fitness function is represented by large set of data points given by query to database. Such data set can consist billions of data records. Thus the presented research is based on idea to use population of small number of individuals which tend to big number of fitness function evaluations. If the training data satisfies condition of stationarity, this fitness function is than evaluated on small data window. Because it is difficult to expect that this data window will contain complete information about modeled system behaviors, floating data window is used. Presented study tests if there are some difference for various overlaps of neighborhood window positions.

2 Floating Data Window

As it was discussed above, this paper discusses problem of GPA using floating data window onto large stationary data. Frequently, small data windows are computationally more efficient than bigger ones, if the used data set contains all relevant properties of system behaviors. If the data are chosen randomly or on the base of any other criteria, they can cause inefficient computation or irrelevant conclusions. Thus for application in the area of large and big data, it his paper the use of floating data window is tested. This approach is based on assumptions that information obtained from previous positions of data window will not be replaced by information from newer data totally and thus that behavior of moving window will be similar to large window ones except increased computational efficiency.

Floating data window can be moved about random or fixed number of data records in each evolutionary cycle of used GP algorithm. The size of the floating data window movement step can be chosen to prevent exceeding number of data array records during evaluation.

It is possible to expect that described algorithm will be applicable to nonstationary data too. Processing of these data typically require modeling of static moments change in the time, changes of model structure. There are at least three different approaches – dividing data records vector into smaller ones and development of separate model in each position, an application on floating window to create separate model in each position too or an application of evolutionary algorithm with long-time adaptability like haploid algorithms are. The problem is, that now there are developed rather haploid (and especially diploid) genetic algorithm [12] than haploid GPAs. Especially, second case represents recursion of above described algorithm. It is possible to represent it as computing of the same algorithm in different positions of big (superior) data window inside which smaller one will move in each step of big one repeatedly.

The following Fig. 1 sketches modification of GPA-ES evolutionary algorithm which allows after each cycle of evolutionary algorithm to move floating data window into new position. This movement causes evaluation of large part of training data vector without need to use evaluation of extreme large number of data points in each evolutionary step.

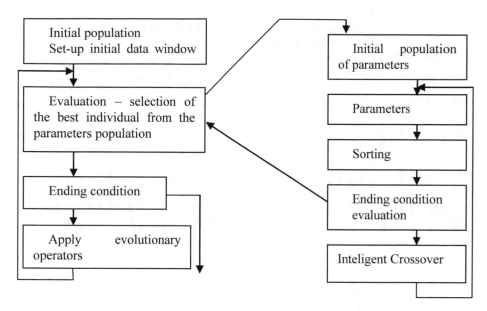

Fig. 1. GPA-ES algorithm structure modified for floating window use.

Floating data window overlap is equivalent to window movement speed because of linear dependency of these parameters – especially if the movement is unidirectional and it is not random. The application of floating data window can be more efficient than repeated evaluation of big stationary one. Thus the main aim of this work is to investigate applicability of small floating window on large data vector. There can be expected that small floating window will be more computationally efficient than single one covering whole training data vector.

3   Experiment Design and Obtained Results

Experiments used to verify applicability of floating data window on large data sets where each position of data window is used maximally once were performed on 4 node cluster. Each node was equipped by 4 processors, each with 12 cores. Amount of RAM was 384 GB per node.

Training data vector was chosen 3 million lines and corresponds to limit of 30000 evolutionary cycles and 64 samples in data window (64 * 30000 < 3000000). As the pseudo random number generator served standard C++ rand() function of GNU compiler. These training data were generated by Lorenz attractor system model in long double (80 bit) precision without additional noise. Experiments were repeated 3000 times with different seed magnitude of used pseudo random number generator influencing genetic programming algorithm. This number of experiments gives error of resulting average number of GPA cycles less than 1% (Figs. 2 and 3).

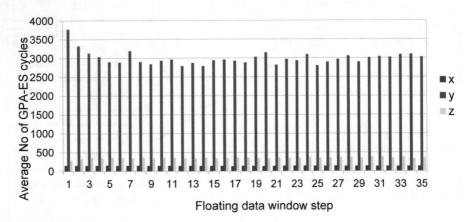

Fig. 2. This figure demonstrates dependency of average numbers of GPA cycles on movement step for floating data window size 32 samples.

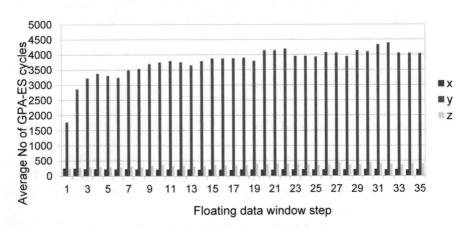

Fig. 3. The same situation for 64 samples floating data window.

For comparison, evaluating of whole data set produced results summarized in the following Table 1:

Table 1. Results of experiments with evaluating of whole training data vector in each GPA cycle.

Window size	x	y	Z
30000	321	1675	193

This table presents results of experiments with large data vector. Above discussed results were computed by significantly large number of fitness function evaluation than above presented results computed using floating data window.

4 Conclusion

This contribution describes motivation of GPA evaluation scheme change to incorporate large amount of data without inefficient evaluation of fitness function in big number of points. Modification of GPA-ES algorithm is described too. Presented simulation experiments point that floating data window concept is applicable for evaluation of large data sets and it is efficient.

Acknowledgement. The work was supported from ERDF/ESF "Co-operation in Applied Research between the University of Pardubice and companies, in the Field of Positioning, Detection and Simulation Technology for Transport Systems (PosiTrans)" (No. CZ.02.1.01/0.0/ 0.0/17_049/0008394).

References

1. Koza, J.R.: Genetic Programming: On the Programming of Computers by Means of Natural Selection. MIT Press, Cambridge (1992)
2. Cartledge, C.H.: How Many VS Are There in Big Data? http://www.clc-ent.com/TBDE/Docs/vs.pdf. Accessed 8 June 2019
3. Verma, G., Verma, V.: Role and applications of genetic algorithm in data mining. Int. J. Comput. Appl. **48**(17), 5–8 (2012)
4. Gandomi, A.H., Sajedi, S., Kiani, B., Huang, Q.: Genetic programming for experimental big data mining: a case study on concrete creep formulation. Autom. Constr. **70**, 89–97 (2016). https://doi.org/10.1016/j.procs.2018.07.264
5. Poli, R., Langdon, W.B., McPhee, N.F., Koza R.J.: A field guide to genetic programming (2008). Published via http://lulu.com and freely available at http://www.gp-field-guide.org. uk. (With contributions by Koza R.J.)
6. Bader-EL-Den, M., Adda, M.: Automatic Programming for Optimization Problems with Big Data. https://www.findaphd.com/phds/project/automatic-programming-for-optimization-problems-with-big-data/?p62613. Accessed 12 Mar 2019
7. Hmida, H., Hamida, S.B., Borgi, A., Rukoz, M.: Scale genetic programming for large data sets: case of Higgs Bosons classification. Procedia Comput. Sci. **126**, 302–311 (2018)
8. Kojecký, L., Zelinka, I., Šaloun, P.: Evolutionary synthesis of automatic classification on astroinformatic big data. Int. J. Parallel Emergent Distrib. Syst. **32**(5), 429–447 (2017). https://doi.org/10.1080/17445760.2016.1194984
9. Andre, D., Koza, J.R.: Parallel genetic programming: a scalable implementation using the transputer network architecture. In: Angeline, P.J., Kinnear Jr., K.E. (eds.) Advances in Genetic Programming, vol. 2. The MIT Press, Cambridge (1996). (Chapter 16)
10. Juille, H., Pollack, J.B.: Parallel genetic programming and fine-grained SIMD architecture. In: Siegel, E.V., Koza, J.R. (eds.) Working Notes for the AAAI Symposium on Genetic Programming, pp. 31–37. MIT, Cambridge (1995)

11. Goldberg, D.E., Smith, R.E.: Nonstationary function optimization using genetic algorithms with dominance and diploidy. In: ICGA, pp. 59–68 (1987)
12. Vekaria, K., Clack, C.: Genetic programming with gene dominance. In: Koza, J.R. (ed.) Late Breaking Papers at the Genetic Programming 1997 Conference, vol. 300. Stanford University Bookstore, Stanford (1997)

Performance Analysis of Collaborative Data Mining vs Context Aware Data Mining in a Practical Scenario for Predicting Air Humidity

Carmen Ana Anton[✉], Anca Avram, Adrian Petrovan, and Oliviu Matei

Electric, Electronic and Computer Engineering Department,
Technical University of Cluj-Napoca, North University Center Baia Mare,
Dr. Victor Babes 62A, 430083 Baia-Mare, Romania
{carmen.anton,adrian.petrovan}@cunbm.utcluj.ro,
anca.avram@ieee.org, oliviu.matei@holisun.com
https://inginerie.utcluj.ro/

Abstract. Predictions in data mining are a difficult process but useful in various areas. The purpose of this article is to make a parallel between the classical data mining process and two new approaches in the process of data mining: collaborative data mining and context-aware data mining. Data gathered from seven meteorological stations in Transylvania served as baseline for the research. Processes for predicting the air humidity were designed and analyzed using the same machine learning algorithms and data. The results obtained prove that collaborative and context-aware data mining approaches bring better results than the standalone approach and highlight some of the algorithms that are more suitable for each approach. The combination of the two notions could be another example of a successful approach for future research.

Keywords: Data mining · Collaborative · Context aware ·
Virtual machine learning

1 Introduction

In a data mining process, the quality of the data used is very important. In reality, such a process does not always have complete and appropriate data, so a new approach is needed. The two new approaches that are the subject of the research in this article are: collaborative data mining (CDM) and context-aware data mining (CADM).

The concept of collaborative data mining, was introduced by Matei et al. in [6], which states that more data sources are used to predict the behavior of one source. The result of the study is that if the sources are correlated, the accuracy raises and if the sources are not correlated, the accuracy decreases.

© Springer Nature Switzerland AG 2019
R. Silhavy et al. (Eds.): CoMeSySo 2019, AISC 1047, pp. 31–40, 2019.
https://doi.org/10.1007/978-3-030-31362-3_5

Context Awareness is a concept that is relatively new and is more and more exploited in the knowledge management research. Dey has provided in [15] a definition of the context that is widely accepted: "Context is any information that can be used to characterize the situation of an entity". Studies performed on this area by Matei et al. in [16] and Avram et al. in [4], prove already that context-aware data mining improves the accuracy of predictions.

In [5] Matei et al. proposes a stratified architecture for data mining process. This is based on simple calculations at the bottom and uses complex algorithms in the upper layer. The proposed multi-layered structure is:

- Local analysis: limited resources, simple calculations.
- Stand-alone complex Data Mining: enough resources, algorithm of virtual machine learning applied, increased complexity.
- Context-aware Data Mining: take into account another data related to source studied.
- Collaborative Data Mining: uses data from another correlated sources, concept use in case of missing data from one source.

The architecture proposed by Matei et al. in [5] is more complex and combines both approaches in a single system. The purpose of this article is to analyze the results obtained when using collaborative data mining (CDM) and context-aware data mining (CADM) approaches, by reporting to the classical data mining approach. The analysis is performed by comparing the results obtained on the same data.

Section 2 presents an overview of the methods used in this article. Section 2.1 describes the collaborative data mining approach, followed by an overview of the context-aware data mining notions in Sect. 2.2. Then, Sect. 2.3 provides information on the experimental setup performed. Section 3 presents the results obtained after running the experiments, while Sect. 4 presents a short overview of the results and future work that could be performed in this area.

2   Methods

2.1   Collaborative Data Mining (CDM)

Data mining collaborative is a method of approaching a machine learning process that involves completing the data of a studied source with data taken from other similar sources. The process offers better results than the one that only uses the data of the studied source. A schematic of the collaborative data mining process can be seen in Fig. 1, where source 2 is used to fill in data and get a prediction for source 1.

The data collaboration system to obtain a prediction was experimented and studied by Anton et al. in [3], comparing its effects on the values obtained in parallel with the process using only the data from a single source. The conclusions state that adapting the algorithms used and optimizing their parameters can achieve much improved results.

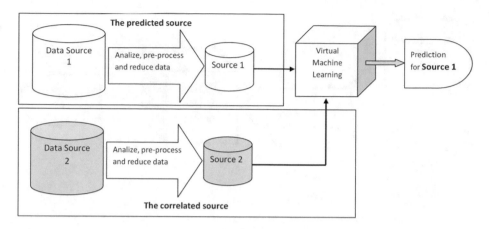

Fig. 1. Proposed of the collaborative data mining process.

Hann et al. in [1] specify that the data we want to analyze through data extraction techniques may be incomplete, noisy and inconsistent, so methods that can cover these discrepancies can yield improved process results. The collaborative method was applied by Stahl et al. in [2] where Pocket Data Mining is a new approach that describes the collaborative exploitation of streaming data in mobile and distributed computing environments. Potential use of mobile data streams were useful in making decisions through exploitation techniques.

Also, in [7], Correia et al. presented a collaborative exploitation framework that allows users with different expertise to analyze data using the results obtained by other researchers. The data mining process has been implemented for the fields of Molecular Biology and Chemoinformatics. As experts from geographically diverse domains use collaborative data exploitation and develop techniques to solve complex problems, the collaborative concept can be used in the case of non-human data sources.

The collaborative data mining process is encountered in experiments that applied the concept for teams of humans that share knowledge and results, such as in [8] by Mladenic et al. or in [9] by Blockeel et al. The concept can be translated into a process of predicting values in which data sources are combined to achieve an increase in the resulting values.

2.2 Context Aware Data Mining (CADM)

Context aware data mining (CADM) has the same steps as the classical data mining approach. Beside these, it comes with a new step consisting in integrating valuable context data in the process. Figure 2 presents an overview of the context-aware data mining process.

Choosing and modeling the context is one of the main concerns when integrating into a context-aware system. Research has been performed on what is the best way to describe the context. It can be modeled in multiple forms, starting

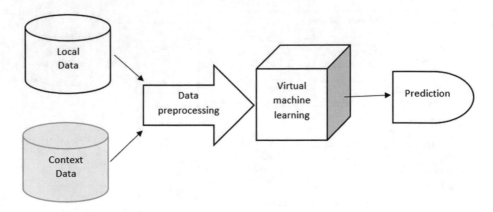

Fig. 2. CADM process overview

from storing information in simple key-value format and going to more complex models like ontology model, object-oriented model or graphical model. Still, the trend of modeling the context in both academic environment and industry seems to be oriented towards using Ontology [10–13].

In the experiments that are the subject of this article, the context was not very complex, hence a simple key-value format was enough for modeling the context. Also, the steps described by Lee et al. in [14], have been followed, namely: (1) the context information was acquired, (2) context was stored, (3) the level of abstraction was analyzed and decided, (4) the context was used in the process for predicting the air humidity.

2.3   Experimental Setup and Implementation

The purpose of this article is to investigate the outcome of applying the two new data mining approaches on the same data. As a baseline, the classical/standalone data mining approach was considered and taken as point of reference. This work comes as a continuation on the investigations done by Anton et al. in [3] and by Avram et al. in [4]. Anton et. al in [3] studied how the application of collaborative data mining techniques, would offer solution for the cases in which one sources do not poses useful data for mining, and the process uses date from another sources correlated, while Avram et al. in [4] analyzed the performance improvements brought by introducing context information in a data mining process, versus the classical approach.

Data Source. Data was downloaded from a public web site[1] that beside offering the service of weather prognosis, it also gives access to the archived meteorological information gathered from weather stations around the globe. Data that served this investigation was collected from seven weather stations, located in

---

[1] https://rp5.ru/.

Podisul Transilvaniei, Romania. The observation interval was chosen between 01.01.2018 and 31.12.2018.

Preprocessing Data. The raw data obtained from the web site contained more entries for each day for a location. These entries had different information gathered from the monitored weather stations (e.g. air temperature, wind direction, wind speed, pressure tendency). The first step that was performed on the obtained data was preprocessing. This consisted in removing unnecessary information (in the context of the experiment) from the initial set of data. The most important information to be considered as source data for the experiment were: the temperature (degrees Celsius) and air humidity (percent) measured at 2 meters above ground.

Data to Be Predicted. The attribute that is the object of study for this article is the air humidity at a specific location. Analysis is performed on predicting the value for this attribute.

Machine Learning Algorithms Chosen. The chosen algorithms for the experiment were: k-Nearest Neighbor model (k-NN), Local Polynomial Regression (LPR), Support Vector Machine (SVM) and Neural Net model (NN). These algorithms were tested both with the collaborative approach, by Anton et al. in [3] and with the context-aware approach by Avram et al. in [4]. The conclusions of the research performed showed that k-NN and LPR plead to better results when applied in a collaborative process, while NN and k-NN lead to the best results in a context-aware process. So, this seem like a good option to start also from the four mentioned algorithms.

Tools and Techniques. In order to model the data and apply the machine learning algorithms, Rapid Miner tool was used. For each location there were created separate processes for the standalone, collaborative and context-aware data mining techniques.

Way of Working. First we ran the process using the standalone data mining approach and registered the results for each algorithm and location. Then we ran the collaborative and context-aware processes and registered the obtained values. Conclusions were drawn based after comparing the results for all three types of techniques.

3   Results

The standalone processes were run and results were registered for each chosen location and machine learning algorithm. The created collaborative processes were run and results were compared with the standalone approach as described in Sect. 3.1. Also, the context-aware processes results were compared with the standalone approach as presented in Sect. 3.2. The value that was monitored in terms of results represents the prediction trend accuracy. This is a metric that measures the average amount of times a regression prediction was able to correctly predict the trend of the regression.

3.1   Collaborative Data Mining (CDM) Results

In the context of the collaborative data mining technique, processes were created and applied for all the pairs of sources among the 7 selected. Thus, for the 4 algorithms, 6 moisture prediction values were obtained for each station, number final for values obtained was 168.

Table 1 presents the obtained results for each algorithm used. The values presented for each weather station are: result for the standalone process; minimal and maximal result, obtained when running the CDM process; and also the average value for each studied source.

Table 1. Collaborative vs standalone data mining results

		S1	S2	S3	S4	S5	S6	S7
k-NN	Standalone	0.601	0.598	0.601	0.596	0.596	0.579	0.576
	Collaborative minim	0.571	0.554	0.565	0.584	0.576	0.571	0.573
	Collaborative maxim	0.648	0.612	0.620	0.632	0.593	0.607	0.618
	Collaborative average	0.612	0.581	0.586	0.611	0.584	0.584	0.587
		S1	S2	S3	S4	S5	S6	S7
SVM	Standalone	0.670	0.634	0.640	0.659	0.659	0.648	0.645
	Collaborative minim	0.651	0.634	0.632	0.665	0.648	0.645	0.626
	Collaborative maxim	0.676	0.654	0.657	0.687	0.662	0.657	0.651
	Collaborative average	0.660	0.647	0.644	0.681	0.653	0.654	0.638
		S1	S2	S3	S4	S5	S6	S7
NN	Standalone	0.643	0.626	0.626	0.645	0.634	0.640	0.665
	Collaborative minim	0.648	0.612	0.634	0.643	0.632	0.626	0.620
	Collaborative maxim	0.665	0.637	0.659	0.668	0.659	0.654	0.654
	Collaborative average	0.657	0.623	0.645	0.655	0.645	0.638	0.638
		S1	S2	S3	S4	S5	S6	S7
LPR	Standalone	0.640	0.573	0.562	0.598	0.551	0.573	0.593
	Collaborative minim	0.568	0.565	0.546	0.596	0.565	0.584	0.568
	Collaborative maxim	0.620	0.618	0.640	0.637	0.626	0.620	0.620
	Collaborative average	0.603	0.590	0.589	0.620	0.601	0.601	0.593

The conclusions that appear in Table 1 are:

- each source has a value obtained collaborative that is higher than standalone one.
- the intervals of variation of the collaborative values are as follows: k-NN [−0.003, 0.047], LPR [−0.020, 0.078], SVM [0.003, 0.028], NN [−0.011, 0.033].

Number of values obtained collaborative compared with standalone can be seen in Table 2. For six of the seven studied sources there is at least one algorithm of those applied that has a 100% success rate, ie all the collaborative values are higher than the standalone ones. Variation depends on the algorithm used and the source that is used as a pair in the collaborative process.

Table 2. Number of higher values obtained collaboratively from standalone

	S1	S2	S3	S4	S5	S6	S7
k-NN	4	1	1	5	0	4	4
LPR	0	5	5	5	6	6	3
SVM	1	6	4	6	1	5	1
NN	6	2	6	5	5	2	0

3.2   Context-Aware Data Mining (CADM) Results

Table 3 presents the CADM results for each algorithm and observed weather station, beside the standalone data mining results.

Table 3. Context-aware vs standalone data mining results

		S1	S2	S3	S4	S5	S6	S7
k-NN	Standalone	0.601	0.598	0.601	0.596	0.596	0.579	0.576
	Context-aware	0.699	0.678	0.703	0.691	0.666	0.705	0.667
		S1	S2	S3	S4	S5	S6	S7
SVM	Standalone	0.670	0.634	0.640	0.659	0.659	0.648	0.645
	Context-aware	0.653	0.625	0.634	0.653	0.626	0.644	0.617
		S1	S2	S3	S4	S5	S6	S7
NN	Standalone	0.643	0.626	0.626	0.645	0.634	0.640	0.665
	Context-aware	0.667	0.633	0.638	0.666	0.637	0.672	0.635
		S1	S2	S3	S4	S5	S6	S7
LPR	Standalone	0.640	0.573	0.562	0.598	0.551	0.573	0.593
	Context-aware	0.652	0.626	0.626	0.631	0.640	0.636	0.620

It can be seen that the context aware approach produces better results for k-NN, NN and LPR. For SVM instead, the resulted values are close to the ones obtained in the standalone process, but none is better than the standalone process. This confirms the same finding done by Avram et al. in [4] where, for predicting the soil moisture, the SVM algorithm did not produce notable improvements. It seems to be a step forward towards confirming the fact that SVM is not a suitable choice to be made when choosing machine learning algorithms for predicting time-series, in a context-aware system.

3.3   CDM vs CADM Results

Table 4 presents a summary of the results for both of the new data mining techniques (collaborative and context-aware) versus the standalone data mining approach. For each machine learning algorithm and technique are presented: the percentage of values higher than in the standalone process and the value range difference of the results obtained using the technique versus standalone.

Table 4. CDM vs CADM results related to standalone data mining

	CDM	Value range for diff CDM-STD	CADM	Value range for diff CADM-STD
k-NN	85.71%	[−0.003, 0.047]	100.00%	[0.070, 0.126]
SVM	100.00%	[0.003, 0.028]	0.00%	[−0.033, −0.004]
NN	85.71%	[−0.011, 0.033]	85.71%	[−0.030, 0.032]
LPR	85.71%	[−0.020, 0.078]	100.00%	[0.012, 0.089]

The conclusions that can be extracted from data in Table 4 are:

- CDM produces improved results for each of the studied machine learning algorithms, while CADM only produces better results for k-NN, NN and LPR algorithms.
- Applying SVM algorithm for predicting air humidity, in CADM produces results closer to the standalone baseline, but none better, while applying CDM produces better results for all situations. This qualifies SVM as a good candidate in a collaborative architecture, while in a context-aware system it does not qualify as an option.
- The most notable improvement (20%) in terms of predicted trend are brought by the CADM process, when using k-NN algorithm.
- There is no significant distinction between the results obtained when using NN or LPR algorithms for either of the techniques.

4   Conclusions

The article studies the results obtained by using the collaborative and context-aware data mining approaches for predicting the air humidity in seven locations and using four different machine learning algorithms.

The research performed showed that both methods bring improvements in the result prediction. The CDM approach provided better results than CADM for NN, while CADM provided better results than CDM when applying k-NN.

A future direction of research could be a combined collaborative and context aware process. Approaches to such a data mining process can be multiple, as follows:

- the same algorithms approached for CDM and CADM in which the prediction of a source is based on context-aware data and data from collaborative sources;
- a combined CDM and CADM process in which a different algorithm is used for each approach, depending on the results obtained in this article. This could be a CDM and CADM data mining process that uses k-NN for context-aware and collaborative SVM.
- different layers of one system, each approached with the most suitable technique, depending on the information available at hand, meaning that if context is available then CADM is used, if there are more sources of information, then it might be that CDM is the best candidate.

References

1. Han, J., Pei, J., Kamber, M.: Data Mining: Concepts and Techniques. Elsevier, Amsterdam (2011)
2. Stahl, F., et al.: Pocket data mining: towards collaborative data mining in mobile computing environments. In: 2010 22nd IEEE International Conference on Tools with Artificial Intelligence, pp. 323–330. IEEE (2010)
3. Anton, C., Matei, O., Avram, A.: Collaborative data mining in agriculture for prediction of soil moisture and temperature. In: Advances in Intelligent Systems and Computing (to appear)
4. Avram, A., Anton, C., Matei, O.: Context-aware data mining vs classical data mining: case study on predicting soil moisture. In: Advances in Intelligent Systems and Computing (to appear)
5. Matei, O., et al.: Multi-layered data mining architecture in the context of internet of things. In: 2017 IEEE 15th International Conference on Industrial Informatics (INDIN), pp. 1193–1198. IEEE (2017)
6. Matei, O., et al.: Collaborative data mining for intelligent home appliances. In: Working Conference on Virtual Enterprises, pp. 313–323. Springer, Cham (2016)
7. Correia, F., Camacho, R., Lopes, J.C.: An architecture for collaborative data mining. In: KDIR 2010-Proceedings of the International Conference on Knowledge Discovery and Information Retrieval (2010)
8. Mladenic, D., et al. (eds.): Data Mining and Decision Support: Integration and Collaboration. Springer, Heidelberg (2003)
9. Blockeel, H., Moyle, S.: Collaborative data mining needs centralised model evaluation (2002)
10. Kotte, O., Elorriaga, A., Stokic, D., Scholze, S.: Context sensitive solution for collaborative decision making on quality assurance in software development processes. In: Intelligent Decision Technologies: Proceedings of the 5th KES International Conference on Intelligent Decision Technologies (KES-IDT 2013), vol. 255, p. 130. IOS Press (2013)

11. Scholze, S., Barata, J., Stokic, D.: Holistic context-sensitivity for run-time optimization of flexible manufacturing systems. Sensors **17**(3), 455 (2017)
12. Scholze, S., Stokic, D., Kotte, O., Barata, J., Di Orio, G., Candido, G.: Reliable self-learning production systems based on context aware services. In: 2013 IEEE International Conference on Systems, Man, and Cybernetics (SMC), pp. 4872–4877. IEEE (2013)
13. Vajirkar, P., Singh, S., Lee, Y.: Context-aware data mining framework for wireless medical application. In: International Conference on Database and Expert Systems Applications, pp. 381–391. Springer, Heidelberg (2003)
14. Lee, S., Chang, J., Lee, S.: Survey and trend analysis of context-aware systems. Inf. Int. Interdisc. J. **14**(2), 527–548 (2011)
15. Dey, A.K.: Understanding and using context. Pers. Ubiquit. Comput. **5**(1), 4–7 (2001)
16. Matei, O., Rusu, T., Bozga, A., Pop-Sitar, P., Anton, C.: Context-aware data mining: embedding external data sources in a machine learning process. In: International Conference on Hybrid Artificial Intelligence Systems, pp. 415–426. Springer, Cham (2017)

Structuring a Multicriteria Model to Optimize the Profitability of a Clinical Laboratory

João Ricardo Capistrano, Plácido Rogerio Pinheiro[(✉)],
Jorcelan Gonçalves da Silva Júnior, and Plínio de Sousa Nogueira

University of Fortaleza, Fortaleza, Brazil
ricardocapistrano3.0@gmail.com,
pliniosousanogueira@gmail.com, jocerlan@gmail.com,
placido@unifor.br

Abstract. Optimizing the profitability of a business is a challenge that involves considering multiple internal and external variables to organizations. Based on a multi-criteria approach, this study aims to propose a model that allows the ordering of significant factors to optimize the profitability of a clinical laboratory. In methodological terms, this study was conducted in two phases. The first phase sought to develop a brainstorming and cognitive mapping in a clinical laboratory located in Acarau, Ceara. The second phase included multi-criteria analysis, based on the MACBETH methodology, from surveys answered by twenty laboratory managers in Ceara state. By employing the M-MACBETH software to analyze data, satisfactory results were obtained as regards the profitability optimization model of laboratories. Therefore, this study concludes that the computational results were promising.

Keywords: Multi-criteria analysis · Clinical laboratory · Macbeth

1 Introduction

Given the economic and political instabilities in Brazil, clinical laboratories have been searching for alternatives to reduce their costs. They aim to remain active without compromising the quality of the services provided to customers, presenting improvements, optimizing internal processes, and investing in technology and continuous training of employees, amid other actions. The current scenario of clinical laboratories in the country is becoming competitive. The optimization of profits; the strict control, which refers to the management of costs; the difficulties encountered in the macroeconomic environment; sectoral regulations; and the market concentration by large groups have created an atmosphere of complexity for the laboratories.

In the face of a situation in which the market determines the cost price, the only way organizations remain in operation is to improve their processes to preserve their profitability. It is of great importance for companies to adopt methodologies, tools, and techniques to achieve the goals constituted by the company's strategy [1]. Nowadays, with technological, scientific, and managerial development, clinical analyses have witnessed changes. The need to adapt to this new reality has led laboratories to seek some solutions. The constant changes, which take place in the economic, political, and

© Springer Nature Switzerland AG 2019
R. Silhavy et al. (Eds.): CoMeSySo 2019, AISC 1047, pp. 41–51, 2019.
https://doi.org/10.1007/978-3-030-31362-3_6

social scenario of Brazil, increase complexity and competition in the sector. This process demands measures so that organizations adopt new standards allowing for adaptation to the changes, constant improvement of their operations, increase in productivity, development of their efficiency, reduction of their costs, and optimization of their decision-making process [2].

The challenge, however, lies in the fact that many managers are not able to define which factors directly affect their business. Given this context, it is challenging to act when there is no clear definition of the decision-making scenario. Thus, it is crucial for decision-makers to outline a structuring model that favors the achievement of the company goals, aiming at its profitability and, therefore, stability.

It is noteworthy that the issue inherent this sector is not revealed, in a proper way, as structured, through the construction of a multi-criteria evaluation prototype, which outlines the ordering of actions. Besides, the profitability of these companies needs to be increased. It is necessary to develop a multi-criteria model that allows structuring, defining, and ordering the concepts and factors that can support management in maximizing profits. With that said, this study asks the following research question: How to prioritize factors that can support the decision-making process of managers to optimize the profitability of a clinical laboratory? The choice of this research topic is justified by the fact that scholarly research on the use of multi-criteria methodologies in clinical laboratories, conducted to support managerial decision-making, is scarce. Indeed, most studies are focused on specific operations and processes due to the global managerial vision.

This study attempts to present a prototype of analysis to improve the effectiveness and the process of decision-making through the application of cognitive mapping. It aims to define the predominant elements, which influence the solution of the problem raised, proposing, with the application of the Macbeth methodology, an ordering of relevant criteria, according to the perception of laboratory managers, when evaluating the optimization of profitability in clinical laboratories. This study hopes that the proposition of a multi-criteria methodology might contribute improving the decision-making process, making an effort to obtain the results sought by the organization efficient, and hence defining the ordering of established criteria, evaluated by laboratory managers. In methodological terms, this study was conducted in two phases. The first phase encompassed a survey of information in a laboratory located in Acarau, Ceara. In this regard, a cognitive mapping was developed, and a selection of 25 criteria was divided into five clusters: Quality, Management, Acquisition and Supply, Technology, and Human Resources. The second phase included multi-criteria analysis, based on the MACBETH methodology, from surveys answered by 20 laboratory managers in Ceara state.

2 Theoretical Reference

2.1 Multi-criteria Support in Decision-Making

By seeking improvements in decision-making, organizations have sought methods that safeguard the development of their decision from acquired knowledge. The decision-making process encompasses several factors and is intrinsically complex. In organizational environments, where one can see a multiplicity of aspects that correlate, this technique becomes obscure. Thus, it is necessary to properly use elaborated techniques for structuring the problems to be solved. In this respect, there is Multi-Criteria Decision Support, or Multi-Criteria Analysis (MCA), a tool that can be widely used in the decision-making of organizations [3]. Over the last four decades, a new approach, called Multiple Criteria Decision Analysis (MCDA), has stood out in scholarship and business. This approach focuses on the study and treatment of decision problems, considering many criteria, which encompass a decision-making process [4]. Most importantly, decision-making is a human activity in which managers, through value judgment, play a significant role in the choice and attractiveness of possibilities for organizational decisions [5].

The MCDA methods insert the decision-maker at the center of the process and provide several techniques to find a solution, considering the particular knowledge involved in the decision, not representing automated methods that constantly cause the same solution for each decision-maker [6].

Accordingly, one can observe that its application is comprehensive, dealing with strategic and tactical decisions in a diversified way, and relying on mathematics, management, computer science, psychology, social sciences, and economics. A large number of multi-criteria methods, including software and applications, are available, as well as the continuous increase of publications related to MCDA [6]. Pointed out that the decision-maker has the challenge of selecting the most appropriate method to support the process since they are not perfect and cannot be applied in all situations. There remains the need for weighing the limitations, peculiarities, eventualities, hypotheses, propositions, and perspectives of each of these methods.

It is noteworthy that the decision-making process affects the results of organizations, consisting of a cognitive process in which a plan of action is chosen to the detriment of others, considering key success factors such as the environment, scenario, and analysis. The decision-making is related to the choice of the appropriate path for companies [7, 19].

Stressed that the choice of the specific method depends on the type of problem being analyzed; the context under study; the defined decision-makers; the procedures of comparison; the alternatives; and the solutions that one aims to obtain [8].

2.2 Management in Clinical Laboratories

Clinical laboratories perform laboratory tests and medical specialty practice called clinical pathology. Their main attribute is to provide data to aid in the diagnosis, prevention, treatment, and the assessment of human health [9]. Pointed out that the development of technology in the field of clinical laboratories has resulted in a considerable increase in the quantity and type of substances that can be analyzed, in the medical opinion, and the taking of therapeutic behaviors [10, 18].

The increasing complexity of clinical laboratory services, leveraged by technological development and the addition of current knowledge, has raised expenditures accordingly [11]. The new technologies have led to the implementation of modern quality processes in laboratories. Nevertheless, such innovations led to an increase in operating costs, which require laboratory managers to improve management [11].

3   Methodology

In methodological terms, this study was conducted in two phases. The first phase included a meeting held with laboratory managers and employees, located in Acaraú (Ceará), to define the management problem. A brainstorming session was held with participants, identifying their key five strategic areas, considered to solve the problem raised.

The ideas were recorded, and a discussion was carried out so that participants could discuss the opinions presented, leading to the definition of common sense about the proposed problem. Following the identification of the strategic areas, a hierarchical tree of value was structured, obtaining managers' key viewpoints. Subsequently, descriptors were defined. The participants were reunited one last time and, based on the problem raised, drew a cognitive map.

The second phase consisted of conducting a survey based on the identified points of view, which included five clusters with five sub-criteria, totalizing 25 criteria. The sample encompassed 20 managers of clinical laboratories in Ceará state. In the evaluation phase, managers and decision-makers were asked to define the degree of importance or attractiveness of the criteria, as proposed in the multi-criteria methodology. It is worthwhile to note that a five-point Likert scale was used to optimize the work, ranging from 1 (no importance) to 5 (extreme importance).

The collected data were treated with M-Macbeth software. Based on the Macbeth methodology, the software established an index, gathering all the criteria of the problem through the analysis and attribution of weights, preferences, and scales, as well as implementing a decision tree and listing a sequence of actions. The definition of the criteria is one of the most delicate parts in the formulation of decision problems, and values are assigned according to a preferred structure of the decision-maker [12, 20].

According to highlights, the proposal to optimize the subjectivity of the decision fits perfectly, by turning the qualitative judgment into an array of judgments and cardinal value scales for the criteria. When using the M-Macbeth parameterization to decide which problem should be treated as a priority, cardinal values and graphs are generated to support the decision-makers [13].

The application of the MACBETH method came from its acceptance by decision-makers, who showed that the issues that were exposed made sense to them. Thus, they were confident in responding to the items. It was necessary to evaluate the acceptance of data, the properties used by the method, as well as to verify whether the result supported the decision-making process. This study also observed and considered ancillary questions such as the existence of computational tools as they allow greater integration with the problem addressed.

4 Results and Analysis

Profits are essential for the survival of a private company. Companies have the prerogative of an endless search for the improvement of their management, aiming at positive financial results that consolidate them. With this perspective in mind, it is of paramount importance to develop a multi-criteria model to optimize the profitability of a clinical laboratory. The model offers managers subsidies to support the decision-making process, and thus the actions to reach the desired goals.

This study took into account highlights, which suggested three phases: structuring, evaluation, and recommendation [14, 16] (Fig. 1).

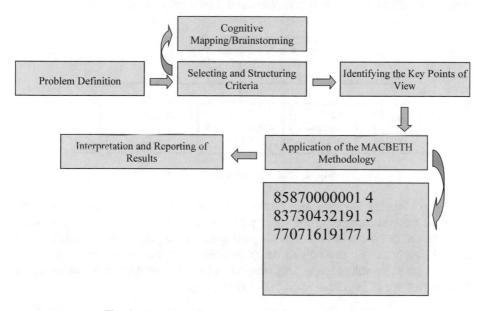

Fig. 1. Multi-criteria support model in decision-making.

Initially, a meeting was held with laboratory managers and employees to define the problem. Subsequently, the key five strategic areas were identified and considered to solve the problem raised. The areas are as follows: Human Resources, Information Technology, Acquisition and Supply, Quality and Management.

After defining the environment in which the problem was established, the next step consisted of identifying the actors involved. In this regard, it was crucial to determine who, in the laboratory, had the power to decide and modify the current situation (the decision-maker); who directly interfered in the process; who was the facilitator; and who suffered the consequences of the decisions taken. Then, it was necessary to observe the support to the decision-maker to identify the information that should be included, starting with the evaluation of the problem. A brainstorming was conducted with the laboratory manager and staff [21, 22]. They were encouraged to speak freely about the problem. Freedom was assessed so that participants could express their opinion and their values according to their specific areas. The next step introduced a series of items to stimulate the emergence of ideas. It is noteworthy that participants had autonomy in answering the items, and brainstorming took place, spontaneously. The facilitator did not make any value judgment on the problem raised. Indeed, the facilitator encouraged the participants to express their ideas promptly, and they had no previous judgments or any coercive pressure from their peers.

These observations led to the identification of the Primary Evaluation Elements (EPA), considered significant to optimize the profitability of the laboratory. The next step consisted of determining the actions that would best achieve the established goals. This study structured a hierarchical tree of value, according to the obtained concepts, which considered the key points of view and the model itself (Fig. 2).

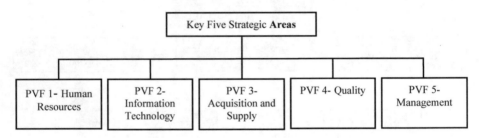

Fig. 2. The hierarchical tree of value.

By considering the structuring of the hierarchical tree of value, the last step involved the construction of the descriptors. Based on the definition of the central problem and the goals to be achieved, participants were once again gathered. In this respect, they were informed about the methodology for the constitution of and stimulated to create cognitive maps. This study used Decision Explorer software to support the construction of cognitive maps [15, 17] (Fig. 3).

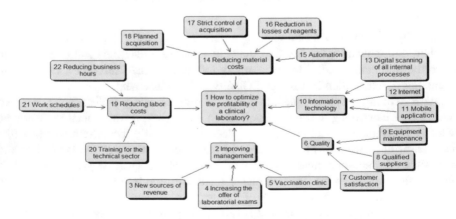

Fig. 3. Elaborated cognitive map (Source: Own elaboration).

Table 1. Weighting of the criteria generated by the MACBETH (Source: Own elaboration based on research data, 2019).

	Current scale	Anchored MACBETH	Basic MACBETH
Continuous education	8.36	8.36	43.00
Access to results and printing online	8.16	8.16	42.00
Identifying the feasibility of new sources of revenue	7.57	7.57	39.00
Digital management of processes	7.38	7.38	38.00
Control of equipment maintenance	6.80	6.80	35.00
Increasing the number of locations which provide laboratory services	6.02	6.02	31.00
Planned acquisition	5.63	5.63	29.00
Quality reagents	5.63	5.63	29.00
Reduction in costs	5.44	5.44	28.00
Incorporating diagnostic imaging service	5.05	5.05	26.00
Awareness of rational use of resources	4.85	4.85	25.00
Automation solutions	4.08	4.08	21.00
Reduction in losses of reagents	3.88	3.88	20.00
Customer loyalty program	3.88	3.88	20.00
Outsourcing of the material transportation	3.69	3.69	19.00
Laboratory application	2.91	2.91	15.00
Human resources	2.72	2.72	14.00
Intermittent working day	1.75	1.75	9.00
Assessment of methodologies	1.75	1.75	9.00
Notification of results	1.55	1.55	8.00
Setting up vaccination clinics	1.55	1.55	8.00
Survey of the prices of reagents	0.59	0.59	3.00
Online Pre-registration	0.39	0.39	2.00
Business hours	0.19	0.19	1.00
Pet examinations	0.19	0.19	1.00
Lower level	0.00	0.00	0.00

The descriptors were checked for each key point of view of the hierarchical tree of value. For the elaboration of each descriptor, this study considered the characteristics of the points of view, as well as the impact levels and the combination between them. Besides, it was established the reference levels, upper and lower, which represented the impact on the company's results. It is noteworthy that the descriptors encompass a structure with two levels of reference impact: the first has a significant effect, and the second has no important impact. The employees and managers pointed out twenty-five qualitative, indirect, and discrete descriptors.

By considering the M-Macbeth software as a multi-criteria tool, the scales were generated and analyzed under the light of the MACBETH method. Therefore, this allowed for using the model in the decision-aid process that employs an additive aggregation function. Given the literature and the context of clinical laboratories, all 25 criteria were registered and grouped into five broader clusters related to the operational and administrative activities of clinical laboratories: Human Resources (5 criteria), Technology (5 criteria), Acquisition and Supply (5 criteria), Quality (5 criteria), and Management (5 criteria). Each researched decision-maker refereed the difference in attractiveness between two actions presented for comparison. According to the answers of the decision-maker, matrices of judgments were elaborated and whose values are used as inputs in the software. Thus, it was verified the consistency of these judgments, and the value function was determined, providing the cardinal measurement of the descriptors.

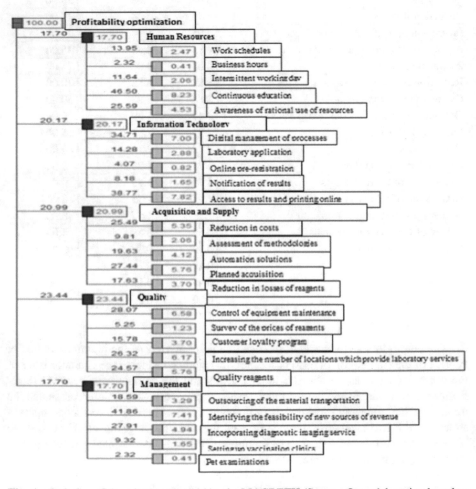

Fig. 4. Ordering of the criteria generated by the MACBETH (Source: Own elaboration based on research data, 2019).

Subsequently to the process of obtaining the matrix and the definition of the local scales (Current scale), the MACBETH proposes a score for each option anchored at the two reference levels (Macbeth basic). Additionally, a range was developed by imposing the values of the reference levels as 100 and 0, respectively (Macbeth transformed), as Table 1 (Fig. 4).

It is worthwhile to note that the most significant sub-criterion in the cluster Human Resources was "continuous education". This is due to the importance of the development of continuous education, which contributes to the improvement of the quality of services and the optimization of routines in an effective way. Such process bolsters a culture of efficiency and enhances the development of the laboratory. As the cluster Information Technology is concerned, the outstanding sub-criterion was "access to results and printing online". Such evidence has to do with the competitive differential for the company as it speeds up the availability of the results to customers. In other words, customers do not need to leave home to obtain the results.

As regards the cluster Acquisition and Supply, the most significant sub-criterion for optimizing profitability was "planned acquisition". In the cluster Quality, it is noteworthy that "control of equipment maintenance," "increasing the number of locations which provide laboratory services", and "quality reagents" were the predominant sub-criteria. It is worth highlighting that the cluster Quality presented the highest score, observing the aggregate result of all clusters. The sub-criterion "identifying the feasibility of new sources of revenue" stood out in the cluster Management.

5 Conclusion

Based on a multi-criteria analysis, this study attempted to propose a model that allows the ordering of significant factors to optimize the profitability of a clinical laboratory, employing the Macbeth methodology. As shown in Sect. 4, the research aim was fulfilled. Following the application of brainstorming and cognitive mapping techniques, this study structured the management problem and managers' key viewpoints, also defining their respective descriptors. The criteria were defined and grouped into five clusters: Human Resources (5 criteria), Information Technology (5 criteria), Acquisition and Supply (5 criteria), Quality (5 criteria), and Management (5 criteria). By employing the Macbeth methodology, findings showed that the Quality cluster is the most relevant among the others and the sub-criteria "continuous education". The latter composes the cluster Human Resources and is the most relevant in this regard.

In terms of research limitations, the sample of this study focuses rather narrowly on 20 laboratory managers located in Ceara state. Future research could investigate such quantitative aspect in more detail, hence applying the Macbeth methodology in other Brazilian states and economic sectors.

References

1. Bana e Costa, C.A., Corte, J.M., de Vansnick, J.C.M.: Int. J. Inf. Technol. Decis. Making **11** (2), 359–387 (2012)
2. Bana e Costa, C.A., Angulo Meza, L., Oliveira, M.D.: The MACBETH method and application in Brazil. Engevista **15**(1), 3–27 (2013)
3. Banxia Software. http://www.banxia.com
4. Farias, F.E.S., Nascente, I.M.C., Lima, M.C.A., Pinheiro, P.R., Fornari, T.: Towards the methodology for the implementation of sales: forecast model to sustain the sales and operations planning process. In: Silhavy, R., Silhavy, P., Prokopova, Z. (eds.) CONFERENCE 2018. LNCS, vol. 859, pp. 296–306. Springer, Cham (2018)
5. Hair, J.F., Black, B., Babin, B., Anderson, R.E., Tatham, R.L.: Multivariate Data Analysis, 6th edn. Bookman, Porto Alegre (2009)
6. Ishizaka, A., Nemery, P.: Multi-criteria Decision Analysis: Methods and Software. Wiley, Chichester (2013)
7. Leite, M.L.S., Pinheiro, P.R., Simão Filho, M., Araújo, M.L.S.: Ordering assertive strategies for corporate travel agencies: verbal decision analysis model. In: Silhavy, R., Silhavy, P., Prokopova, Z. (eds.) Advances in Intelligent Systems and Computing, vol. 859, pp. 374–384. Springer, Cham (2018)
8. Manzo, B.F., Brito, M.J.M., Corrêa, A.R.: Implications of the hospital accreditation process in the daily life of health professionals. Mag. Nursing USP São Paulo **46**(2), 388–394 (2012)
9. Mendes, M.E.: Evaluation of the implementation of a quality system in a public clinical laboratory. Master thesis, Master Program in Applied in Medicina, University of São Paulo (1998)
10. Morais, D.C., Almeida, A.T.: Modelo de decisão em grupo para gerenciar perdas de água. Rev. Oper. Res. **26**(3), 567–584 (2006)
11. Pinheiro, R.P., Souza, G.G.C., Castro, A.K.A.: Structuring the multicriteria problem for newspaper production. Operacional Res. **28**(2), 203–216 (2008)
12. Plebani, M.: Charting the course of medical laboratories in a changing environment. Clinica Chimica Acta **319**(2), 87–100 (2002)
13. Plebani, M.: Appropriateness in programs for continuous quality improvement in clinical laboratories. Clinica Chemical Acta **333**(2), 131–139 (2003)
14. Silva, C.F.G., Nery, A., Pinheiro, P.R.: A multicriteria model in information technology infrastructure problems. Procedia Comput. Sci. **91**, 642–651 (2016)
15. Silva, C.F.G., Pinheiro, P.R., Barreira, O.: Multicriteria problem structuring for the prioritization of information technology infrastructure problems. In: Silhavy, R., Silhavy, P., Kopova, Z.P. (eds.) Applied Computational Intelligence and Mathematical Methods, vol. 662, pp. 326–337. Springer, Cham (2017)
16. da Silva, O.B., Holanda, R., Pinheiro, P.R.: A decision model for change management based on multicriteria methodology. In: 8th International Workshop on Business-driven IT Management (BDIM), Proceedings of the 8th International Workshop on Business-driven IT Management (BDIM), pp. 1241–1244 (2013)
17. Pinheiro, P.R., de Souza, G.G.C.: A multicriteria model for production of a newspaper. In: The 17th International Conference on Multiple Criteria Decision Analysis. The 17th International Conference on Multiple Criteria Decision Analysis. British Columbia: Simon Fraser University, vol. 17, pp. 315–325 (2004)
18. Tamanini, I., Castro, A.K.A., Pinheiro, P.R., Pinheiro, M.C.D.: Verbal decision analysis applied on the optimization of Alzheimer's disease diagnosis: a study case based on neuroimaging. Adv. Exp. Med. Biol. **696**(7), 555–564 (2011)

19. Nunes, L.C., Pinheiro, P.R., Pequeno, T.C., Pinheiro, M.C.D.: Toward an applied to the diagnosis of psychological disorders. Adv. Exp. Med. Biol. **696**(7), 23–31 (2011)
20. Simão Filho, M., Pinheiro, P.R., Albuquerque, A.B.: Task allocation in distributed software development aided by verbal decision analysis. In: Silhavy, R., Senkerik, R., Oplatkova, Z. K., Silhavy, P., Prokopova, Z. (eds.) Advances in Intelligent Systems and Computing, vol. 465, 1st edn, pp. 127–137. Springer, Berlin (2016)
21. Barbosa, P.A.M., Pinheiro, P.R., de Vasconcelos Silveira, F.R., Simão Filho, M.: Applying verbal analysis of decision to prioritize software requirement considering the stability of the requirement. In: Advances in Intelligent Systems and Computing, vol. 575, 1st edn, pp. 416–426. Springer, Cham (2017)
22. Vasconcelos, M.F., Pinheiro, P.R., Simao Filho, M.: A multicriteria model applied to the choice of a competitive strategy for the printed newspaper. In: Silhavy, R., Silhavy, P., Prokopova, Z. (eds.) Cybernetics Approaches In Intelligent Systems, 1edn, vol. 1, pp. 206–215. Springer, Cham (2017)

Discharge of Lithium–Oxygen Power Source: Effect of Active Layer Thickness and Current Density on Overall Characteristics of Positive Electrode

Yurii G. Chirkov[1], Oleg V. Korchagin[1], Vladimir N. Andreev[1], Vera A. Bogdanovskaya[1(✉)], and Viktor I. Rostokin[2]

[1] A.N. Frumkin Institute of Physical Chemistry and Electrochemistry, Russian Academy of Sciences, Moscow, Russia
bogd@elchem.ac.ru
[2] National Research Nuclear University (Moscow Engineering Physics Institute), Moscow, Russia

Abstract. A distinctive feature of discharge of the lithium–oxygen power source (LOPS) with nonaqueous electrolyte is the filling of the positive electrode pores by lithium peroxide that is not soluble in the electrolyte and is characterized by low conductivity. Generally, the cathodic discharge process can be carried out only in a comparatively thin, several tens of micrometers, porous layer bordering on the gas phase. Therefore, the capacity per 1 cm^2 of the outer cathode surface proves to be small. In this connection, the problem arises of developing more advanced LOPS and providing efficient performance of the active layers of the positive electrode at an increase in their thickness to achieve higher overall characteristics. In this work, the authors obtain experimental dependences of the positive electrode capacity on the active layer thickness and various current density values. Theoretical analysis of the obtained experimental data is performed. Here, the issues that are of considerable interest of the LOPS discharge theory are discussed.

Keywords: Lithium–oxygen battery cathode · Discharge ·
Charging · Computer simulation · Monoporous cathode theory ·
Oxygen consumption constant k

1 Introduction

The concept of lithium-oxygen (lithium-air) batteries was proposed in the second half of the last century in the United States. However, the first successful lithium-air current source with aprotic electrolyte was tested only in 1996, and its cyclability was demonstrated only 10 years later. Thus, the study of this type of current sources, the object is extremely complex, in fact, has just begun. Here it is also appropriate to note that the creation of new power sources in general is much slower compared, for example, with the development of electronics or vehicles, for which they are largely created. Thus, fuel cells have been developed for more than 150 years, lithium-ion systems were studied for 35 years before the first commercialization in 1991.

© Springer Nature Switzerland AG 2019
R. Silhavy et al. (Eds.): CoMeSySo 2019, AISC 1047, pp. 52–69, 2019.
https://doi.org/10.1007/978-3-030-31362-3_7

One of the most promising trends for development of lithium power sources (LPS) is development of lithium–oxygen (air) batteries (LOB) and primary cells (LOC), with their power density much exceeding the characteristics of state–of–the–art lithium–ion batteries [1]. However, the implementation of theoretical advantages of the Li–O_2 system is complicated due to a number of factors, among which the most important ones are transport and ohmic losses on the positive electrode.

The characteristic feature of the discharge process of lithium–oxygen power sources (LOPS) with nonaqueous electrolyte is the filling of the pores of the positive electrode insoluble in the electrolyte and nonconducting lithium peroxide (Li_2O_2) [2, 3], the final reaction product in the course of oxygen reduction [4, 5]:

$$O_2 + \bar{e} \rightarrow O_2^-$$
$$O_2^- + Li^+ \rightarrow LiO_2 \tag{1}$$
$$2LiO_2 \rightarrow Li_2O_2 + O_2$$

Lithium peroxide can be oxidized by application of anodic current, which allows considering LOPS with nonaqueous electrolyte not only as a primary cell, but also as a promising battery. When LOBs are charged, oxygen is generated on the anode. Namely, lithium peroxide accumulated in the positive electrode (cathode) pores is decomposed with formation of lithium ions, oxygen molecules, and electrons:

$$Li_2O_2 \rightarrow 2Li^+ + O_2 + 2e \tag{2}$$

At present, most researchers relate the transition from primary cells to LOB to application of aprotic electrolytes, which would provide the stability of the lithium anode. As opposed to aqueous electrolytes that allow preserving an extended system of gas pores in the oxygen electrode, the aprotic electrolyte wholly floods the electrode pore space. Under these conditions, supply of O_2 to active surface centers accessible for electrons is limited by the diffusion coefficient of oxygen, its solubility, and also solvent viscosity and composition of electrolyte [6].

Ohmic losses of the Li–O_2 system are due not only to the resistance of the electrolyte, but also to the extremely low conductivity of lithium peroxide deposited in the electrode in the course of the discharge. Accumulation of peroxide also results in complication of oxygen transport into the bulk of the positive electrode.

The capacity of LOB determined by the Li_2O_2 amount accumulated during the discharge is the main criterion that serves for comparative assessment of active materials. Most of the authors are using the capacity value expressed in mAh per mass of the material. Here, thin active layers are generally discussed, which provides high utilization of the electrode pore space and obtaining the capacity of up to 50000 mAh g^{-1} [7]. However, a more important characteristic is the capacity per geometric surface area of the electrode. At present, the best parameters of LPS with nonaqueous electrolyte are ~ 30 mAh cm^{-2} [8], which is considerably inferior to the characteristics of primary LOCs with aqueous electrolyte equipped with a protected lithium anode [9].

One of the central problems of enhancing the capacity of LPS is optimization of the active material loading on the positive electrode. In [8–14], the effect of the positive electrode thickness on the capacity is studied, among other issues. Here, while the general trend is the decrease in the capacity expressed in mAh g^{-1}, controversial results have been obtained at an increase in the electrode thickness for the capacity per geometric electrode surface area. According to [14], the latter indicator reaches the maximum values at the active material loading of 0.96 mg cm^{-2} and remains practically unchanged at the further increase in the loading. On the contrary, the authors of [12] observed a pronounced growth of the capacity (mAh cm^{-2}) at an increase in the loading from 5 to 15 mg cm^{-2} and then a 3-fold decrease in the capacity at an increase in the loading to 25 mg cm^{-2}. The data of [8] are close to the results of [12], but they are characterized by a significant scatter in a wide range of the active material loadings. The observed data scatter is probably related to the structure of the material used for the positive electrode active layer (AL).

This paper is dedicated to analysis of processes in LOB on the positive electrode in the course of discharge. The processes preventing supply of oxygen into the cathode pores are studied using computer simulation techniques. An attempt is made to explain phenomena observed as a result of variation of the cathode AL thickness and current density in LOBs.

2   Experimental

2.1   Electrochemical Measurements

The active material of the positive electrode used in the experiments was the PtCo catalytic system formed on the surface of carbon nanotubes subjected to alkaline pretreatment (CNT-T$_{NaOH}$ [15]). The choice of the PtCo catalyst is based on the results of paper [16] that established its high activity in the oxygen reduction reaction in aqueous media. The chosen support was the CNT-T$_{NaOH}$ material. It provides high LOB discharge capacity according to the data of the previous work [15].

The effect of the active material content on the LOB characteristics was studied in "Swagelok"-type cell with electrodes with the diameter of 12 mm. The AL based on the PtCo/CNT-T$_{NaOH}$ catalyst was formed on the surface of a 39BC gas diffusion layer (GDL) (Sigracet) by spreading the suspension of the catalyst, binder (PVDF), and solvent (1-methyl-2-pyrrolidone) using a brush on microporous GDL at the temperature of 110–120 °C. The PVDF/carbon material ratio was 1/4. The formed electrode was dried in a vacuum drying box for at least 12 h at 100 °C.

The electrolytes were prepared using the LiClO$_4$ salt (battery grade, dry, 99.99%, Sigma-Aldrich) and DMSO solvent (anhydrous, ≥ 99.9%, Sigma-Aldrich).

The cell was assembled in a deaerated dry box according to the technique described earlier [15]. The negative electrode was lithium foil; the separator was filter paper. In the course of the assembly, the positive electrode was turned with its active layer to the separator.

The electrodes with the different catalyst loading (0.2 to 1.0 mg cm^{-2} by carbon material that corresponds to the AL thickness from $\sim$15 to 55 μm) were compared by performing deep discharge of the cell to the voltage of 2 V in the galvanostatic mode. Before the tests, the system was purged by oxygen (especially pure grade) for 10–15 min.

Figure 1 presents the data characterizing the effect of the catalyst amount applied on the positive electrode and current density on the LOC parameters. Two experiments were performed for each catalyst loading. They included measurement of LOB discharge curves in the constant current density mode as per carbon support mass $(210\,\text{mA}\,\text{g}_C^{-1})$ or geometric surface area of the electrode (0.112 mAh cm^{-2}).

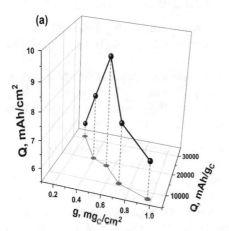 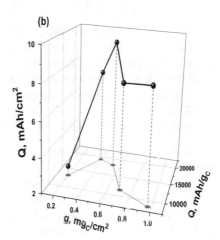

Fig. 1. Dependence of discharge capacity of LOB (Q) on the amount of the PtCo/CNT-T$_{\text{NaOH}}$ catalyst on the electrode (mg$_C$/cm^2). The characteristics are obtained at the constant current density per carbon material mass (210 mA/g$_C$) (a) and per geometric electrode surface area (0.112 mA/cm^2) (b).

One can see that in the first case Q expressed in mAh g$_C^{-1}$ decreases at an increase in the catalyst amount. Q [mAh cm^{-2}] manifests a maximum corresponding to the catalyst loading of 0.53 mg$_C$ cm^{-2}. In case of tests in the constant current density mode, both $Q\left[\text{mAh g}_C^{-1}\right]$ and Q [mAh cm^{-2}] calculated per geometric surface area of the electrode (0.112 mA cm^{-2}) pass a maximum. Here, the maximum $Q(\text{mAh g}_C^{-1})$ is reached at the loading of 0.39 mg$_C$ cm^{-2} and the maximum Q (mAh cm^{-2}) is observed at the loading of 0.53 mg$_C$ cm^{-2}.

The data presented in Fig. 1 pose a number of issues requiring discussion. The following part of this work presents the theoretical analysis of the obtained experimental data.

2.2   Analysis of Experimental Data

A review of both experimental and theoretical works devoted to the discharge process at the cathode of lithium-oxygen current sources indicates the presence of two most characteristic regularities: the cathode capacity decreases with increasing thickness of the active layer and increases with decreasing current density. However, there are no explanations of these phenomena in the literature. The purpose of our study is to attempt to give a physical interpretation of these effects.

The structure of our article is as follows. At first, the results of experiments were presented, which clearly demonstrate how the thickness of the active layer and the current density affect the cathode capacitance. Now begins the presentation of the theoretical section of this article. It is not directly related to the above experimental section of the article, being quite independent, and actually splits into two parts.

The theory of cathode discharge of lithium–oxygen power sources should answer the questions arising in the course of analysis of the data of Fig. 1:

1. Why is the allowable thickness of the cathode AL so small, tens of micrometers? In other words, why cannot gas–phase oxygen penetrate deep into AL? This results in the low values of the cathode capacity.
2. Why can a gradual increase in the cathode AL thickness result at first in a certain increase in the obtained capacity, but then inevitably causes its decrease?
3. Why, when we want to increase the capacity, not the current density should be increased, but, vice versa, its value must be wherever possible decreased?

Before answering the above questions, let us first consider the state-of-the-art understanding of the role of the cathode AL structure in the discharge.

Numerous experimental and theoretical studies form a basis for the general concept of the optimum porous structure of the positive electrode AL. Of special interest are theoretical works [17, 18] developing a new multiscale approach to model studies. They show that the positive electrode characteristics in the course of discharge are affected not by separate pore groups, but by the whole microstructure of the cathode pore space. It is directly stated in [18] that porosity must be increased up to the overall specific porosity of $3.5–4.5 \ 10^6 \ cm^{-1}$. Here, the pore size distribution must contain two peaks in the pore size range of 7–8 and 33–110 nm.

Thus, one can point out the main general requirements to the structure of the cathode AL pores:

1. The presence of cathode macropores with the radii of several hundreds of nm. These pores are practically not "occluded" by lithium peroxide and therefore they can be a basis for formation of a transport channel for oxygen supply into the cathode AL. Thus, prerequisites are created for a more complete use of the AL porous space, which would allow increasing the cathode capacity value.
2. The presence of cathode mesopores that are characterized by high specific surface area required for the first stage of the electrochemical reaction of lithium peroxide formation (1).

However, meeting these two requirements is insufficient for providing efficient supply of oxygen molecules. A large pore that is safe from "occlusion" by Li_2O_2 is required that must be in contact with the neighboring large pores. In other words, a cluster of large pores must be developed and each of these pores must be able to receive oxygen from the neighboring pore and pass it on. But for this, the concentration of large pores in the active layers must be high enough.

If the number of large pores in the cathode AL is low and small pores prevail, they "encompass" large pores and stop the movement of oxygen molecules across the chains of large pores. As a result of this, oxygen cannot penetrate deep into the cathode AL.

Is the necessary condition of existence of a sufficiently large concentration of large pores fulfilled? Most probably is not present. At any rate, there is no information in the literature to this effect.

Now, let us pass to search for the answer to the **first** of the questions posed in this section. Let us do this using the computer simulation technique.

At first, let us visualize the active layer of the cathode as a set of small regions, so that either large or small pores were predominant in each such region. All these regions are cubic. Thus, the model of the biporous cathode AL structure that we chose appears. It is presented in Fig. 2: the model of equally sized cubes of two types. The black and light cubes are randomly mixed in it.

Fig. 2. Model cube of a biporous cathode AL. Black cubes contain large, "occlusion–proof" pores, through which oxygen is supplied; light cubes are grains with small pores, in which lithium peroxide is formed.

Take a detailed look at rather arbitrary diffusion coefficients of oxygen molecules for each cube type. Here, we assume that diffusion coefficient D_l of oxygen in a cube with large pores is much larger than diffusion coefficient D_s in a cube with small pores. Let us calculate how the effective diffusion coefficient D^* of oxygen in a mixture of black and light cubes depends on fraction ω of cubes with large pores ($0 \leq \omega \leq 1$).

The results of such a calculation of efficient diffusion coefficient D^* of oxygen are presented in Fig. 3. One can see that the efficient diffusion coefficient grows at an increase in parameter ω from the obvious value of $D^* = D_s$ (at $\omega = 0$, there are only light cubes with small pores in the model cube) to the value of $D^* = D_l$ (at $\omega = 1$, there are only black cubes with large pores in the model cube). Ratio D_s/D_l of diffusion coefficients was varied from the value of 10 (curve 1) to 100 (curve 2).

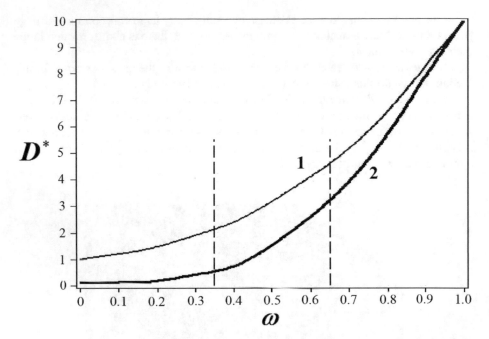

Fig. 3. Dependence of efficient diffusion coefficient D^* of oxygen molecules in the electrolyte on fraction ω of cubes with large pores in the cathode AL. The ratios of the arbitrary values of the diffusion coefficients are (1) 10 and (2) 100.

Now characterize the conditions, at which a percolation cluster consisting only of large pores can be formed in the cathode AL. The percolation theory [19] allows calculating the value of percolation threshold ω^*: the critical moment, at which a percolation cluster consisting of cubes with large pores appears in the system. Paper [20] presents the characteristic equation allowing find the value of percolation threshold for systems of rectangular lattices (including a square planar lattice, a 3D cubic lattice, and their counterparts in n-dimensional space with any n). Here is this equation:

$$2\,(d-1)\gamma^2\omega^3 + (2d-3)\gamma\omega - 1 = 0, \tag{3}$$

where d is the space dimension, γ is the fraction of bonds in the "link problem", ω is the fraction of bonds in the "node problem". In our case, we are dealing with a cubic lattice (Fig. 2), so that $d = 3$ must be assumed in Eq. (3). Further, $\gamma = 1$ must be set in Eq. (3). As a result, instead of characteristic Eq. (3), we obtain the following equation:

$$4\omega^3 + 3\omega - 1 = 0. \tag{4}$$

In this Eq. (4), the only real solution obtained for percolation threshold ω^* is:

$$\omega* = \left[\left(1 + 2^{1/2}\right)^{1/3} + \left(1 - 2^{1/2}\right)^{1/3}\right]/2 = 0.298. \tag{5}$$

Thus, a "black" percolation cluster is formed in the model cube if the concentration of black cubes becomes approximately 30%. If this condition is not fulfilled, then light cubes with small pores can block the bonds between black cubes in the model cube in the range of $0 \leq \omega \leq 0.3$.

Besides, one must point out that, as shown in [21], to realize a *complete* percolation cluster of black cubes with large pores their concentration ω must be somewhat increased: from 0.3 (condition (5)) to the concentration of about 0.35. In Fig. 3, the position of this boundary ($\omega = 0.35$) is indicated by the first dashed vertical line.

Secondly, it is obvious from symmetry considerations that at $\omega > 0.65$ (on the right of the second dashed vertical line in Fig. 3), the percolation cluster of light cubes with small pores providing transport of lithium ions cases to exist in the cathode AL. Thus, the optimum conditions for the functioning of the cathode AL must be looked for in the range of the values of parameters ω, where both percolation cluster ("black" and "light") exist simultaneously. This condition is fulfilled only in the range of $0.35 < \omega < 0.65$ (Fig. 3).

The percolation cluster of black cubes with large pores for $\omega = 0.35$ is shown in Fig. 4. It is important that such a cluster can supply oxygen molecules into the cathode AL to any depth. However, the concentration of large pores in real ALs of LOPS cathodes is apparently not high enough ($\omega < 0.35$). In this range of values of parameter ω, according to the data of Fig. 3, efficient diffusion coefficient $D*$ of oxygen molecules is as yet small, so that oxygen, while moving through large lithium peroxide–free pores (black cubes), can penetrate only a very thin surface layer of the electrode. This is probably the cause of the very low cathode discharge capacity.

All the above considerations together provide the answer to the first of the questions formulated in the beginning of this paper section.

Consider the **second** issue and discuss the character of the dependence of the discharge capacity of the LOPS cathode on the thickness and also appearance here of local capacity maximums (the data of Fig. 1). Such maximums and the overall decrease in the cathode capacity at an increase in the AL thickness are mentioned in many studies. It is shown in [13] that the capacity measured at an increase in the cathode active layer thickness in the series of 19.7, 65.5, 219.2 μm, forms a decreasing sequence of values: 2550, 1580, 350 mAh g^{-1}. Such facts are rather easily explained. The region, in which Li_2O_2 is indeed formed, is narrow and adjoins the plane separating the cathode and the region, from which oxygen is supplied. The supply of lithium ions to this surface layer is naturally ever more complicated at the same current density at an increase in the cathode AL. Lithium ions must additionally move not only in the interelectrode space, but also, in part, via small AL pores, also somewhat "littered" by insoluble Li_2O_2. This results in a decrease in the concentration of lithium ions in the region in which Li_2O_2 is formed. Therefore, capacity decreases at an increase in the cathode AL thickness.

Fig. 4. Arbitrary section of the model cube (Fig. 2) by a plane parallel to its front and rear surfaces. The cubes with large pores are marked black; the cubes with small pores are marked gray. The fraction of black cubes is $\omega = 0.35$; the fraction of gray cubes is $\omega = 0.65$. The white color in the figure indicates isolated black and gray cubes not included into the two percolation clusters.

The additional remark concerning an opportunity of concentration of lithium ions decrease on an interface the cathode/oxygen. There is no doubt that there is a certain characteristic region at the cathode/oxygen boundary, in which only the process of lithium peroxide formation takes place. If the thickness of the cathode exceeds the size of this characteristic region, then there is already a ballast, non-working part of the cathode. This section of the cathode AL has an increased resistance to the movement of ions. Thus, in the active part of the AL, where the formation of lithium peroxide occurs, there is an inevitable decrease in both the current density and the concentration of lithium ions.

One can see from the experimental data (Fig. 1) that capacity maximums are observed. In case of a very thin cathode AL, a slight increase in its thickness obviously does not result in a drastic decrease in the concentration of lithium ions and even allow somewhat increasing the capacity. However, as already pointed out above, when a certain critical thickness is reached, the concentration of lithium ions in AL proves to be insufficient. And then, after the maximum is reached, the corresponding decrease in the capacity value starts. All this explains the nature of the local extremum in the dependence of the capacity value on the cathode AL thickness.

3 Computer Simulations

3.1 Relationship of Cathode Capacity with Current Density

The computer simulation technique we used to elucidate the causes for an increase in the cathode capacity at a decrease in current density and, thus, consider the **third** of the above issues. Let us use the simplest monoporous model of the porous structure in the modeling of the cathode AL. Let us assume that there is a set of tortuous and nonintersecting capillaries with constant radius r_{po} starting on the front cathode surface bordering on the gas phase. The active layer thickness is L, its initial porosity is g_0.

Consider the processes of Li_2O_2 formation in an individual pore, as all pores are in fact identical. Assume that the concentration of lithium ions in the electrolyte filling all pores cathode AL remains practically constant [22]. As follows from the above, this can be provided for formation of thin ALs with a negligible resistance for supply of lithium ions. Then, following the example of [23, 24], let us assume the concentration of oxygen in the pores $a(x, t)$ to be the main value determining the discharge process on the cathode.

When a pore is filled by Li_2O_2, free volume fraction g and real pore radius r_p is changed in each pore section. Then one can form a system of equations that describe processes occurring in a monoporous cathode AL in the course of discharge of a lithium–oxygen power source. First of all, let us take into account that effective diffusion coefficient D of oxygen in the electrolyte is not constant; it is proportional to porosity g. Then we have the following obvious relationships:

$$D = D * g/\tau, \tau = g^{-1/2} \tag{6}$$

Here, $D*$ is the true diffusion coefficient of oxygen, τ is the pore tortuosity. Movement in the pore of oxygen dissolved in the electrolyte is described by a diffusion equation with oxygen consumption in formation of lithium peroxide:

$$\frac{\partial(ga)}{\partial t_p} = \frac{\partial}{\partial x}\left(D\frac{\partial a}{\partial x}\right) + R(a, g, r_p) \tag{7}$$

Here, $R(a, g, r_p)$ is the consumption of oxygen as a result of formation of Li_2O_2, the final product of reaction chain (1) that is precipitated in the given pore section. It is also characterized by porosity g and pore radius r_p. It is desirable to know how exactly the consumption of oxygen depends on the type of the electrolyte and regularities of the pore structure of the material used for the cathode AL. However, we do not have such information yet. Thus, it is usually suggested to characterize the complex process of oxygen consumption by a single integral parameter: oxygen consumption constant k. Let us assume that

$$R(a, g, r_p) = -2g \cdot k \cdot a(x, t_p)/r_p. \tag{8}$$

In the right part of condition (8), $2 g/r_p$ is the specific surface area, on which the first stage of process (1) occurs, $a(x, t_p)$ is the concentration of oxygen in the given section of pore x at the given time moment t_p, and k plays the role of the proportionality factor with the dimension of cm/s. Let us also take into account that

$$\frac{\partial g}{\partial t_p} = R(a, g, r_p) \frac{M}{\rho} \tag{9}$$

Equation (9) describes the change in cathode porosity g in time, M/ρ is the molar volume. Here, M, g/mol, is the molar mass and ρ, g/cm^3 is the density of Li_2O_2.

In the course of discharge in the pores of the cathode AL, oxygen is supplied from the front surface into the cathode pore space. It is reduced with addition of an electron on the pore walls and then it interacts with lithium ions. The formed final product Li_2O_2 is precipitated in the pores and gradually fills their initial volume. For definiteness, let us assume that the discharge process (the process of oxygen supply into the cathode pores) is over when only 10% of the initial of pore volume on the front AL surface remain unfilled.

Thus, there are in fact two equations with three variables: a, g, and r_p. It is, however, obvious that the last two quantities are related as follows:

$$(g/g_0)^{3/2} = (r_p/r_{po})^2, \tag{10}$$

Here, g_0 is the initial porosity, r_{po} is the initial pore radius. Further assume that in the beginning of the discharge process, at $t = 0$, the whole pore is free of the reaction product and of oxygen dissolved in the electrolyte. Then, in the reduced coordinate range of $0 \le z = x/L < 1$, where L is the cathode AL thickness:

$$r_p/r_{po} = 1 \text{ and } a(z,t) = 0. \tag{11}$$

Here, we have the following condition on the front cathode AL surface for the given concentration c in the pore:

$$c = a/a^* = 1 \text{ at } z = 0, \tag{12}$$

where a^* is the solubility of oxygen in the electrolyte. Assume the following condition at the rear side of the active layer:

$$\partial c/\partial z = 0 \text{ at } z = 1. \tag{13}$$

Putting system of equations and conditions (3)–(6) in a nondimensional form will be effective for carrying out specific numerical calculations. Thus, we set

$$t = t_p D[g_0]^{1/2}/L^2, r = r_p/r_{po}, z = x/L, c = c_{O2}/c_{O2,0}, \tag{14}$$

where t_p is current time.

Then, at $0 \le r, z$ and $c \le 1$, expressions (3)–(6) can be rewritten as a system of two equations in two main variables c and r:

$$dr^{4/3}c/dt = d(r^2 dc/dz) - \beta r^{1/3}c \qquad (15)$$

$$dr/dt = -\gamma c \qquad (16)$$

where the parameter

$$\beta = k[2L^2/D(g_o)^{1/2}r_{po}], \gamma = k[(3L^2 c_{O2,0}/2D(g_o)^{1/2}r_{po})(M_{Li2O2}/\rho_{Li2O2})] \qquad (17)$$

The final total of calculations is estimation of variation in time of the value of overall current $I(t)$, A cm^{-2}, of the LOPS, which should be calculated in accordance with the formula

$$I(t) = k[4Fg_0 c_{O2,0}]L/r_{po}] \int (\text{from } z = 0 \text{ to } z = 1)[r^{1/3}(z,t)c(z,t)]dz \qquad (18)$$

The final value of specific capacity of the cathode C, C cm^{-2}, can be calculated as by the formula

$$C = \int (\text{from t} = 0 \text{ to } t = t^*)I(\tau)d\tau, \qquad (19)$$

where the reduced time is

$$t^* = T/t_{po} \qquad (20)$$

where T is the time of ending of discharge (at this moment, the pore throats become "overgrown" with lithium peroxide by 90%) and the discharge time scale is

$$t_{po} = L^2/D(g_o)^{1/2} = L^2/(g_0)^2 D_{02} \qquad (21)$$

The values presented in Table 1 were chosen for calculations.

Table 1. Parameters of the experimental model used for calculations

| Parameter | r_o | L | g_o | M_{Li2O2} | ρ_{Li2O2} |
	cm 10^7	cm		g mol^{-1}	g cm^{-3}
Value	2.00	0.07	0.73	45.8768	2.3

The solvent was tetraglyme. Further, the value of parameter k was varied. The value of k was estimated in [25]. For example, for such a solvent as propylene carbonate (PC), the value of $k = 1.025 \times 10^{-8}$ cm s^{-1} was obtained.

3.2   Method of Calculations

The differential equations in partial derivatives (12)–(13) were solved numerically using the Maple software [27–29]. In order to separate out the desired variables $c(x,t)$ and $r(x,t)$, these equations were preliminarily replaced by the equivalent system of three equation:

$$\frac{\partial}{\partial t}c(z,t) = r(z,t)^{2/3}\frac{\partial^2}{\partial x^2}c(z,t) + \frac{2}{r(z,t)^{1/3}}\cdot\frac{\partial}{\partial z}r(z,t)\cdot\frac{\partial}{\partial z}c(z,t) + \frac{c(z,t)\cdot\left(\beta - \frac{4}{3}\gamma\cdot c(z,t)\right)}{r(z,t)};$$
$$\frac{\partial}{\partial z}r(z,t) = v(z,t);$$
$$\frac{\partial}{\partial t}r(z,t) = -\gamma\cdot c(z,t);$$

with the boundary conditions:

$$c(0,t) = 1 \qquad \frac{\partial}{\partial z}c(1,t) = 0 \qquad \frac{\partial}{\partial z}v(1,t) = 0;$$

and initial conditions:

$$c(z,0) = \begin{cases} 1 & z = 0 \\ 0 & 0 < z \le 1, \ r(z,0) = 1 \end{cases}$$

It should be noted that system of Eqs. (12)–(13) was repeatedly used and discussed in detail by many authors (see, for example, [22, 23]). Our work differs in that we varied one of the main parameters of the system under investigation, the oxygen consumption constant k. This enabled us to show the phenomenon of an increase of the cathode capacity with decreasing parameter k, and to explain the physical factors that promote an increase in capacitance. Below it will be shown that the variation of parameter k is symbate to the variation of the current density i (they either increase or decrease concurrently). Therefore, the cathode capacity increases with decreasing current density.

3.3   Results of Calculation of Cathode Characteristics

The physical meaning of oxygen consumption constant k in expression (8) is the fraction of oxygen molecules converted into the superoxide anion O_2^- that can react further with lithium ions and ultimately form lithium peroxide. It is easy to understand that parameter k is in fact proportional to current density. In any case, an increase in current density can cause an increase in k. Indeed, in case of zero supply of lithium ions into the pore ($i = 0$), the process of formation of lithium peroxide molecules (1), obviously, must stop. Here, $k = 0$ should be assumed. Thus, constant k indeed proves to be proportional to current density i, mA cm^{-2}.

Indeed, it is rather difficult to characterize the exact relationship between k and i. One must take into account many factors: the thickness of cathode AL, its pore radius, parameters of the solvent, current density i. But it seems probable that current density will change in the same direction as parameter k. Here, an issue that is not quite obvious should be pointed out: it turns out that a decrease in current density i and therefore in parameter k does not cause a decrease in the cathode capacity, as could be expected, but, on the contrary, results in its growth. This feature of current density has been repeatedly confirmed experimentally in many works. It is shown in [6] that a decrease in the current density from 0.5 to 0.1 mA cm^{-2} causes a gradual increase in the cathode capacity.

A detailed analysis of the effect of parameter k on the overall characteristics of a discharging cathode is carried out in [26]. This study indicates the causes of the increase in the cathode discharge capacity at a decrease in parameter k. Let us consider these causes at a specific example. Let us calculate the cathode capacity according to the above system of Eqs. (6)–(10). The change in cathode capacity C, C cm^{-2} with variation of parameter k is shown in the Table 2. It contains the calculated dependences of specific capacity C, C cm^{-2}, of the cathode on the value of parameter k for the chosen initial pore radius $r_{po} = 40$ nm. This parameter was varied in the range from 10^{-6} to 10^{-9} cm s^{-1}. One can see that the cathode capacity grows at a decrease in parameter k and reaches very high values: hundreds of C cm^{-2}. It must be explained why the cathode capacity increases at a decrease in k (and therefore in current density i).

Table 2. Dependence of specific cathode capacity on oxygen consumption constant k

log k	−6	−7	−8	−9
C, C cm^{-2}	0.08	37.5	100	275

Table 3 shows how the distribution of lithium peroxide in the pores of the cathode AL changes through the reduced active layer depth z at a decrease in parameter k at the end of discharge. Let us point out that the degree of filling of any given pore section by lithium peroxide in the monoporous cathode model at the end of the cathode discharge process corresponds to the value of reduced radius $r = r_p/r_{po}$ equal to the ratio of radius r_p at the end of the cathode discharge to initial pore radius r_{po}. Obviously, the condition of $r = 1$ means that Li$_2$O$_2$ is completely absent in this pore section. On the contrary, at $r < 1$ the pore section is already partially filled by Li$_2$O$_2$. Parameter r assumes the lowest value in the pore opening, at $x = 0$.

Table 3. Distribution of lithium peroxide through reduced depth z of cathode AL at the end of discharge for some values of parameter k

log k	$z = x/L$				
	0.2	0.4	0.6	0.8	1.0
−7	0.95	0.99	1.0	1.0	1.0
−8	0.7	0.87	0.93	0.97	0.98
−9	0.4	0.53	0.6	0.63	0.64

Table 3 shows that only the narrowest pore region near its opening, at the boundary with the gas phase, is "overgrown" by lithium peroxide at $k = 10^{-7}$ cm s^{-1}. Li$_2$O$_2$ is practically absent in the rest of the pore ($z > 0.4$). Starting with $z = 0.6$, reduced radius $r = r_p/r_{p0}$ becomes identical to unity, which points to the absence of Li_2O_2 in this region.

At a decrease in parameter k, the Li_2O_2 formation range starts increasing. At the lowest value of $k = 10^{-9}$ cm s^{-1}, the process of Li_2O_2 formation already encompasses the whole pore (up to the coordinate of $x = L$). Therefore, the possibility of decreasing the values of parameter k or, which is in itself analogous to the possibility of decreasing the value of current density i, can be a desirable effect. It implies an increase in the discharge capacity. However, there is a negative effect. A decrease in parameter k or current density i cause a significant decrease in the duration of the cathode discharge process.

The character of growth of the discharge duration as dependent on parameter k for a monoporous structure with pore radius $r_{p0} = 40$ nm and 2 nm is presented in Table 4.

Table 4. Dependence of discharge duration T of a LOPS cathode on parameter k at two pore radii (given in the brackets).

k, cm s^{-1}	10^{-7}	10^{-8}	10^{-9}
T, s (40 nm)	667	1300	11333
T, s (2 nm)	50	100	500

Thus, a well known experimental fact is confirmed: a decrease in current density i results in a fast growth of the time required for discharge. When a cathode in LOPS discharges, a decrease in current density results in manifestation of two factors causing an increase in specific cathode capacity: the duration of this process that increases at a decrease in the discharge current density and the growing penetration depth of the process of Li_2O_2 formation in the pores. It is these conditions that cause an increase in capacity at a decrease in current density.

Table 4 also demonstrates the character of growth in the discharge time on the value of parameter k for the structure with pore radius $r_{p0} = 2$ nm. One can see how an increase in the pore radius (transition from $r_{p0} = 2$ nm to $r_{p0} = 40$ nm) causes an increase in the cathode discharge time. This also allows assuming that the macropores in fact fail to be "occluded" by Li_2O_2, and they can supply oxygen to the mesopore openings.

4   Conclusions

This work presents experimental data characterizing the amount of the active material on the positive LOB electrode on the discharge capacity (Q) and also these data are analyzed theoretically using the computer simulation technique. The active material used is the PtCo/CNT-T$_{NaOH}$ catalyst. It is found that an increase in the amount of the supported catalyst from 0.20 to 0.97 mg$_C$ cm^{-2} (per carbon material mass) results in a decrease in Q expressed in mAh g$_C^{-1}$. The dependence of Q on current density is of similar character. Here, Q [mAh cm^{-2}] manifests a maximum corresponding to the catalyst loading of 0.53 mg$_C$ cm^{-2} and increase in the loading.

Theoretical modeling of the work of the cathode AL with a biporous pore (macropores not filled with Li$_2$O$_2$ and mesopores in which it is accumulated) is performed. It is shown that effective diffusion coefficient $D*$ of oxygen is extremely small at the low mesopores concentration, which is observed in practice. Formation of a percolation cluster providing oxygen transport through the whole AL depth is realized at macropore fraction $\omega \geq 0.35$. However, in actual practice, the macropore fraction is lower than 0.35 and thus the effective diffusion coefficient of oxygen is low. This results in the fact that penetration of oxygen molecules into AL is possible only to a small depth, due to which small overall cathode discharge parameters are obtained. Besides, it is shown that the extremely dependence of the positive electrode capacity on the active layer thickness (mass) is related not only to supply of oxygen into the reaction zone, but also of lithium ions, the concentration of which at the electrode/oxygen interface is the lower, the higher the AL thickness.

Computer simulation is used to study the effect of oxygen consumption constant k on the discharge reaction on the basis of the model of a monoporous cathode AL. Parameter k characterizes the number of oxygen molecules reaching the pore surface and capable of reacting with lithium ions with formation of Li$_2$O$_2$. It is shown that parameter k is almost proportional to current density i, mA cm^{-2}. The decrease in the value of k or, which is, in fact, the same, of current density i can provide high LOB capacity owing to expansion of the region of lithium peroxide formation in AL of the positive electrode and an increase in discharge time T. This conclusion corresponds to the character of the experimental dependence of discharge capacity on current density.

Acknowledgement. The work was performed with support of Ministry of Science and Higher Education of Russian Federation.

References

1. Tarasevich, M.R., Andreev, V.N., Korchagin, O.V., Tripachev, O.V.: Lithium-oxygen (air) batteries (state-of-the-art and perspectives). Prot. Met. Phys. Chem. Surf. **53**, 1–48 (2017)
2. Tran, C., Yang, X.-Q., Qu, D.: Investigation of the gas-diffusion-electrode used as lithium/air cathode in non-aqueous electrolyte and the importance of carbon material porosity. J. Power Sources **195**, 2057–2063 (2010)
3. Yang, X.-H., He, P., Xia, Y.-Y.: Preparation of mesocellular carbon foam and its application for lithium/air battery. Electrochem. Commun. **11**, 1127–1130 (2009)
4. Laoire, C.O., Mukerjee, S., Abraham, K.M., Plichta, E.J., Hendrickson, M.A.: Elucidating the mechanism of oxygen reduction for lithium-air battery applications. J. Phys. Chem. C **113**, 20127–20134 (2009)
5. Laoire, C.O., Mukerjee, S., Abraham, K.M., Plichta, E.J., Hendrickson, M.A.: Influence of nonaqueous solvents on the electrochemistry of oxygen in the rechargeable lithium − air battery. J. Phys. Chem. C **114**, 9178–9186 (2010)

6. Mohazabrad, F., Wang, F., Li, X.: Experimental studies of salt concentration in electrolyte on the performance of Li-O$_2$ batteries at various current densities. J. Electrochem. Soc. **163**, A2623–A2627 (2016)

7. Liu, T., Leskes, M., Yu, W., Moore, A.J., Zhou, L., Bayley, P.M., Kim, G., Grey, C.P.: Cycling Li-O2 batteries via LiOH formation and decomposition. Science **350**, 530–533 (2015)

8. Nomura, A., Ito, K., Kubo, Y.: CNT sheet air electrode for the development of ultra-high cell capacity in lithium-air batteries. Sci. Rep. **7**, 45596 (2017)

9. Imanishi, N., Luntz, A.C., Bruce, P.G. (eds.): The Lithium-Air Battery. Fundamentals. Springer, New York (2014)

10. Abraham, K.M., Jiang, Z.: A polymer electrolyte-based rechargeable lithium/oxygen battery. J. Electrochem. Soc. **143**, 1–5 (1996)

11. Beattie, S.D., Manolescu, D.M., Blair, S.L.: High-capacity lithium-air cathodes. J. Electrochem. Soc. **156**, A44–A47 (2009)

12. Xiao, J., Wang, D., Xu, W., Wang, D., Williford, R.E., Liu, J., Zhang, J.-G.: Optimization of air electrode for Li/air batteries. J. Electrochem. Soc. **157**, A487–A492 (2010)

13. Zhang, G.Q., Zheng, J.P., Liang, R., Zhang, C., Wang, B., Hendrickson, M., Plichta, E.J.: Lithium–air batteries using SWNT/CNF buckypapers as air electrodes. J Electrochem Soc **157**, A953–A956 (2010)

14. Landa-Medrano, I., Pinedo, R., Ruiz de Larramendi, I., Ortiz-Vitoriano, N., Rojo, T.: Monitoring the location of cathode-reactions in Li-O$_2$ batteries. J. Electrochem. Soc. **162**, A3126–A3132 (2015)

15. Bogdanovskaya, V.A., Korchagin, O.V., Tarasevich, M.R., Andreev, V.N., Nizhnikovskii, E.A., Radina, M.V., Tripachev, O.V., Emets, V.V.: Mesoporous nanostructured materials for the positive electrode of a lithium-oxygen battery. Prot. Met. Phys. Chem. Surf. **54**, 373–388 (2018)

16. Bogdanovskaya, V.A., Kol'tsova, E.M., Tarasevich, M.R., Radina, M.V., Zhutaeva, G.V., Kuzov, A.V., Gavrilova, N.N.: Highly active and stable catalysts based on nanotubes and modified platinum for fuel cells. Russ. J. Electrochem. **52**, 723–734 (2016)

17. Bao, J., Hu, W., Bhattacharya, P., Stewart, M., Zhang, J.-G., Pan, W.: Discharge performance of Li-O$_2$ batteries using a multiscale modeling approach. J. Phys. Chem. C **119**, 14851–14860 (2015)

18. Pan, W., Yang, X., Bao, J., Wang, M.: Optimizing discharge capacity of Li-O$_2$ batteries dy design of air-electrode porous structure: multifidelity modeling and optimization. J. Electrochem. Soc. **164**, E3499–E3511 (2017)

19. Tarasevich, Y.Y.: Percolation: Theory, Applications, Algorithms. Editorial URSS, Moscow (2011)

20. Chirkov, Y.G.: Theory of porous electrodes: the percolation and a calculation of percolation lines. Russ. J. Electrochem. **35**, 1281–1290 (1999)

21. Chirkov, Y.G., Rostokin, V.I., Skundin, A.M.: Computer modeling of positive electrode operation in lithium-ion battery: model of equal-sized grains, percolation calculations. Russ. J. Electrochem. **47**, 71–83 (2011)

22. Sandhu, S.S., Fellner, J.P., Brutchen, G.W.: Diffusion-limited model for a lithium/air battery with an organic electrolyte. J. Power Sources **164**, 365–371 (2007)

23. Read, J., Mutolo, K., Ervin, M., Behl, W., Wolfenstine, J., Driedger, A., Foster, D.: Oxygen transport properties of organic electrolytes and performance of lithium/oxygen battery. J. Electrochem. Soc. **150**, A1351–A1356 (2003)

24. Dabrowski, T., Struck, A., Fenske, D., Maaβ, P., Ciacchi, L.C.: Optimization of catalytically active sites positioning in porous cathodes of lithium/air batteries filled with different electrolytes. J. Electrochem. Soc. **162**, A2796–A2804 (2015)
25. Bogdanovskaya, V.A., Andreev, V.N., Chirkov, Y., Rostokin, V.I., Emets, V.V., Korchagin, O.V., Tripachev, O.V.: Effect of positive electrode structure on process of discharge of lithium-oxygen (air) power source. theory of monoporous cathode. Prot. Met. Phys. Chem. Surf. **54**, 1015–1025 (2018)
26. Chirkov, Y., Andreev, V.N., Rostokin, V.I., Bogdanovskaya, V.A.: Discharge of lithium-oxygen power source: monoporous cathode theory and role of constant of oxygen consumption process. Altern. Energy Ecol. (ISJAEE) **4–6**, 95–107 (2018)
27. Dyakonov, V.P.: Maple 10/11/12/13/14 in Mathematical Calculations (Russian Edition). DMK-Press, Moscow (2011)
28. Davis, J.H.: Differential Equations with Maple: An Interactive Approach. Birkhauser, Boston (2001)
29. Edwards, C.H., Penny, D.E.: Differential Equations and Boundary Value Problems: Computing and Modeling, 3rd edn. Moscow (2008)

Comprehensive Framework for Classification of Abnormalities in Brain MRI Using Neural Network

S. Harish[1(✉)] and G. F. Ali Ahammed[2]

[1] VTU, Belagavi, India
harishsrinivasaiah@gmail.com
[2] VTU PG-Centre, Mysuru, India
aliahammed78@gmail.com

Abstract. Precise identification of the abnormalities in brain is one of the challenging mechanisms to deal with in clinical diagnosis process. Irrespective of presence of various sophisticated diagnosis system in present time, there are frequent reporting of error-prone diagnosis even by the medical standards. Review of existing system toward detection of brain MRI shows that there are very less quantum of work being carried out towards emphasizing over classification process. This lead to formulate a novel framework in proposed system for assisting in classification process. The proposed system offers a simple and yet robust segmentation and classification process in multiple level which is further boosted by adopting Artificial Neural Network. The study outcome of the proposed system shows that it excels better accuracy in contrast to existing learning methods that are frequently used by researchers.

Keywords: Brain MRI · Segmentation · Detection · Classification · Benign · Malignant · Magnetic resonance imaging

1 Introduction

Magnetic Resonance Imaging (MRI) is one of the dominant medium of diagnosing the abnormalities present in human body [1]. Brain is considered as one of the essential as well as complex human organ where diagnosis of the abnormalities are bit challenging [2]. The challenge factor involved in diagnosis through brain MRI is the presence of various artifacts e.g. noise, inferior contrast, absence of ridges, frequent adoption of the supervised techniques, etc. Such problems are anticipated to be solved by medical image processing techniques. At present, there has been various archives of the literatures towards focusing on proper identification of abnormalities present in brain as well as there are certain degree of work towards solving classification problems also [3–6]; however, they are also associated with significant amount of problems too which are quite hard to solve. A closer look into the quantity of the research work towards brain MRI shows that there are more number of work towards detection problem and few work focusing on the classification problem. The prime reason behind this problem is that existing system doesn't offer extensive and, comprehensive extraction of

© Springer Nature Switzerland AG 2019
R. Silhavy et al. (Eds.): CoMeSySo 2019, AISC 1047, pp. 70–80, 2019.
https://doi.org/10.1007/978-3-030-31362-3_8

significant features [7]. This causes the system to deplete important set of clinical information which is predominantly required for making decision of the presence of abnormalities present in the brain MRI image. Apart from this, existing segmentation process are too straight forward and less work are found to be extensive [8, 9]. Existing segmentation techniques are more focused on background factor and not much on the foreground, which is also one of the possible reason impacting on the accuracy. Therefore, this manuscript presents a framework which offers a comprehensive segmentation that acts as a compliment towards feature extraction too. Section 2 discusses about the existing research work followed by problem identification in Sect. 3. Section 4 discusses about proposed methodology followed by elaborated discussion of algorithm implementation in Sect. 5. Comparative analysis of accomplished result is discussed under Sect. 6 followed by conclusion in Sect. 7.

2   Related Work

This section is the extension of the brief review of the study towards brain MRI [10]. Most recently, the work of Mallick et al. [11] have presented a classification scheme using neural network and wavelet based approach over brain MRI. Shao et al. [12] have carried out segmentation process in order to identify specific region of brain for facilitating classification process. Study using segmentation was also carried out by Wang et al. [13] where the authors have used convolution neural network for better accuracy. Explicit discussion of the decomposition-based approaches was carried out by Gudigar et al. [14] for investigation classification performance for multiple classes of abnormalities of brain. Literature has also witnessed usage of feature extraction along with learning approach method for assisting classification of brain as seen in the work of Gumaei et al. [15]. Segmentation process for specific disease condition of brain was seen in work of Liu et al. [16] where linearized kernel was used for classification. Similar direction of work was also carried out by Wang et al. [17]. Study towards identifying complicated state of disease of brain was discussed by Huang et al. [18] where the authors have used kernel-based learning approach for assisting in classification. Kermi et al. [19] have presented multi-stage segmentation method over three dimensional input image of brain MRI. Usage of random forest over the active contours with multiple patches has been presented in the work of Ma et al. [20] for enhancing the segmentation technique. The implementation of linearized discriminant analysis is found in Wang et al. [21] for functional MRI which brings significant classification accuracy. Yuan et al. [22] have discussed about multi-center brain MRI classification by using convolutional neural networks (CNN) approach and obtained 92% of classification accuracy for large MRI dataset. A work of Zhan et al. [23] have discussed a Glioma segmentation mechanisms by using multiple classifier based collaborative training. A unique work of Liu et al. [24] has presented the hierarchical brain networks for classification of structural MRI by using schizophrenia based method. Similar kind of research with different approaches are found in Kaur et al. [25], Kasobov et al. [26, 27], Aemananzas et al. [28], Liu et al. [29, 30] etc. From the extensive survey analysis, the research problem is formed with current state of art in the research domain. The next section briefs about problem description.

3 Problem Description

After reviewing the existing system, it can be seen that existing system towards classification is much prone to usage of one type of segmentation technique which is absolutely not comprehensive. Such forms of mechanism are actually focused much on detection approach and bit less on the classification process. In order to implement classification process, it is essential to ensure that there is comprehensive number of discrete features without which the accuracy score cannot be improved. Unfortunately, usage of the sophisticated optimization technique in existing approaches are only focused on achieving accuracy without considering various comprehensive information for improving the decision making system as an outcome. Hence, there is a need of developing a framework that focuses on comprehensive optimization towards enhancing the classification performance of brain MRI.

4 Proposed Methodology

The current work is the continuation of the prior work [32, 33] which has introduced an unique enhancement modeling towards assisting the identification of tumor in bran MRI[P-2][P-3]. The present work introduces a framework that is capable of performing non-conventional segmentation of the brain MRI image. The adopted scheme is shown in Fig. 1.

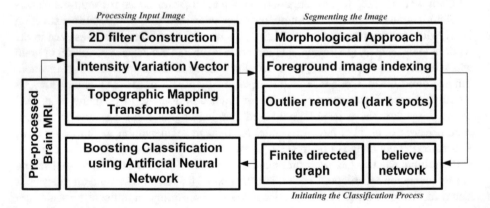

Fig. 1. Undertaken scheme of proposed classification

The input image is basically a pre-processed image obtained from [31] which undergoes 4 different rounds of operation viz. processing input image, segmenting image initiating the classification process, and applying neural network. Each processes are very much sequential which leads to explore more features which are numerically distinct for assisting in decision making process while performing training using neural network. The proposed system also introduces a unique segmentation process which facilitates in both foreground and background marking that is further subjected to believe network and finite directed graph before applying training using artificial neural network. The next section elaborates about system implementation.

5 System Implementation

The complete implementation of the proposed system is designed and developed for accomplishing the target of superior classification performance of the brain MRI image. The core strategy for the implementation process is basically to apply different forms of suitable processes that perform two essential task i.e. segmentation and classification. The accuracy of the classification process is boosted by applying Artificial Neural Network. The complete discussion of the system implementation is carried out on the basis of following sequential stages of operations:

(i) Processing Input Image: The system takes the input image of brain MRI followed by digitization of it in order to retain in the form of Matrix. This stage of operation also converts the input image to grayscale for easing the further processing steps. Once the grayscale input image is considered than it is subject next process for obtaining intensity variation vector after a two dimensional filter is constructed. A multi-dimensional image filtering process is applied on the grayscale image to obtain two dimensional vectors which is further used for constructing intensity variation vector. The obtained vector is than subjected to topographic mapping transformation which represents the input grayscale image in the form of topographical map where information about the brightness as well as ridges of the image is obtained. This process is the beginning of primary segmentation process which is capable of identifying the regions with primary possibility of tumor. Another advantage of this transformation process is that it colors all the significant regions differently to classify different regions of abnormalities.

(ii) Segmenting the Image: Different from any existing segmentation approach, the proposed system performs very different series of approaches which actually complements the segmentation process for better classification performance. For that purpose, the system constructs elements of morphological structure with a shape of disk which will accelerate the process of dilation and erosion in morphological operation. Basically such elements are a two dimensional matrix with neighboring elements of binary value which is meant for considering the *true* pixel while discarding the *false* pixel elements. A reference pixel which is at the center is considered to carry out identification of *true* and *false* pixels. Such forms of segmentation are possible for both binarized as well as grayscale brain MRI images. This operation results in generation of mapped version of input image as well as morphed eroded image. The obtained morphed eroded image is then subjected to reconstruction process with respect to grayscale input image. Finally, the operation results in marking of the foreground image and elimination of the dark spots to remove any possibilities of outliers. The morphological operation is further continued by performing morphological dilation operation over the reconstructed image with respect to the structural elements from the prior processing steps. The obtained dilated image is now subjected to image reconstruction process followed by computation of the complement of recently obtained dilated image. This operation results in transforming the darker region of brain MRI image into lighter one while the lighter region of brain MRI image into darker one. Finally, the maximum region is obtained for assisting the segmentation process.

(iii) Initiating the Classification Process: The next process is to apply the image binarization process to the recently obtained compliment image with respect to the gray thresholding of it. The obtained binarized matrix is now subjected to the computation process for obtaining the Euclidean distance between the binary images obtained. Further, the initial segmentation process using topographic mapping transformation is applied on the binarized image to further obtained more segmented image and thereby this process assists in marking the background. For better classification process it is necessary to control the intensity factor too at the end process. This process alters the image intensity with an aid of reconstruction of morphological elements with respect to obtained intensity variation vector and concatenation of marked background and compliment image. The obtained segmented image is than further subjected to topographic mapping transform. This process results in further segmentation process thereby offering better classified outcomes. The next step of this process is to apply finite directed graph without a form of involvement of cycles and with more inclusion of edges and vertex for better classification assistance. This process is further followed by constructing a believe network which is used for facilitating the decision making process in the presence of various variable sequences. As all the variables are encoded in the belief network therefore it can successfully control any form of missing data. Another advantage of this process is that it assists in good prediction over statistical data. This process is followed up by obtaining the entire region corresponding to the belief network followed by applying histogram on the top of it to perform region appending operation.

(iv) Boosting Classification using Artificial Neural Network: The proposed system doesn't directly apply Artificial Neural Network but it chooses to incorporate precise inputs to the neuron before starting training operation. For this purpose, the proposed system carries out feature extraction process where all the statistical features can be predicted. The proposed system uses all the frequently used descriptive statistical parameters over the brain MRI image for computing the features. Incorporating artificial Neural Network offers significant advantage in terms of extracting the inference from the large set of data as well as vague data. Therefore, better pattern extraction as well as identification of significant trends can be investigated. Therefore, the obtained information can be used for constructing an information structure using Artificial Neural Network. Such newly formulated structure comprises of massive quantity of highly interconnected elements in order to solve classification problem. The training operation is carried out by considering the prior steps of input image processing, segmentation, and initial classification operation. The proposed system uses real-valued function in order to construct an activation function in Artificial Neural Network. This adoption of training method offers various benefits e.g. forecasting of time-series as well as function approximation. This training operation involves iterating the learning process till elite classification information is obtained by the process. Finally, after the completion of the training operation, the proposed system can successfully perform classification that if the obtained image has presence of tumor or there is absence of tumors. Therefore, an effective classification process is obtained and a precise classification process is obtained. Figure 2 highlights the process flow of proposed system implementation.

6 Results Discussion

This section discusses about the outcomes obtain after implementing the proposed system discussed in prior section. Scripted in MATLAB, the proposed system targets assessing the classification performance of brain MRI image. Following are the brief of database used, result analysis strategy, visual outcomes, and numerical outcomes.

6.1 About Database

The study considers the input image to be brain MRI data from standard dataset [31–33]. The dataset consist of brain MRI data of more than 1000 subjects with 2168 sessions of magnetic resonance, and 1608 Positron Emission Tomography session. The dataset were captured considering various clinical test state for the subject. The proposed study has been tested with more than 5000 brain MRI data with variable sizes of the data. The dataset is also accompanied by the ground truth value for the purpose of the model validation.

6.2 Analysis Strategy

As the proposed study aims for addressing classification problems considering the case study of brain MRI image, therefore, accuracy is the primary performance parameters considered for the analysis. Apart from accuracy, the proposed study is also assessed considering processing time for the classification process. This adoption will also ensure analysis of the complexity involved in the learning process.

6.3 Visual Results

The visual outcomes of the proposed system for one sample brain MRI image are shown in Fig. 3. After taking the input image (Fig. 3(a)), Intensity Variation vector (IVV) (as shown in Fig. 3(b)) is obtained followed by obtaining image for Topographic Mapping transformation (TMT) (as shown in Fig. 3(c)). The foreground in indexed (Fig. 3(d)) followed by elimination of the dark spots (Fig. 3(e)) for resisting outliers followed by clearing all the edges (Fig. 3(f)).

This process is followed by TMT-based segmentation process which offers a comprehensive color based classification over the discrete regions of brain MRI image (Fig. 3(g)) and this operation is also followed by indexing different regions with specific index in numbers. Finally, after applying training using Artificial Neural Network, the proposed system can be extended in future to performs identification of the tumor region as well as predicts if the tumor is malignant or benign.

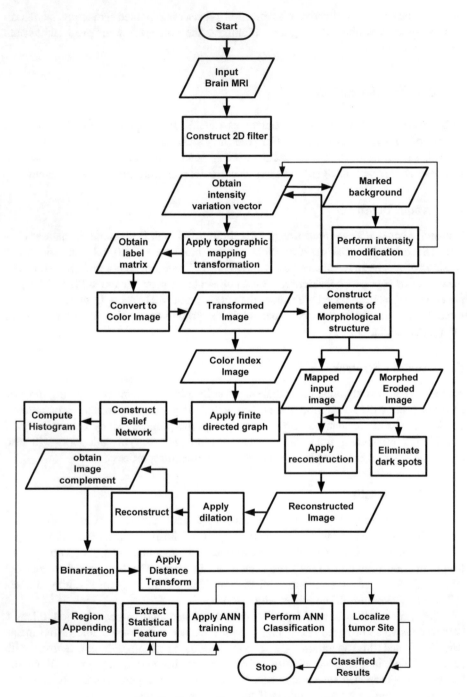

Fig. 2. Process flow of proposed system

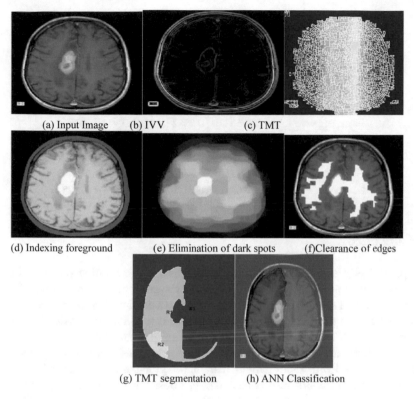

(a) Input Image   (b) IVV   (c) TMT

(d) Indexing foreground   (e) Elimination of dark spots   (f)Clearance of edges

(g) TMT segmentation   (h) ANN Classification

Fig. 3. Visual outcomes

6.4   Numerical Results

In order to evaluate the effectiveness of the proposed system, the outcomes of it are compared with the other approaches of training e.g. support vector machine, self-organizing map, feed-forward, etc. Similar set of images (set of images with abnormalities of tumor and normal regions) are considered for comparative analysis. The study also considers various parameters for computing accuracy e.g. True Positive, True Negative, False Positive, False Negative, Sensitivity/ Recall Rate, Specificity, Precision, F1-Score, etc. Table 1 highlights the numerical outcomes of comparative analysis.

From the outcome, it can be seen that proposed system offers better accuracy score as seen the precision and F1-score values. The complete processing time of the algorithm is found to be 0.28871 s while the range of processing time of existing system varies between 0.42883–1.6883 s. Hence, it can be seen that adopted artificial neural network has a significant positive impact on the classification performance for brain MRI image irrespective of the type and complexity present within the image. Apart from this, there is no buffer stored as the complete process works on run-time stating negligibly less spatial complexity.

Table 1. Comparative analysis of results

Accuracy parameters	Existing system		Proposed system	
	Normal	Abnormal	Normal	Abnormal
True positive	0.36629	0.78221	0.65116	0.88739
True negative	0.19778	0.46213	0.99992	0.2544
False positive	0.76882	0.92721	0.5512	0.74555
False negative	0.66828	0.29824	0.34599	0.10794
Sensitivity/Recall rate	0.26536	0.81299	0.65302	0.89155
Specificity	0.76559	0.48712	0.99995	0.25442
Precision	0.87656	0.27769	0.99992	0.543442
F1-Score	0.67998	0.24192	0.79007	0.67528

7   Conclusion

This paper has presented a mechanism that assists in performing classification. The significant contribution/novelty in the proposed system arc (i) the proposed system introduces a novel mechanism of intensity variation vector which assist in identification of foreground and background intensities associated with the brain MRI, (ii) introduction of topographic mapping transformation technique also assists in proper identification as well as indexing of foreground image which significantly assists in identifying the region inflicted with abnormalities with higher granularity, (iii) and adoption of the Artificial Neural Network also contributes to boost the classification process of the brain MRI even over the dynamic situation.

References

1. Landini, L., Positano, V., Santarelli, M.: Advanced Image Processing in Magnetic Resonance Imaging. CRC Press, Boca Raton (2018)
2. Wang, S.-H., Zhang, Y.-D., Dong, Z., Phillips, P.: Pathological Brain Detection. Springer, Singapore (2018)
3. Despotović, I., Goossens, B., Philips, W.: MRI segmentation of the human brain: challenges, methods, and applications. Hindawi-Comput. Math. Methods Med. 1–23 (2015). Article ID 450341
4. Sugimori, H.: Classification of computed tomography images in different slice positions using deep learning. Hindawi-J. Healthcare Eng. 1–9 (2018). Article ID 1753480
5. Xiao, Z., et al.: Brain MR image classification for Alzheimer's disease diagnosis based on multifeature fusion. Hindawi-Comput. Math. Methods Med. 1–13 (2017). Article ID 1952373
6. Bahadure, N.B., Ray, A.K., Thethi, H.P.: Image analysis for MRI based brain tumor detection and feature extraction using biologically inspired BWT and SVM. Hindawi-Int. J. Biomed. Imaging 1–12 (2017). Article ID 9749108
7. Saman, S., Narayanan, S.J.: Survey on brain tumor segmentation and feature extraction of MR images. Int. J. Multimedia Inf. Retrieval **8**, 1–21 (2018)

8. Liu, J., et al.: A survey of MRI-based brain tumor segmentation methods. Tsinghua Sci. Technol. **19**, 578–595 (2014). https://doi.org/10.1109/tst.2014.6961028
9. Hiralal, R., Menon, H.P.: A survey of brain MRI image segmentation methods and the issues involved. In: The International Symposium on Intelligent Systems Technologies and Applications, pp. 245–259. Springer, Cham (2016)
10. Harish, S., Ahammed, G.A.A., Banu, R.: An extensive research survey on brain MRI enhancement, segmentation and classification. In: 2017 International Conference on Electrical, Electronics, Communication, Computer, and Optimization Techniques (ICEEC-COT), Mysuru, pp. 1–8 (2017)
11. Mallick, P.K., Ryu, S.H., Satapathy, S.K., Mishra, S., Nguyen, G.N., Tiwari, P.: Brain MRI image classification for cancer detection using deep wavelet autoencoder-based deep neural network. IEEE Access **7**, 46278–46287 (2019)
12. Shao, Y., et al.: Hippocampal segmentation from longitudinal infant brain mr images via classification-guided boundary regression. IEEE Access **7**, 33728–33740 (2019)
13. Wang, L., Xie, C., Zeng, N.: RP-Net: a 3D convolutional neural network for brain segmentation from magnetic resonance imaging. IEEE Access **7**, 39670–39679 (2019)
14. Gudigar, A., et al.: Automated categorization of multi-class brain abnormalities using decomposition techniques with MRI images: a comparative study. IEEE Access **7**, 28498–28509 (2019)
15. Gumaei, A., Hassan, M.M., Hassan, M.R., Alelaiwi, A., Fortino, G.: A hybrid feature extraction method with regularized extreme learning machine for brain tumor classification. IEEE Access **7**, 36266–36273 (2019)
16. Liu, Y,, Wei, Y, Wang, C.: Subcortical brain segmentation based on atlas registration and linearized kernel sparse representative classifier. IEEE Access **7**, 31547–31557 (2019)
17. Wang, M., et al.: Graph-kernel based structured feature selection for brain disease classification using functional connectivity networks. IEEE Access **7**, 35001–35011 (2019)
18. Huang, J., Zhu, Q., Hao, X., Shi, X., Gao, S., Xu, X., Zhang, D.: Identifying resting-state multi-frequency biomarkers via tree-guided group sparse learning for schizophrenia classification. IEEE J. Biomed. Health Inf. **23**(1), 342–350 (2018)
19. Kermi, A., et al.: Fully automated brain tumour segmentation system in 3D-MRI using symmetry analysis of brain and level sets. IET Image Process. **12**(11), 1964–1971 (2018)
20. Ma, C., Luo, G., Wang, K.: Concatenated and connected random forests with multiscale patch driven active contour model for automated brain tumor segmentation of MR images. IEEE Trans. Med. Imaging **37**(8), 1943–1954 (2018)
21. Wang, Z., Zheng, Y., Zhu, D.C., Bozoki, A.C., Li, T.: Classification of Alzheimer's disease, mild cognitive impairment and normal control subjects using resting-state fMRI based network connectivity analysis. IEEE J. Translational Eng. Health Med. **6**, 1–9 (2018). https://doi.org/10.1109/jtehm.2018.2874887
22. Yuan, L., Wei, X., Shen, H., Zeng, L., Hu, D.: Multi-center brain imaging classification using a novel 3D CNN approach. IEEE Access **6**, 49925–49934 (2018)
23. Zhan, T., et al.: A glioma segmentation method using CoTraining and superpixel-based spatial and clinical constraints. IEEE Access **6**, 57113–57122 (2018)
24. Liu, J., Li, M., Pan, Y., Wu, F., Chen, X., Wang, J.: Classification of schizophrenia based on individual hierarchical brain networks constructed from structural MRI images. IEEE Trans. Nanobiosci. **16**(7), 600–608 (2017)
25. Kaur, T., Saini, B.S., Gupta, S.: Quantitative metric for MR brain tumour grade classification using sample space density measure of analytic intrinsic mode function representation. IET Image Process. **11**(8), 620–632 (2017)

26. Kasabov, N.K., Doborjeh, M.G., Doborjeh, Z.G.: Mapping, learning, visualization, classification, and understanding of fMRI data in the NeuCube evolving spatiotemporal data machine of spiking neural networks. IEEE Trans. Neural Netw. Learn. Syst. **28**(4), 887–899 (2017)
27. Kasabov, N., Zhou, L., Doborjeh, M.G., Doborjeh, Z.G., Yang, J.: New algorithms for encoding, learning and classification of fMRI data in a spiking neural network architecture: a case on modeling and understanding of dynamic cognitive processes. IEEE Trans. Cogn. Dev. Syst. **9**(4), 293–303 (2017)
28. Armañanzas, R., Iglesias, M., Morales, D.A., Alonso-Nanclares, L.: Voxel-based diagnosis of alzheimer's disease using classifier ensembles. IEEE J. Biomed. Health Inf. **21**(3), 778–784 (2017)
29. Liu, J., Li, M., Lan, W., Wu, F., Pan, Y., Wang, J.: Classification of Alzheimer's disease using whole brain hierarchical network. IEEE/ACM Trans. Comput. Biol. Bioinform. **15**(2), 624–632 (2018)
30. Liu, M., Zhang, J., Adeli, E., Shen, D.: Joint classification and regression via deep multi-task multi-channel learning for Alzheimer's disease diagnosis. IEEE Trans. Biomed. Eng. **66**(5), 1195–1206 (2019)
31. Openfrmri. https://openfmri.org/dataset/. Accessed 20 April 2019
32. Harish, S., Ahammed, G.A.: BrainMRI enhancement as a pre-processing: an evaluation framework using optimal gamma, homographic and DWT based methods (2019). https://doi.org/10.1007/978-3-030-00184-1_27
33. Harish, S., Ali Ahammed, G.F.: Integrated modelling approach for enhancing brain MRI with flexible pre-processing capability. Int. J. Electr. Comput. Eng. (IJECE) **9**(4), 2416–2424 (2019). https://doi.org/10.11591/ijece.v9i4.pp2416-2424. ISSN: 2088-8708

Application of Value Set Concept to Ellipsoidal Polynomial Families with Multilinear Uncertainty Structure

Radek Matušů[1]([⊠])[iD] and Bilal Şenol[2][iD]

[1] Centre for Security, Information and Advanced Technologies (CEBIA–Tech),
Faculty of Applied Informatics, Tomas Bata University in Zlín,
nám. T. G. Masaryka, 5555, 760 01 Zlín, Czech Republic
`rmatusu@utb.cz`
[2] Department of Computer Engineering, Faculty of Engineering,
Inonu University, 44280 Malatya, Turkey
`bilal.senol@inonu.edu.tr`

Abstract. The contribution intends to present the application of the value set concept to the ellipsoidal polynomial families with multilinear uncertainty structure. It is a follow-up to the previously published work, where the ellipsoidal polynomial families with affine linear uncertainty structure were studied. In the first parts of this paper, the basic terms related to the robustness under parametric uncertainty (e.g., uncertainty structure, uncertainty bounding set, family, and value set) are briefly recalled, with the accent on the ellipsoidal polynomial families. Subsequently, the non-convex value sets of the illustrative ellipsoidal polynomial family with multilinear uncertainty structure are plotted and analyzed. It is shown that the boundaries of the value set need not to mapped only from the boundaries in the parameter space but possibly also from the internal points.

Keywords: Value set · Ellipsoidal uncertainty · Spherical uncertainty · Family of polynomials · Multilinear uncertainty

1 Introduction

The systems with uncertain parameters represent a natural and comprehensible tool for the mathematical description of real-life objects with potentially complicated behavior by means of LTI models. Such systems are frequently given by so-called families of plants or families of polynomials, and so the robustness of these families have been studied for several decades [1, 2]. The family is determined by two main factors – the uncertainty structure, and the uncertainty bounding set.

The possible structures of uncertainty include the independent one, the affine linear one, the multilinear one, and the non-linear one (e.g., the polynomial or completely general ones). Besides, the special cases are represented, e.g., by the single parameter uncertainty or by the uncertain quasipolynomials [1, 3, 4]. This contribution focuses on

R. Silhavy et al. (Eds.): CoMeSySo 2019, AISC 1047, pp. 81–89, 2019.
https://doi.org/10.1007/978-3-030-31362-3_9

the case of the multilinear uncertainty structure, and it is considered to be a follow-up to the previously published work [5], in which the affine linear uncertainty structure was investigated.

The uncertainty bounding set is supposed to be defined by a ball in a norm [1]. The far most common case utilizes L_∞ norm (a box). The other possible approaches use either L_2 norm (a sphere or an ellipsoid) or L_1 norm (a diamond). This contribution assumes the uncertainty bounding set given by using L_2 norm. The corresponding families of polynomials are called as the ellipsoidal [5–8] or spherical [1, 9–13] polynomial families. Strictly speaking, some works, e.g. [1, 9], use the term "spherical polynomial family" solely for a family with independent uncertainty structure and uncertainty bounding set in the shape of an ellipsoid. In such a case, it can be considered as the analogy to the standard interval polynomial. However, the term "ellipsoidal/spherical polynomial family" is used more generally here, i.e., for any uncertainty structure. Despite the fact that the ellipsoidal/spherical uncertainty is not researched so often as the classical "box" uncertainty, there is still a range of works dealing with the ellipsoidal/spherical uncertainty and related problems, see e.g. [5–18].

Obviously, the most crucial property of all control loops is their stability, and thus the closed-loop control systems under uncertainty are requested to be robustly stable, which means they remain stable for all possible members from the assumed family. A number of methods for robust stability analysis of SISO systems were published, and their proper choice depends primarily on the uncertainty structure and the uncertainty bounding set [1–4]. A popular method is based on the value set concept and zero exclusion condition [1]. It represents a straightforward but efficient technique that is applicable to many cases, and that can be especially advantageous for complicated uncertainty structures with the lack of other robust stability analysis methods [4, 19].

The contribution is intended as a follow-up to the previously published paper [5]. The current work concentrates on the value sets of the ellipsoidal polynomial families with multilinear uncertainty structure, while the former paper [5] was focused on a less general affine linear uncertainty structure. Besides, other preliminary results have been already presented e.g. in [12, 13], but these works were aimed only at the families with the independent uncertainty structure, i.e., the "spherical version" of the interval polynomials, and they used the Polynomial Toolbox for Matlab for plotting the value sets [9]. Nevertheless, this contribution deals with more general case represented by the ellipsoidal polynomial families with the multilinear structure of uncertainty. It is shown that for the multilinear case, the value set is generally a non-convex shape and, moreover, that the boundaries of the value set in the complex plane can be mapped not only from the boundaries in the parameter space but possibly also from the internal points in the parameter space.

2   Uncertainty Structure

As mentioned above, the families of systems with parametric uncertainty are given by the uncertainty structure and by the uncertainty bounding set. Suppose the uncertain polynomial:

$$p(s,q) = \sum_{i=0}^{n} \rho_i(q)s^i \tag{1}$$

where q is a vector of real uncertain parameters and $\rho_i(q)$ are coefficient functions. The form of $\rho_i(q)$ is crucial as it determines the uncertainty structure and consequently it predetermines the choice of suitable robust stability analysis method. The brief classification is stated in the second paragraph of the Introduction. This work is aimed at the multilinear uncertainty structure, which means that if all but one uncertain parameters are fixed, then $\rho_i(q)$ is affine linear in the remaining (non-fixed) parameter. Practically speaking, the coefficients can contain the product of uncertain parameters [1, 3, 4, 20].

3 Uncertainty Bounding Set

The uncertainty bounding set Q is supposed as a ball in a norm. The typical shapes of Q are a box (for L_∞ norm), a sphere or an ellipsoid (for L_2 norm or weighted L_2 norm), and a diamond (for L_1 norm). The L_2 norm case is:

$$\|q\|_2 = \sqrt{\sum_{i=1}^{n} q_i^2} \tag{2}$$

or:

$$\|q\|_{2,w} = \sqrt{q^T W q} \tag{3}$$

where $q \in \mathbb{R}^k$ and $W = diag\left(w_1^2, w_2^2, \ldots, w_k^2\right)$ is a positive definite symmetric matrix (weighting matrix) of size $k \times k$. Provided that $r \geq 0$ and $q^0 \in \mathbb{R}^k$, the ellipsoid (in $\mathbb{R}^k$) centered at q^0 can be expressed via the inequality:

$$\left(q - q^0\right)^T W \left(q - q^0\right) \leq r^2 \tag{4}$$

or equivalently:

$$\left\|q - q^0\right\|_{2,W} \leq r \tag{5}$$

For example, the special case of the ellipse of uncertain parameters can be readily visualized for $k = 2$ (two-dimensional space) [1, 12, 13].

4 Ellipsoidal Polynomial Family

In this paper, the family of polynomials is given by [1, 3, 4]:

$$P = \{p(s,q) : q \in Q\} \tag{6}$$

where $p(s,q)$ has the multilinear uncertainty structure, and Q is an ellipsoid. Therefore, the resulting family of polynomials (6) is called the ellipsoidal polynomial family.

5 Value Sets

The concept of the value sets combined with the utilization of the zero exclusion condition offers a universal tool for robust stability analysis [1, 3, 4].

Assume a family of polynomials (6). As defined in [1], its value set at frequency $\omega \in \mathbb{R}$ is given by:

$$p(j\omega, Q) = \{p(j\omega, q) : q \in Q\} \tag{7}$$

and so $p(j\omega, Q)$ is the image of Q under $p(j\omega, \cdot)$. The value set can be practically constructed by substituting s for $j\omega$ and letting q range over Q. The examples of the typical shapes of the value sets for the polynomial families with various uncertainty structures and Q in the shape of a box are shown e.g. in [3, 4].

From the viewpoint of robust stability testing, the value sets are plotted for selected samples of non-negative frequencies and subsequently the zero exclusion condition fulfillment is checked. More information on the parametric uncertainty, the value sets, and the robust stability investigation can be found e.g. in [1, 3, 4].

The formulas for computation of elliptical value sets of a spherical polynomial family with independent uncertainty structure were derived in [1] on the basis of the proof of the Soh-Berger-Dabke theorem [15].

In this paper, the value sets for an ellipsoidal polynomial family with multilinear uncertainty structure are numerically calculated and plotted.

6 Illustrative Example

Assume a polynomial family given by the multilinear uncertainty structure:

$$p(s,q) = s^3 + (4q_1q_2 + 2)s^2 + (q_1q_2 + q_1 + q_2 + 5)s + (q_1 + q_2 + 3) \tag{8}$$

combined with two types of uncertainty bounding sets:

(a) Weighted L_2 norm (ellipsoid – ellipse):

$$\begin{aligned} \|q - q^0\|_{2,W} &\leq 1 \\ q^0 &= [0, 0] \\ W &= \begin{bmatrix} w_1^2 & 0 \\ 0 & w_2^2 \end{bmatrix} = \begin{bmatrix} 1 & 0 \\ 0 & 2 \end{bmatrix} \end{aligned} \tag{9}$$

that is equivalently:

$$\begin{aligned}&\left(q_1 - q_1^0\right)^2 + 2\left(q_2 - q_2^0\right)^2 \leq 1\\&q^0 = [0,0]\end{aligned} \qquad (10)$$

(b) L_∞ norm (box – rectangle):

$$\begin{aligned}&q_1 \in [-1, 1]\\&q_2 \in [-0.5, 0.5]\end{aligned} \qquad (11)$$

The elliptical uncertainty bounding set (9) is visualized in Fig. 1, which shows not only the boundary (blue curve) but also 1000 randomly chosen internal points. Moreover, the rectangle formed by the displayed axes in Fig. 1 coincides with the rectangular uncertainty bounding set (11).

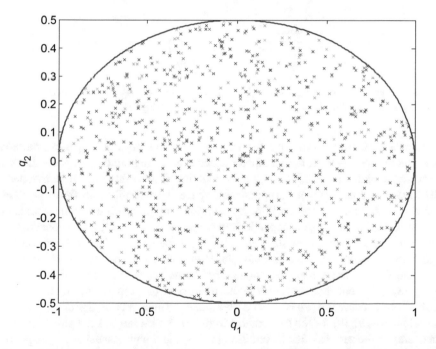

Fig. 1. Elliptical uncertainty bounding set (9) with 1000 randomly chosen internal points.

The value set of the ellipsoidal polynomial family with multilinear uncertainty structure (8) and the uncertainty bounding set (9) for the frequency $\omega = 1$ is plotted in Fig. 2. This value set is comprised of its "boundaries", which are related to the boundaries of the elliptical uncertainty bounding set (9) from Fig. 1, and the images of 1000 randomly selected points within this uncertainty bounding set.

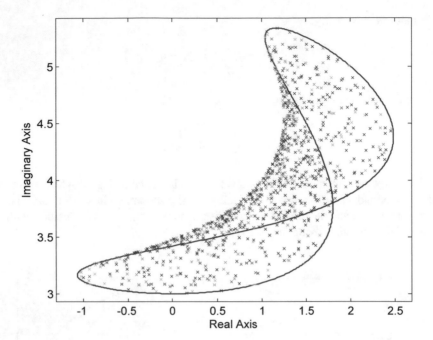

Fig. 2. The value set of the family (8), (9) for $\omega = 1$ with the images of 1000 randomly chosen internal points from the elliptical uncertainty bounding set (Fig. 1).

As can be seen, not only that the value set is a non-convex shape, but moreover, some internal points from the parameter space create the boundary of the value set in the complex plane. In other words, the boundaries of the value set can be mapped not only from the boundaries in the parameter space but possibly also from the internal points. This can happen for multilinear (and more complicated) uncertainty structures, but not for affine linear or independent uncertainty structures [5]. Obviously, it generally complicates the robust stability analysis of the polynomial family, because the edge-based tests are not directly available. However, the value set concept is valid.

Then, Fig. 3 shows both the "boundary" value set of the ellipsoidal type of the family (8), (9) (solid blue curve) and the traditional box type of the family (8), (11) (dashed black curve) for $\omega = 1$. The "boundary" value sets of the ellipsoidal polynomial family (8), (9) for the frequencies from 0 to 3 with step 0.1 are plotted in Fig. 4. According to the zero exclusion condition [1, 3, 4], and with respect to the shape of the full versus "boundary" value sets (Fig. 2), the "boundary" value sets imply the robust stability of the family (8), (9).

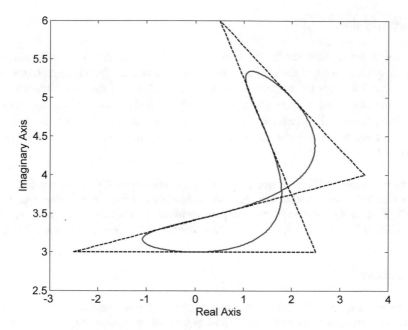

Fig. 3. Comparison of the "boundary" value sets of the family (8), (9) (solid blue curve) and of the family (8), (11) (dashed black curve) for $\omega = 1$.

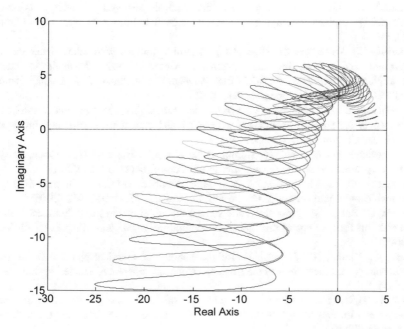

Fig. 4. The "boundary" value sets of the ellipsoidal polynomial family (8), (9) for $\omega = 0 : 0.1 : 3$.

7 Conclusion

This contribution dealt with the ellipsoidal polynomial families with multilinear uncertainty structure and their value sets. After recalling the fundamental terms, it was demonstrated that for the multilinear case, the boundaries of the non-convex value set in the complex plane need not be mapped only from the boundaries in the parameter space but possibly also from the internal points in the parameter space. Moreover, the obtained value set was compared with the standard "box" version of the polynomial family.

Acknowledgments. This work was supported by the Ministry of Education, Youth and Sports of the Czech Republic within the National Sustainability Programme project No. LO1303 (MSMT-7778/2014) and also by the European Regional Development Fund under the project CEBIA-Tech No. CZ.1.05/2.1.00/03.0089. This assistance is very gratefully acknowledged.

References

1. Barmish, B.R.: New Tools for Robustness of Linear Systems. Macmillan, New York (1994)
2. Bhattacharyya, S.P.: Robust control under parametric uncertainty: an overview and recent results. Ann. Rev. Control **44**, 45–77 (2017)
3. Matušů, R., Prokop, R.: Graphical analysis of robust stability for systems with parametric uncertainty: an overview. Trans. Inst. Meas. Control **33**(2), 274–290 (2011)
4. Matušů, R., Prokop, R.: Robust stability analysis for systems with real parametric uncertainty: implementation of graphical tests in Matlab. Int. J. Circ. Syst. Signal Process. **7** (1), 26–33 (2013)
5. Matušů, R.: Value sets of ellipsoidal polynomial families with affine linear uncertainty structure. In: Cybernetics and Automation Control Theory Methods in Intelligent Algorithms. Proceedings of CSOC 2019. Advances in Intelligent Systems and Computing, vol. 986, pp. 255–263. Springer, Cham (2019)
6. Sadeghzadeh, A., Momeni, H.: Fixed-order robust H∞ control and control-oriented uncertainty set shaping for systems with ellipsoidal parametric uncertainty. Int. J. Robust Nonlinear Control **21**(6), 648–665 (2011)
7. Sadeghzadeh, A., Momeni, H., Karimi, A.: Fixed-order H∞ controller design for systems with ellipsoidal parametric uncertainty. Int. J. Control **84**(1), 57–65 (2011)
8. Sadeghzadeh, A., Momeni, H.: Robust output feedback control for discrete-time systems with ellipsoidal uncertainty. IMA J. Math. Control Inf. **33**(4), 911–932 (2016)
9. Hurák, Z., Šebek, M.: New tools for spherical uncertain systems in polynomial toolbox for Matlab. In: Proceedings of the Technical Computing Prague 2000, Prague, Czech Republic (2000)
10. Tesi, A., Vicino, A., Villoresi, F.: Robust stability of spherical plants with unstructured uncertainty. In: Proceedings of the American Control Conference, Seattle, Washington, USA (1995)
11. Polyak, B.T., Shcherbakov, P.S.: Random spherical uncertainty in estimation and robustness. In: Proceedings of the 39th IEEE Conference on Decision and Control, Sydney, Australia (2000)

12. Matušů, R., Prokop, R.: Robust stability analysis for families of spherical polynomials. In: Intelligent Systems in Cybernetics and Automation Theory. Proceedings of CSOC 2015. Advances in Intelligent Systems and Computing, vol. 348, pp. 57–65, Springer, Cham (2015)
13. Matušů, R.: Spherical families of polynomials: a graphical approach to robust stability analysis. Int. J. Circ. Syst. Signal Process. **10**, 326–332 (2016)
14. Chen, J., Niculescu, S.-I., Fu, P.: Robust stability of quasi-polynomials: frequency-sweeping conditions and vertex tests. IEEE Trans. Autom. Control **53**(5), 1219–1234 (2008)
15. Soh, C.B., Berger, C.S., Dabke, K.P.: On the stability properties of polynomials with perturbed coefficients. IEEE Trans. Autom. Control **30**(10), 1033–1036 (1985)
16. Barmish, B.R., Tempo, R.: On the spectral set for a family of polynomials. IEEE Trans. Autom. Control **36**(1), 111–115 (1991)
17. Tsypkin, Y.Z., Polyak, B.T.: Frequency domain criteria for l^p-robust stability of continuous linear systems. IEEE Trans. Autom. Control **36**(12), 1464–1469 (1991)
18. Biernacki, R.M., Hwang, H., Bhattacharyya, S.P.: Robust stability with structured real parameter perturbations. IEEE Trans. Autom. Control **32**(6), 495–506 (1987)
19. Matušů, R., Pekař, R.: Robust stability of thermal control systems with uncertain parameters: the graphical analysis examples. Appl. Therm. Eng. **125**, 1157–1163 (2017)
20. Matušů, R., Şenol, B., Pekař, L.: Robust stability of fractional order polynomials with complicated uncertainty structure. PLoS ONE **12**(6), e0180274 (2017)

Comparison of Methods for Time Series Data Analysis for Further Use of Machine Learning Algorithms

Andrea Nemethova[✉], Dmitrii Borkin, and German Michalconok

Faculty of Materials Science and Technology in Trnava,
Institute of Applied Informatics, Automation and Mechatronics,
Slovak University of Technology in Bratislava, Bratislava, Slovakia
{andrea.peterkova,dmitrii.borkin,
german.michalconok}@stuba.sk

Abstract. The aim of this paper is to cover the initial stage of data analysis of time series data. In our research we are working with real world data obtained from a thermal plant. The main objective of this paper is to analyze the data and to discover potential trends. We also present the comparison of various time series data smoothing methods. This stage of research is necessary in order to continue with applying machine learning algorithms in next stage. After brief introduction we introduce the nature and character of given data. In following chapters of this paper we introduce used methods and present the results. The whole data analysis was performed in Phyton.

Keywords: Data analysis · Forecast · Exponential smoothing

1 Introduction

Using of data analysis methods are nowadays used extensively. The main reason for this is, that many processes (industrial, financial, medical, etc.) are generating huge amounts of data every year. These data are collected, stored and have great knowledge potential. This, many times, hidden knowledge can help to optimize processes, minimize costs or maximize the revenue. There are many methods, which are helping to unveil the knowledge from the data. It highly depends on the nature of the data and also on the objective of the analysis, which method to choose. In our paper we are dealing with the time series data. Time series data is a series (or sequence) of data points, which are indexed in time order. We can also say, that the time series data is a series of discrete-time data. This kind of data is usually analyzed to discover the characteristics of the data like for example trends. After discovering trends in time series data a prediction model can be designed. Based on this model we can predict future values of the time series based on previous values. These models are often based on machine learning methods and algorithms. This paper focuses on the analysis of the real world time series data and comparison of trend-finding methods. The used methods are based on traditional statistical analysis and are described in further sections of this paper.

© Springer Nature Switzerland AG 2019
R. Silhavy et al. (Eds.): CoMeSySo 2019, AISC 1047, pp. 90–99, 2019.
https://doi.org/10.1007/978-3-030-31362-3_10

2 Background and Methods

First step in data analysis using data mining methods is to understand the data. In our paper we are dealing with time series type of data, which are describing thermal output power in megawatts (MW). This data were acquired from a local heating plant. For the

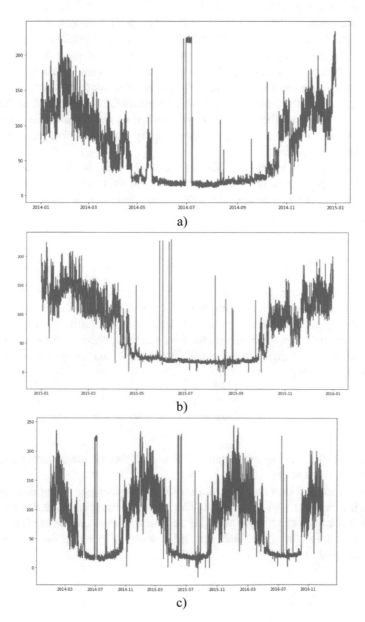

a)

b)

c)

Fig. 1. The overall overview of the given data: (a) the first year of measurements (b) the second year of measurement (c) the whole time of measurement

purposes of our research the data were from the years 2014 to the 2016. The data were obtained in their raw form as an export to a .xls files. Data were divided into three separate tables/files according to the corresponding year. Each table consisted of approximately 50000 rows of raw data. The dataset has 2 parameters. First, there was the parameter describing the date and time of the measurement. Second parameter was the actual thermal output power in MW.

After analyzing the data structure of given data set it is needed to search the actual data set for any kind of data errors. Data from any kind of industrial processes are prone to errors. After analyzing the dataset for potential errors, we have discovered that the most common error was missing data error. This was caused mainly by temporary sensor failures. We have decided to remove the empty records and also to combine all three files into one consistent data set. Figure 1a shows the histogram of the first year of measurements, where on the x axis there is time and on the y axis is the actual thermal output power in MW. The Fig. 1b shows exactly the same histogram, but for the second year of measurement.

After combining the separate files of given data set we have obtained overall overview of the given data. On the Fig. 1c there is the overview through the whole time of measurement. As it can be seen from this figure, there is a clear and visible trending in given data.

3 Moving Average

The first statistical data analysis method after data preprocessing stage was the moving average. This method produces moving average values, which are simplifying the overall data image. With the use of this method we can see trending and seasoning in the data set much clearer. With this method we can also predict next value based on the previous values of the time series data. Following formula shows the method of calculation of the moving average.

$$\widehat{y}_t = \frac{1}{k} \sum_{n=0}^{k-1} y_t - n \tag{1}$$

Where $\widehat{y}_t$ is the moving mean value, k is the number of actual values and y_t is the actual value of time series.

Figure 2a and b shows the computation of moving average on our actual data. The red line represents the moving average. The blue and orange lines represent the upper and lower boundary, which were computed by the 2 times sigma rule. Figure 2a shows the moving average for whole data set and b shows a close up for a time frame of approximately one month.

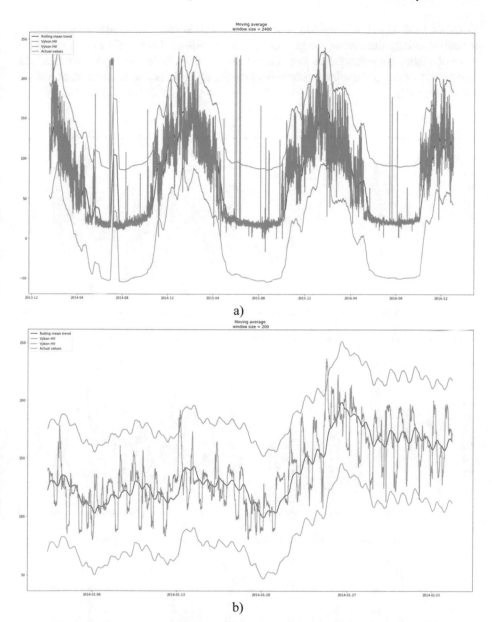

Fig. 2. The use of moving average method: (a) (b) the deeper look at the data

4   Exponential Smoothing

Exponential smoothing is a powerful method also called time series forecasting method. It is used for univariate data that could extend to support data with some systematic trend. The methods of exponential smoothing forecasting have in common fact that prediction is a weighted sum of past observations. The model explicitly uses an exponentially decreasing weight for past observations. There exist three main types of exponential smoothing time series forecasting methods. In our paper, we deal with two types: single and double exponential smoothing. Lets take a closer look at each of these methods.

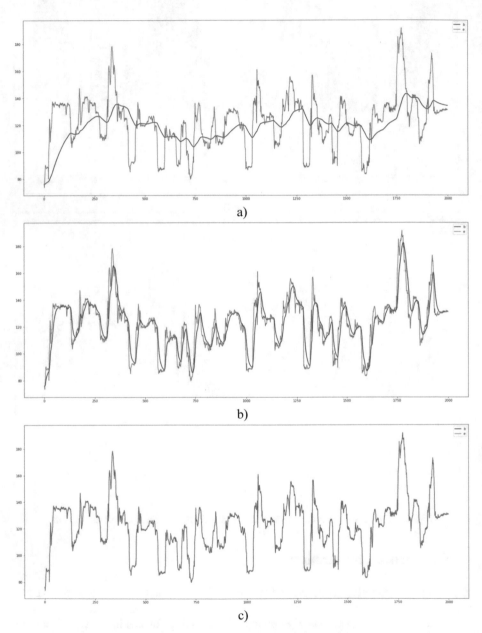

Fig. 3. The single exponential smoothing: (a) alpha = 0.01, (b) alpha = 0.1, (c) alpha = 0.9

4.1 Single Exponential Smoothing

Single exponential smoothing also called simple exponential smoothing is a time series forecasting method for univariate data without any trend or even seasonality. This method requires a smoothing factor. This single parameter is called coefficient alpha. The alpha parameter can be set to a value between 0 and 1. Large values of parameter alpha mean that the model focuses mainly to the most recent past observations. When using a small value of parameter alpha, more of the history is taken into account when making a forecast.

Values close to 1 demonstrate fast learning, which mean that only the most recent values influence the forecasts. Opposite values close to 0 indicates slow learning that means that past observations have a large influence on the forecast (Fig. 3).

$$\widehat{y}_t = \propto * y_t + (1 - \propto) * \widehat{y}_{t-1} \tag{2}$$

4.2 Double Exponential Smoothing

Double exponential Smoothing is explicitly adding a support for trends in the univariate time series. This method uses two parameters. The alpha parameter is also use to control smoothing factor, but an additional smoothing factor is added to control the decay of the influence of the change in trend. This additional parameter is called beta. This method supports trends that change in this two different ways:

- Additive trend when trend is linear,
- Multiplicative when trend is exponential.

For multi-step forecasts, forecasts with longer range, the trend can continue on unrealistically. Than it could be useful to dampen the trend over the time, which mean reducing size of the trend over further time steps down to a straight line (Fig. 4).

$$l_x = \propto y_x + (1 - \propto) * (l_{x-1} + b_{x-1}) \tag{3}$$

$$b_x = \beta(l_x - l_{x-1}) + (1 - \beta) * b_{x-1} \tag{4}$$

$$\widehat{y}_{x-1} = l_x + b_x \tag{5}$$

4.3 Triple Exponential Smoothing

Holt-Winters model is also called a triple exponential smoothing. This type of smoothing is a way to model three aspects of the time series: a typical value (average), a slope (trend) over time, and a cyclical repeating pattern (seasonality). In our research, we use this type of smoothing to encode values from the past and use them to forecast the future values (Fig. 5).

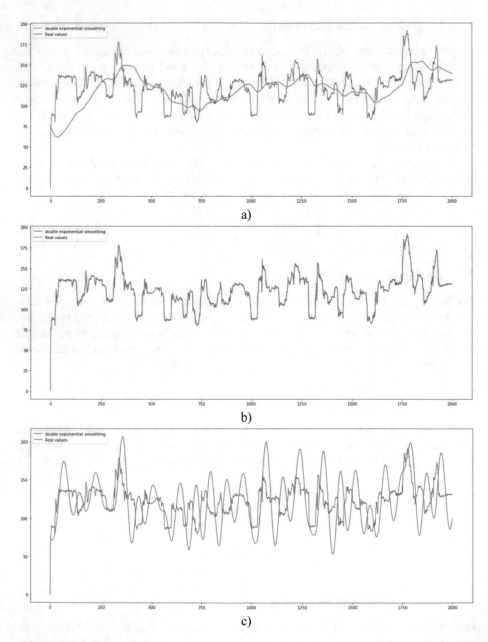

Fig. 4. The double exponential smoothing: (a) alpha = 0.01, beta = 0.01, (b) alpha = 0.5, beta = 0.01, (c) alpha = 0.01, beta = 0.5

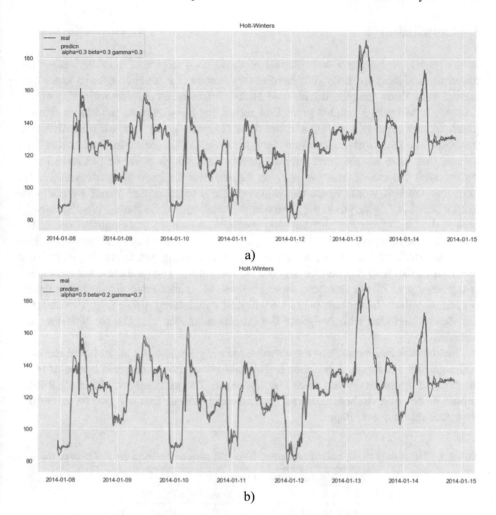

Fig. 5. The Holt-Winters method: (a) alpha = 0.3, beta = 0.3, gamma = 0.3 (b) alpha = 0.5, beta = 0.2, gamma = 0.7

$$l_x = \propto (y_x - s_x - L) + (1 - \alpha)(l_{x-1} + b_{x-1}) \tag{3}$$

$$b_x = \beta(l_x - l_{x-1}) + (1 - \beta) * b_{x-1} \tag{4}$$

$$s_x = \gamma(y_x - l_x) + (1 - \gamma)s_{x-L} \tag{5}$$

$$\widehat{y}_{x+m} = l_x + mb_x + s_{x-L+(m-1)modL} \tag{6}$$

5   Conclusion

In this paper we have covered classical methods used to predict the time series data, based on statistical methods. We have used methods like moving average and exponential smoothing (simple, double and Holt-Winters method). The main aim of our research is to build a reliable prediction model for the described process of thermal plant. In this stage of our research we have analysed given data with the statistical methods. The results of this analysis show trending and seasoning in our data set. That means, that there are recurrent patterns in this data, which could be predicted in the future, with the use of machine learning algorithms. We have also discovered spots with unusual behaviour, where the values of the thermal power output were significantly different (higher) than they were in that particular time frame. We also see the potential of using the machine learning methods to analyse these unusual spots. We have used the described methods also to predict future values based on the values from the past. However the main disadvantage of predicting future values with these methods is, that they can predict only the nearest future. The prediction system, which is the main goal of our research, should be able to predict more distant future development of values. In our next paper we will deal with adding supplement parameters in our dataset and also with designing the prediction model with the machine learning methods.

From our initial research of the machine learning methods we have chosen methods and algorithms mentioned in Table 1. This table also shows the performance of these methods in the means of the mean absolute error. The parameters of these algorithms should be however further optimized for best performance. This will also be part of our next research and next paper.

Table 1. The results of the trained and tested SVM for parameters from the first case of training

Mean absolute error		
Model	Train [%]	Test [%]
Linear model	8.95	10.64
Elastic net	8.97	10.66
Decision tree regression	9.36	11.98
Random forest regression	4.27	11.69
Linear SVR	39.54	25.29

Acknowledgments. This publication is the result of implementation of the project: "UNIVERSITY SCIENTIFIC PARK: CAMPUS MTF STU - CAMBO" (ITMS: 26220220179) supported by the Research & Development Operational Program funded by the EFRR.

This publication is the result of implementation of the project VEGA 1/0673/15: "Knowledge discovery for hierarchical control of technological and production processes" supported by the VEGA.

References

1. Gardner, J.R., Everette, S.: Exponential smoothing: the state of the art. J. Forecasting **4**(1), 1–28 (1985)
2. Hyndman, R., et al.: Forecasting with exponential smoothing: the state space approach. Springer, Heidelberg (2008)
3. Trigg, D.W., Leach, A.G.: Exponential smoothing with an adaptive response rate. J. Oper. Res. Soc. **18**(1), 53–59 (1967)
4. Michalakes, J., et al.: The weather research and forecast model: software architecture and performance. In: Use of High Performance Computing in Meteorology, pp. 156–168 (2005)
5. Michalakes, J., et al.: Development of a next-generation regional weather research and forecast model. In: Developments in Teracomputing, pp. 269–276 (2001)
6. Chatfield, C., Yar, M.: Holt-Winters forecasting: some practical issues. J. Roy. Stat. Soc. Ser. D (Stat.) **37**(2), 129–140 (1988)
7. Chatfield, C.: The Holt-winters forecasting procedure. J. Roy. Stat. Soc. Ser. D (Stat.) **27**(3), 264–279 (1978)

Missing Data in Collaborative Data Mining

Carmen Ana Anton$^{(\boxtimes)}$, Oliviu Matei, and Anca Avram

Electric, Electronic and Computer Engineering Department,
Technical University of Cluj-Napoca, North University Center Baia Mare,
Dr. Victor Babes 62A, 430083 Baia-Mare, Romania
carmen.anton@cunbm.utcluj.ro, oliviu.matei@holisun.com, anca.avram@ieee.org
https://inginerie.utcluj.ro/

Abstract. Incomplete data in the data mining process is a common case. In collaborative data mining, treating this case can bring better performance and results, so the study of missing data from the perspective of the collaborative data mining approach is addressed in this article. Completing missing data for research sources have been done in two ways: using the average from other sources or adding an operator from the RapidMiner application to the processes. Both approaches have generated good results and can be considered as viable alternatives in collaborative data mining processes.

Keywords: Data mining · Collaborative · Missing data

1 Introduction

In a data mining process we need complete data to get the most accurate results. The lack of data implies the approach of solutions that offer results close to those with complete data and with as little decrease in the accuracy of the results. In a collaborative data mining (CDM) process, addressing the lack of data could be done quite easily, considering that for predicting a source we can use data from another source.

Matei et al. proposes in [1] a stratified architecture for data mining process, that can be the starting point for initiating a CDM process applied for missing data. We can use the idea define by Matei et al. in [2], which says that for predictions of one source we can use data from another sources.

In many research, in terms of temperature prediction, a number of virtual machine learning algorithms has approached, with better results than traditional statistical methods. Thus, Radhika and Shashi in [13], present a study of the application of Support Vector Machines (SVM) compared with Multi-Layer Perceptron (MLP) for weather prediction. The subject was approached, also by Matei et al. in [3], by Avram et al. in [6], by Anton et al. in [4,5] and was tested with different algorithms, such as k-Nearest Neighbor model (KNN), Local polynomial regression (LPR), Neural Net model (NN) and Support Vector Machine (SVM) and for collaborative data mining approach.

© Springer Nature Switzerland AG 2019
R. Silhavy et al. (Eds.): CoMeSySo 2019, AISC 1047, pp. 100–109, 2019.
https://doi.org/10.1007/978-3-030-31362-3_11

The problem of missing values for some attributes is important for the process of data mining and for statistical reasoning. In both ways exist diverse theoretical methods to deal with it and were described by Imielinski and Lipski in [11] and in [12].

Multiple research has been done in determining the proper method for filling out the missing data. An example is reported by Grzymala-Busse et al. in [9], where nine different approaches to missing attribute values are presented and compared. From that research shows that the method of applying the C4.5 algorithm and the method of ignoring the examples with missing values of the attributes are the best ones. Other approaches have been studied by Kotsiantis et al. in [10], where some incorrect values are removed and treated as "missing data" using different interpolation techniques and filling methods.

2   Methods

2.1   Collaborative Data Mining with Missing Data

The data collaboration system use to obtain a prediction, was experimented and studied by Anton et al. in [4] and in [5], comparing its effects on the values obtained in parallel with the process using only the data from a single source. The conclusions state that adapting the algorithms used and optimizing their parameters can achieve much improved results.

Grzymala-BusseWitold and Grzymala-Busse in [8], divided the methods to handle the problem of missing values in two categories:

- sequential methods (called also preprocessing methods)
- parallel methods (methods in which taken into account during the main process of acquiring knowledge).

In this research, we have tested both methods, choosing for the first case, replacing for missing attributes with values calculated by the arithmetic average of other numerical attributes, and we call this method cMDa (collaborative Missing Data average). For the second approach, from the parallel methods, we have selected the mode to include in the data mining process a sub-process which completes the missing data with the values from a learning algorithm, naming the method cMDimv (collaborative Missing Data Impute missing values).

2.2   Preparation of Data Sources

Data for the research was downloaded from a public web site[1]. Weather stations are from the Transylvania Plateau. The study continues the research conduct by Anton et al. in [5] where it is defined the collaborative approaches that use multiple sources. The 12 meteorological stations, from that study, were also used in this article to address missing data problem, and the source data for

---

[1] https://rp5.ru/.

the experiment were the temperature (degrees Celsius) measured at 2 m above ground. The process are implement in Rapid Miner and we want to predict the air temperature at a source location. The time frame for the study is February 1, 2010 to April 30, 2010. The structure of the data used in the research is composed of temperatures expressed in degrees Celsius recorded by meteorological stations for each day of the specified period.

2.3   Missing Data - cMDaa Method

As is mentioned by Zhang et al. in [14] data preparation is a very important step in a data mining process. Therefore, the methods used for missing data should be tailored to the situation and found the ones with the highest success rate. One of the problem, in data mining, is the lack of data. In this article, we propose to approach two variants of completing the missing data and studying their impact on CDM processes.

For the first stage of the study, five meteorological stations were selected for prediction process (S1 to S5) and these were investigated in various stages of missing data. Consequently for the five stations, data was deleted in different percentages, namely: 25%, 50%, 75% and were included in collaborative data mining process with 3 sources. For the collaborative pair sources from the process was use another 7 sources. The data mining processes were applied differently as follows:

- standalone processes with missing data, 3 variants (std - standalone Missing Data): std25 (25% missing data), std50 (50% missing data), std75 (75% missing data);
- collaborative processes with 3 stations, 3 variants (cMDaa - collaborative Missing Data arithmetic average): cMDaa25, cMDaa50, cMDaa75. For these processes, at the source that had missing data, attributes were filled in with the value representing the arithmetic average of the temperature from the other sources in the process. We obtained for each source 21 values.

For each source in the study we obtained results for standalone process and collaborative one. The results of the S5 source for the stand-alone process, for each case of missing data in part, were: 0.388 (std25), 0.212 (std50), 0.106 (std75). Instead, for the collaborative process approached for each of the specified cases, the results can be observed in the Fig. 1. It can be seen in the Fig. 1 that all the values obtained in the missing data processes (MD25, MD50, MD75) are smaller than those in which the collaborative method was approached with the filling in the missing data with arithmetic average (MD25a, MD50a, MD75a). For all studied sources results were similar to S5 source.

The values obtained with the collaborative process (with the addition of the average temperature of other sources from the process) are higher than those obtained with the standalone process, in a percent of 100%. Depending on the missing data volume, the differences between the collaborative value and the standalone value are in the following ranges:

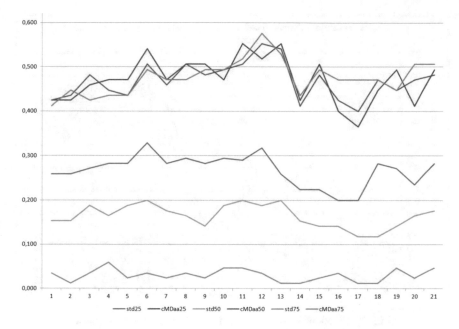

Fig. 1. Results obtained for S5 in CDM process with missing data.

- Missing data 25%: [0.165, 0.330]
- Missing data 50%: [0.247, 0.424]
- Missing data 75%: [0.376, 0.565]

It is obvious that, there is a significant increase in the values obtained in collaborative mode, compared with the standalone ones, which leads to a better prediction if missing data are replaced with another value, in this case, the average value of the sources in the process.

2.4 Results - cMDaa Method

For all 5 studied sources, the values obtained in collaborative processes with 25%, 50% and 75% missing data are higher than the value obtained in the standalone mode of the source. The basic conclusion is that using data from other stations significantly improves the accuracy of the collaborative process results. In the Fig. 2 were used the following notations: S1–S5 data sources, the OY axis is represented by the predictions values obtained for each station and the OX axis represents the types of processes applied to the stations with the minimum and maximum values obtained, such as:

- Minimum and maximum values for collaborative process with sources with complete data: c_min, c_max;
- Minimum and maximum values for collaborative process for sources with incomplete data in different percentages: c25_min, c25_max, c50_min, c50_max, c75_min, c75_max;

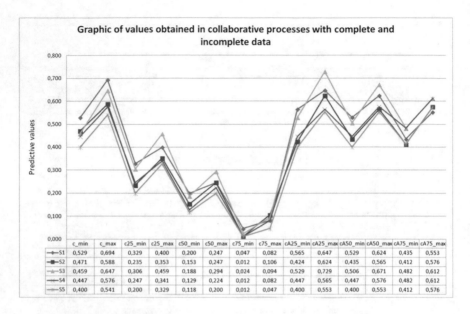

Fig. 2. Chart of values obtained in collaborative processes.

- Minimum and maximum values for a collaborative process for sources with incomplete data but with filling them with arithmetic average (also, in different percentages): cA25_min, cA25_max, cA50_min, cA50_max, cA75_min, cA75_max.

The observations that reveal from the Fig. 2 arc: the values obtained in collaborative processes with incomplete data are the lowest, which was to be expected, the values obtained with collaborative processes with the data filled in with the arithmetic mean are about the same as those of the sources that were predicted with their initial data, some stations (S2, S3) recorded higher values in the processes where they had incomplete data completed with arithmetic mean from other sources.

2.5   Missing Data - cMDimv Method

The second approach (cMDimv - collaborative Missing Data impute missing value) involved the inclusion in the collaborative process of a specific operator from RapidMiner (Impute Missing Value = IMV) to fill in the missing data with data generated based on an algorithm. This operator estimates the values for the missing values of the selected attributes. This estimate is made by applying a learned model that is a sub-process that uses a particular algorithm, in our case k-NN. Because in the previous stage the results were promising and the values obtained in collaborative mode were higher, the current stage made the forecasts for seven sources (ST1 to ST7), obtaining the visible results in the Table 1 for standalone process and in Table 2 for collaborative one. The standalone prediction process was applied to seven stations as follows:

- Standalone value for a 100% complete source
- Standalone value for a source that lacks 50% of the data;
- Standalone value for a source that lacks 50% of the data, but which have been completed with the operator IMV.

Table 1. Values for standalone process

Source	A_C	VP_CD	VP_ID	
ST1	0.812	0.482	IMV50	0.529
			MD50	0.282
ST2	0.839	0.459	IMV50	0.447
			MD50	0.212
ST3	0.607	0.494	IMV50	0.424
			MD50	0.212
ST4	0.841	0.482	IMV50	0.471
			MD50	0.259
ST5	0.578	0.518	IMV50	0.565
			MD50	0.294
ST6	0.836	0.494	IMV50	0.424
			MD50	0.224
ST7	0.829	0.482	IMV50	0.459
			MD50	0.247

In the Table 1 the abbreviations is such as: A_C represent the average of correlations between sources, VP_CD represent values of predictions for the standalone process with complete data, VP_ID are the values of predictions for the standalone process with incomplete data in percent of 50%, IMV50 represent the standalone process with 50% missing data and using the operator Impute Missing Data from RapidMiner and MD50 represent the standalone process with missing data in percent of 50%.

It is noted that all IMV50 values are higher than MD50 values, and their growth is in the range of [0.200, 0.247]. Sources ST1 and ST5 have even higher values obtained with the IMV50 than those obtained with a process that takes into account the full and initial data of the stations.

The collaborative prediction process was applied to 7 stations as follows:

- Collaborative value with 3 sources (1 predictive and 2 alternatives) using the IMV operator (50% missing data). (imv)
- Collaborative value with 3 sources (1 predictive and 2 alternatives) using the average temperature from pairs in collaborative (50% missing data). (aa)

The imv and aa columns in the Table 2 represent the collaborative process for the source with missing data, using the operator Impute Missing Data from Rapid-Miner application (imv) and the collaborative process for source with missing data, filling out using the average of data from the sources of the process (aa).

Table 2. Values for collaborative process

	ST1		ST2		ST3		ST4		ST5		ST6		ST7	
	imv	aa	imv	aa	imv	aa	imv	aa	imv	aa	imv	aa	imv	aa
1,2					0.624	0.553	0.412	0.435	0.565	0.541	0.412	0.447	0.388	0.412
1,3			0.541	0.494			0.400	0.471	0.471	0.506	0.365	0.435	0.388	0.459
1,4			0.529	0.482	0.659	0.588			0.600	0.612	0.400	0.471	0.400	0.471
1,5			0.565	0.471	0.612	0.565	0.447	0.471			0.412	0.435	0.412	0.435
1,7			0.659	0.471	0.647	0.565	0.471	0.459	0.600	0.588			0.447	0.447
1,8			0.553	0.447	0.682	0.541	0.435	0.447	0.647	0.588	0.376	0.388		
2,3	0.506	0.518					0.482	0.529	0.553	0.553	0.459	0.494	0.471	0.506
2,4	0.541	0.529			0.6	0.576			0.612	0.576	0.447	0.482	0.435	0.471
2,5	0.576	0.494			0.612	0.565	0.482	0.435			0.471	0.424	0.494	0.459
2,6	0.565	0.541			0.612	0.588	0.471	0.541	0.635	0.635			0.459	0.529
2,7	0.6	0.541			0.635	0.624	0.447	0.506	0.659	0.635	0.471	0.541		
3,4	0.518	0.518	0.647	0.529					0.529	0.541	0.494	0.471	0.518	0.506
3,5	0.459	0.494	0.471	0.506			0.447	0.482			0.447	0.471	0.447	0.482
3,6	0.553	0.600	0.647	0.576			0.529	0.565	0.635	0.671			0.471	0.518
3,7	0.541	0.600	0.600	0.565			0.482	0.565	0.541	0.588	0.447	0.518		
4,5	0.624	0.518	0.612	0.471	0.635	0.529					0.518	0.435	0.541	0.482
4,6	0.553	0.565	0.647	0.541	0.671	0.600			0.671	0.647			0.482	0.565
4,7	0.624	0.518	0.576	0.494	0.694	0.576			0.612	0.565	0.412	0.424		
5,6	0.624	0.576	0.659	0.506	0.624	0.576	0.482	0.494					0.459	0.471
5,7	0.6	0.518	0.612	0.506	0.624	0.541	0.565	0.506			0.529	0.447		
6,7	0.541	0.494	0.647	0.541	0.659	0.553	0.459	0.506	0.647	0.612				

In Fig. 3 there is a parallel of the graphs of values obtained with the two methods discussed in this article.

2.6   Results - cMDimv Method

A more detailed analysis of CDM processes reveals the fact that with this approach we can obtain very good results and is being a useful and viable method for the case of lack of data for a source. A summary of the values obtained with collaborative data mining can also be seen in the Table 3 where the case I is the number of values obtained in collaboration with cMDimv higher than those obtained with collaborative cMDaa, case II is the number of collaborative cMDimv values higher than IMV50 standalone and case III is the number of values obtained in collaboration with cMDaa method greater than standalone process with incomplete data.

Applying collaborative processes to sources with missing data generated a series of conclusions in favor of this approach. Values obtained collaborative with arithmetic average (cMDaa) were 100% higher than standalone. This was to be

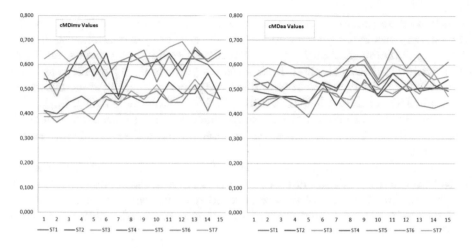

Fig. 3. Parallel between the values obtained with the cMDimv and cMDaa methods.

Table 3. Result for a collaborative process with IMV operator

Source of data	ST1	ST2	ST3	ST4	ST5	ST6	ST7
Case I	10	14	15	12	7	4	4
Case II	13	15	15	8	11	9	8
Case III	15	15	15	15	15	15	15

expected and is a strengthening of the belief that the notion of collaboration can be used successfully in the absence of data. Processes with IMV Operator and a collaborative approach (cMDimv) produce higher values than those approached with the standalone application of the operator IMV. There are two sources that have all the higher collaborative values (ST2, ST3) than standalone. For the other sources, percentages ranging from 86.66% to 53.33% were obtained. A comparison between cMDimv and cMDaa highlights that: cMDimv has very good results if the source has an average correlation of less than 0.650 and a standalone predictive value of less than 0.500 (ST3). Sources with the lowest values obtained in the standalone process with no data (ST2, ST3) obtained the best values at collaborative cMDimv or cMDma.

3   Conclusions

Fayyad et al. in [7] stated that missing and noisy data is a very big problem in data mining, and is needed to include in the process more sophisticated statistical strategies to identify incomplete attributes and complete them.

In this article, we have demonstrated that, in case of lack of data in a collaborative data mining process, their completion can be done successfully with both a sequential and a parallel approach. The difference between the two, in a collaborative process, can be represented by the level of correlations between the participants in the process. As seen in the case of a correlation around 0.500 with the cMDimv method it is obtained higher values (with more than 0.100) than with the cMDaa method. Both methods produce better results than those in which the missing values are not filled, with increases situated as follows: for method cMDaa interval is [0.095, 0.165], and for CMDimv is [0.094, 0.200].

The research has shown that both methods bring significant improvements to the result prediction and offer two approaches to addressing missing data in a CDM process.

The next research in the cases of lack of data should focus on bringing new data into the data extraction process starting from the idea stated by Matei et al. in [15]. Another view of research can be studying the correlations between the sources involved in the process and the results obtained in different types of CDM processes.

References

1. Matei, O., et al.: Multi-layered data mining architecture in the context of Internet of Things. In: 2017 IEEE 15th International Conference on Industrial Informatics (INDIN), pp. 1193–1198. IEEE (2017)
2. Matei, O., et al.: Collaborative data mining for intelligent home appliances. In: Working Conference on Virtual Enterprises, pp. 313–323. Springer, Cham (2016)
3. Matei, O., et al.: A data mining system for real time soil moisture prediction. Procedia Eng. 181, 837–844 (2017)
4. Anton, C., Matei, O., Avram, A.: Collaborative Data Mining in Agriculture for Prediction of Soil Moisture and Temperature. Advances in Intelligent Systems and Computing (to appear)
5. Anton, C., Matei, O., Avram, A.: Use of Multiple Data Sources in Collaborative Data Mining. Computational Methods in Systems and Software (to appear)
6. Avram, A., et al.: Context-aware data mining vs classical data mining: case study on predicting soil moisture. In: International Workshop on Soft Computing Models in Industrial and Environmental Applications, pp. 199–208. Springer, Cham (2019)
7. Fayyad, U.M., et al.: Knowledge discovery and data mining: towards a unifying framework. In: KDD, pp. 82–88 (1996)
8. Grzymala-busse, J.W., Grzymala-busse, W.J.: Handling missing attribute values. In: Data Mining and Knowledge Discovery Handbook, pp. 33–51. Springer, Boston (2009)
9. Grzymala-busse, J.W., Hu, M.: A comparison of several approaches to missing attribute values in data mining. In: International Conference on Rough Sets and Current Trends in Computing, pp. 378–385. Springer, Heidelberg (2000)
10. Kotsiantis, S., et al.: Filling missing temperature values in weather data banks. In: 2006 2nd IET International Conference on Intelligent Environments-IE 2006. IET, pp. 327–334 (2006)
11. Imielinski, T., Lipski Jr., W.: Incomplete information in relational databases. In: Readings in Artificial Intelligence and Databases, pp. 342–360. Morgan Kaufmann (1989)
12. Imielinski, T., Lipski Jr., W.: Incomplete information and dependencies in relational databases. In: ACM SIGMOD Record, pp. 178–184. ACM (1983)

13. Radhika, Y., Shashi, M.: Atmospheric temperature prediction using support vector machines. Int. J. Comput. Theory Eng. **1**(1), 55 (2009)
14. Zhang, S., Zhang, C., Yang, Q.: Data preparation for data mining. Appl. Artif. Intell. **17**(5–6), 375–381 (2003)
15. Matei, O., et al.: Context-aware data mining: embedding external data sources in a machine learning process. In: International Conference on Hybrid Artificial Intelligence Systems, pp. 415–426. Springer, Cham (2017)

A Survey on Program Analysis and Transformation

Ahmed Maghawry[1]([⊠]), Mohamed Kholief[1], Yasser Omar[1],
and Rania Hodhod[2]

[1] Department of Computer Science, College of Computers
and Information Systems, Arab Academy for Science, Technology and Maritime
Transport (AASTMT), Alexandria, Egypt
ahmed.mg.mohamed@gmail.com, kholief@gmail.com,
dr_yaser_omar@yahoo.com
[2] TSYS School of Computer Science,
Columbus State University, Columbus, GA, USA
hodhod_rania@columbusstate.edu

Abstract. Program transformation is a process in which an input program is transformed into another program that achieves a specific goal. Such transformation is done by applying a sequence of transformation rules on the input program to generate another program as the output. Such transformations can be done manually with human intervention (software developer) or automatically by a transformation program applying a transformation algorithm. Automating this process has been of an interest to a myriad of researchers in the past years. Several researches were done to automatically find good transformation sequences to achieve different transformation goals including program optimization and test case generation. The most popular techniques used are search-based meta-heuristic algorithms including genetic algorithm. In this paper, we will survey previous works that used genetic algorithm to achieve optimization goals under the umbrella of program transformation.

Keywords: Program analysis · Program transformation · Genetic algorithms

1 Introduction

Program transformation has been proven to be of a great benefit in different applications such as reverse engineering as well as compiler optimization and program comprehension, in such applications, we define a transformation algorithm to be a sequence of transformation steps (rules) applied on a targeted input program; it's usually constructed manually by software developers for each transformation goal. Finding good transformation sequences automatically is a very hard process that requires human intervention [1]. The problem of finding the optimal transformation sequence that serves a specific optimization goal has been categorized as a search problem and has been tackled by using several search techniques such as random search; hill climbing as well as search based meta-heuristic algorithms, such as genetic algorithms [5]. Program transformation has been shown to be a very useful technique

© Springer Nature Switzerland AG 2019
R. Silhavy et al. (Eds.): CoMeSySo 2019, AISC 1047, pp. 110–124, 2019.
https://doi.org/10.1007/978-3-030-31362-3_12

for search-based software testing using evolutionary techniques [13, 14], program optimization [9] as well as software development support like migrating one program from its original environment to another environment, in addition to test case and test data generation [10–12].

On the other hand, software testing is a technique applied to gain the consumer's confidence in the software; it is usually done manually by software testers. Testing any software system is a laborious wide scale task that is costly and time consuming [15, 16]; it can eat up to 50% of software development resources [17]. No much progress of automating the verification and validation of software through dynamic testing has been achieved [2]. Generally, the goal of software testing is to execute a set of minimal number of test cases to reveal as many faults as possible. Software testing automation is not a straight forward process as for many years; many researchers have proposed different methods to generate test cases/data automatically using different methods to develop test cases/data generators [18–23]. This survey paper aims to review works that focused on using program transformation to achieve program optimization and software development support, especially automatic test case/data generation.

In the next section, the definition of program transformation as a large concept will be reviewed as well as software testing definition and fundamentals, in addition to providing a background on the different application of genetic algorithms in program transformation. The rest of the paper is organized as follows, Sect. 3 focuses on program transformation purposes, Sect. 4 highlights the challenges facing program transformation and genetic algorithms, then Sect. 5 reviews related previous work, and finally conclusions are presented in Sect. 6.

2 Background

2.1 Program Transformation

Program transformation is the process concerning changing a targeted program into another program, in which case we alter the program's syntax but at the same time leaving its semantics unchanged. The process of program transformation has been applied to various software engineering disciplines, such as program synthesis and optimization [24, 25], refactoring and reverse engineering [26], program comprehension [27], and software maintenance [28]. Program transformation has been proven to be a beneficial supporting technology for software engineering branches like software testing [14, 29].

In program transformation, a transformation sequence that consists of a set of simple indivisible transformation rules called axioms are applied to a program's source code. These axioms preserve the semantic equivalence of both the input and output source code; as long as each axiom preserves the semantic equivalence of both the input and the output source code, then a transformation sequence that consists of a set of such axioms is considered to be semantic equivalence preserver. In most of the program transformation systems currently in use today, the order in which the transformation rules are applied to the input program is pre-determined by the designers of the transformation engine [1]; such prefixing is far from being generic and dependant on the input program as it assumes a prior knowledge of the input program. "The effectiveness of a given transformation is also dependent on the order in which the transforms occur." [1]. For example, consider two sequences S1, S2 with the same transformation rules but in different order S1: [T1, T2, T3] S2: [T3, T1, T2] where [Tn] is a single transformation rule that could possibly produce different results for a given source program according to its order of execution. Cooper et al. [1] has described this as "interplay, where a transformation may create opportunities for other transformations and similarly may even eliminate those opportunities".

2.2 Software Testing

Software testing is a part of program analysis and it can be described as conducting an investigative process that provides the customer with details about the quality and performance of the targeted software product or service. It can also provide an objective and independent view of the targeted software in order to allow the business to assess and comprehend the risks of software implementation [2]. Software testing techniques include the process of executing the targeted program with the intent to find errors or other defects as well as verifying that the software product is fit for use, serves its purpose(s), and meets one or more of its targeted properties [2]. These properties can be assumed as an indicator to whether the component or the system under test meets the original requirements that were used as a guide to its design and development. It can also indicate whether the system responds, as expected, to all kinds of inputs, performs its functions within an acceptable time span, sufficiently usable, suitable to its targeted environment and achieves the desired targets of its stakeholders. Figure 1 shows the testing phase within the software development cycle.

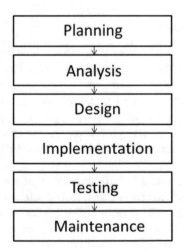

Fig. 1. Testing as a part of software development life cycle.

Software testing requires a software tester to design and execute test cases usually in a manual fashion, which is a laborious and time consuming work [2]. This necessitates the automation of this process to benefit the overall software development life cycle. However, no much progress to automate the validation and verification of software in software engineering through dynamic testing has been achieved [2]. The automatic test data generation and design remains a manual activity although automating software testing can reduce the cost of software developing significantly and facilitate an early test preparation as well as increase the speed of test runs, in addition to increasing the confidence of the testing result(s). While different methods to generate test data automatically have been proposed over the years [18–23], the process of software testing automation is still not straight forward [2].

Applying artificial intelligence (AI) techniques in the vast field software engineering is a promising area of research that brings together the benefits from both domains. Software engineering activities are knowledge intensive and human centred [2], which intrigued researchers to use AI to investigate the fitness of search algorithms like simulated annealing, genetic algorithm and ant colony optimization as a more fit alternative for test data generator developing [18, 19]. Using evolutionary techniques, researchers have made a significant progress in developing test data generators that are based on genetic algorithm [20–23, 30] using previously developed test data generators techniques [31–35].

2.3   Genetic Algorithms

A genetic algorithm is a heuristic search method used in computing. It mimics the process of natural evolution and is used to find the optimal solution for search problems based on the theory of natural selection and evolutionary biology [36]. The development of this method was inspired by the concepts natural selection and genetic dynamics [5]. The basic principles of genetic algorithms were first constructed by Holland [36] and are well described in [37, 38]. As a global optimization algorithm, the performance quality of a genetic algorithm relies on the technique of balancing the two conflicting objectives in which the algorithms explores the search space for better solutions while exploit the best solutions found so far. The main advantage of genetic algorithms comes from their capacity to optimally combine both exploitation and exploration [36]. Figure 2 shows the basic flow of a genetic algorithm. The following steps outline the basic genetic algorithm [4]:

1. **[Start]** Generate random population of individuals, where each individual is a possible solution.
2. **[Fitness Evaluation]** Evaluate the fitness of each individual in the population
3. **[New population]** Generate a new population:
    i. **[Selection]** Select two individuals from population according to their fitness
    ii. **[Crossover]** According to the probability of crossover, parents are crossed over to form new offspring; otherwise, offspring is same as parents.
    iii. **[Mutation]** According to probability of mutation, a gene is mutated within an individual.
    iv. **[Accepting]** Insert new individuals in new population.

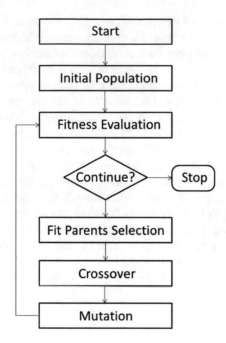

Fig. 2. Basic genetic algorithm flow.

4. **[Replace]** Use new population for another run of the algorithm.
5. **[Test]** If end condition is satisfied, **stop**, and return the best solution in current population.
6. **[Loop]** Go to second step.

2.4   Program Transformation with Genetic Algorithm

Fatiregun, Harman and Hierons explored the efficiency of using search techniques, such as random search, hill climbing, and genetic algorithms for program transformation [1]. The results from their work show that genetic algorithms outperformed the other search methods when tested on small synthetic program transformation problems. The researchers reformulated the transformation problem as a search problem for optimization, providing evidence that genetic algorithms can be used to evolve good transformation sequence. They employed 22 different transformations numbered serially from 1 to 22 where each individual solution is represented as a sequence of 20 transformation rules as follows: TSn:{10,14,1,1,8,16,14,4,22,15,20,4,8,20,9,2,12,20, 9,21}.

3   Challenges

3.1   Program Transformation Challenges

The main challenge that appears in program transformation is the fact that it involves the use of a sequence of transformation rules that are usually made by hand through a laborious and notoriously hard process that requires intensive human intervention [1]. Automating this process while keeping the output source code semantically equivalent to the input source code is a challenge. A semantics-preserving transformation algorithm must be used to achieve this process while preserving the integrity of the output program.

3.2   Genetic Algorithm Challenges

A genetic algorithm is theoretically supposed to be achieving a perfect utilization between each space exploration and search space exploitation. However, although this perfect utilization may be theoretically true for a genetic algorithm, many problems might exist in practice that raises a number of challenges. Holland assumed the following: a. the population size is infinite b. the evaluation function reflects the suitability of a solution accurately and c. the genes interactions are minimum [41]. However, in practice, the population size is finite; this affects the sampling ability of the genetic algorithm and accordingly, affects its performance. Moreover, a genetic algorithm may also punish individuals that do not pass the fitness function evaluation, even if such individuals contain good genes; those individuals will be dropped because the genetic algorithm fundamentally searches for good chromosomes not good genes. In such a process, good genes might be lost in the evolutionary process only because they existed in the wrong chromosome.

4   Achieving Program Transformation and Software Testing Using Genetic Algorithms

This section reviews previous works related to the use of genetic algorithms to support software testing and the generation of software testing cases/data. It also aims to look at the different implementations of genetic algorithms encountered in the literature including hybrid genetic algorithms. The main goal of this section is to shed light on the works done to deal with program analysis and transformation problems as search problems using genetic algorithm.

4.1   Using Genetic Algorithms to Support Software Testing

Software testing is a part of the software engineering process to validate and verify whether a software meets its requirements and specifications. Software testing is knowledge intensive and human centred [2]; usually done manually by human generated test cases as a result of limited operation speed, high cost in terms of time, and limited resources. A properly and carefully generated test case/data is important to effectively test the software. Manual testing and automated testing as testing techniques has been employed over the past decades [42, 43], however, automated testing has been proven to be more efficient and effective based on previous research [44] due to its ability to effectively and efficiently maximize test coverage, detect more errors, increase text execution, decrease cost and improve the quality of software.

The maintenance and improvement process of software has become very difficult due to the recent technological advancement and complexity of software [6, 8], test cases suffering duplicity, and the encounter of unskilled and wrong examinations. Automation oriented techniques can lead to cost savings. Artificial intelligence techniques have been used for artificial engineering in this field [2]. For example, search algorithms including simulated annealing; genetic algorithm and ant colony optimization were used as possible alternatives to develop test data generators [19, 28].

With the importance of having reliable and secure software systems, techniques have been used to avoid mistakes in software development and to make it as reliable as possible. For example, a streamlining investigation about using genetic algorithm to produce test cases that are significantly reliable was introduced in [7]. The overall purpose of this research was to reduce the cost of test cases generation in terms of consumed time and effort, in addition to providing good quality software and helping boost the software development lifecycle.

Shrivastav and Kim proposed a genetic algorithm based technique for test data generation that attempts to search for all possible paths in the targeted program to find test errors. "The proposed algorithm works on the control flow graph (CFG) that operates in the independent route for the new set of statements or conditions" [2]. During testing, every independent path should be visited at least once.

Data flow testing was explored by Chiragis who worked on genetic algorithm based automatic test data generation for data flow testing as well as functional testing [39], while Eyal and Kandel worked on the development of evolutionary based techniques to improve the efficiency of the test cases [40].

As it is impossible to effectively test software extensively with human testing, genetic algorithms can help because they operate on large scale search spaces choosing the best of numerous solutions, while adjusting their behaviour toward better solutions. The work done by Reeves investigated the use of GA with small population [46] while other studies involved the use of large population [47, 48]. Based on this research, Diaz-Gomez and Hougen proposed a technique to help evaluate diversity of fixed length population of chromosomes [6] in which a suitable fitness function -defined in relation to the testing goals- is used to evaluate these test cases. This helps determine whether this fitness function can be used for further generation of test cases. Then a test case is executed, and related feedback information is collected. Test cases with high fitness value are selected and added into the best test cases pool, while test cases with low fitness are dismissed and neglected for further generation or can be combined and mutated to produce new test cases (offspring). The offspring are evaluated against the fitness value. Such process continues until a stopping condition is achieved.

In another research, Kosindrdecha and Daengdej presented various test case generation processes and techniques, such as random testing technique, goal-oriented techniques, specification-based techniques, and sketch diagram-based techniques [50]. The results of this work show several weak points related to the used test case generation techniques. For example, they lacked the ability to identify and reserve the critical domain requirements in the test case generation process. Techniques used for automating test case generation are inefficient because of the limited resources and the fact that most of the automatically generated test cases can be corrupted. In another research, GA has been proven to be an effective technique in test case generation to detect errors in complex systems [49] when compared to other test case generation techniques investigated by [3, 50].

In [49], Brendt and Watkins proposed two new methods to help improve the performance of GA including the use of neural networks. The first method involved replacing the real target system execution to avoid expensive execution costs for evaluating test cases. The second method involved improving and enhancing the major genetic operators, such as selection, crossover and mutation. The results from this work show that the application of genetic algorithm as an optimization technique to automatically generate test cases helps detect more errors in the input programs as it uses the best possible test case. Results from [49] are aligned with the results mentioned earlier in this section where genetic algorithm was found to outperform other automated test case generation techniques such as random testing and also increases the number of test cases thereby improving efficiency.

Exhaustive software testing techniques are not always recommended as they become tough for even medium sized programs. Usually, only parts of a program can be tested that may not necessarily be the error rich parts. As a result, a more selective approach was developed that can focus on those parts that are most critical, so these parts can have a higher priority when tested, hence increase the testing efficiency. The researchers in [2] demonstrated that it is possible to apply GA techniques to find the most critical/error prone paths in a software construct to improve software testing efficiency. In this work, the researchers attempted to optimize software testing efficiency by identifying the most critical path clusters in a program. This was achieved by developing a variable length genetic algorithm that optimizes and selects the software path clusters, which are weighted according to the criticality of the path. The use of GA helps to refine effort and cost estimation in the testing phase.

4.2 Different Implementations of GAs to Support Software Testing

Basic software testing is done in the same environment it has been designed for. It aims to examine the runtime quality and quantity of a targeted software to the maximum limits. However, software testing can be done in foreign environments to explore its scalability [15, 19]. The testing process is expected to output correct results, but, unfortunately, not all defects can be identified at the same time.

Evolutionary testing employs a meta-heuristic search technique, GA, to convert the task of test case generation into an optimization problem [3, 20, 32, 51]. Evolutionary testing searches for the optimal test parameters that satisfy a predefined test criterion. This test criterion relies on the use of a fitness function that measures how well each of the auto generated optimization parameters satisfies the given test criterion [52, 53]. In [3], a collection of different types of Gas have been run on different tools and analyzed in terms of their performance. All these algorithms have the basic core of evolutionary testing but have different cost functions. Test case generation using GA was implemented using three different languages in [3], an implementation in Ruby started with randomly generating test cases to form the initial population, evaluate the fitness function f(x) for each individual of test cases in the population. The following steps will be repeated until a specific number of child test cases are generated. A pair of test cases (individuals) is selected form the current population where the probability of selection is an increasing fitness function. Selection is done with replacement, which means that the same pair of test cases can be selected more than once.

With the probability of cross over Pc, the pair is crossed over at a randomly chosen point to form two new test cases or off spring. With the probability of mutation Pm the two test cases are mutated. The technique was applied and results showed that the mutation rate has a great impact on the average fitness of the genetic algorithm during testing such that, smaller the rate, better the fitness function value will be [3]. Another implementation in C++ is presented with the following Pseudo-code:

```
choose initial_population:
evaluate individual_fitness function
determine population's_averagefitness_function
Repeat
          select best_case individuals to reproduce;
          mate_pairs at random;
          apply crossover_operator;
          apply mutation_operator;
          evaluate Individual fitness;
          determine population's averagefitness;
```

The results in [3] showed that software testing using genetic algorithms became efficient even with increasing number of test cases. Finally, a GA implementation using Matlab was implemented where a genetic algorithm and a random testing method were compared and a specified analysis of the best fitness has been evaluated [3]. In order to successfully compare a genetic algorithm against a pure random method, a total of 150 test cases were generated and tested by both methods [22]. The results show that the average response time of test cases created by genetic algorithm is much more efficient than that of the random method [3], which proves that software testing using GA allows for more efficient software testing despite increasing number of test cases. On the other hand, in random testing methods, since data points are independent on time, it becomes inefficient as code becomes complex.

4.3   Hybrid Genetic Algorithms

Hybrid genetic algorithms have been an interesting field of research that gained significant interest in recent years and are continuously used to solve real-world problems. In [4] it was shown that the concept of hybridizing a genetic algorithm is one possible way to build a strong GA that solves hard problems with less time consumed and higher accuracy than usual without the need of any form of human intervention. This sub-section reviews various forms of integrating GAs and different search and optimization techniques.

Holland assumptions mentioned earlier in the Genetic Algorithms Challenges section collides with the practical facts that a population size is finite, which will affect the sampling ability of a GA. The difficulty encountered by a GA to find the best region in a search space then locate the best solution is related to the genetic algorithm's inability to make fine moves in the neighbourhood of current solutions [57]. A GA can usually locate a region where the global optimum exists but may usually consume long time to find the exact local optimum in the convergence region [60].

A hybrid genetic algorithm enhances search capabilities and can also improve the solution's quality and efficiency if combined with problem specific methods as local method. GA with a local search method can improve the performance of the search process to find the global optimum. Time consumed to find the global optimum can be targeted for reduction if a local search method and local knowledge is employed to accelerate locating promising regions in the search space and global optimum. Balancing global and local search must be carefully designed as for the frequency of global search and the duration of local search.

By incorporating a local search method with a GA, the used local search technique can ensure a fair representation of the different search areas in the search space by sampling their local optimal [56] and can also reduce the possibility of premature convergence. It can also help overcome most of the obstacles that appear as a result of finite population size; it can introduce new genes that combats genetic drift problem [54, 55], accelerates search toward global optimum [58], and improve the exploitation ability of the search algorithm without downgrading its ability to explore the search space [58]. If global exploration and local exploitation capabilities of the GA was perfectly balanced, the algorithm will be able to easily produce solutions with high accuracy [59].

4.4   Using Genetic Algorithms to Evolve Transformation Sequences

Program transformation has been defined as the process of changing a targeted input program into another, altering the input program's syntax while maintaining its semantics unchanged [1]. This process has been used in a variety of software engineering braches such as program synthesis and optimizations [24], refactoring and reverse engineering [24], and program comprehension and software maintenance [27].

Transformation has been also proven to be a useful supporting technology for search-based software testing using evolutionary search techniques [14, 29]. Program transformation is beneficial in a variety of applications such as program comprehension, reverse engineering and compiler optimization. In such applications transformation algorithms are manually designed for each transformation target. A transformation algorithm can be defines a series of transformation rules to be applied on a given program.

Finding good transformation sequences automatically that serve a specific goal is notoriously hard and requires so much human intervention. This sub-section shows how search-based meta-heuristic algorithms can be utilized to automate the problem of finding good transformation sequences that may reduce the program size.

Random search, hill climbing, and genetic algorithm were used in [1] to reduce the program size of given programs and then compare them against each other. For a given transformation sequence, its effectiveness depends on the order in which the transform rules occur. Cooper et al. described this as interplay, where a transformation rule may create opportunities for other transformation rules and similarly may eliminate opportunities [27]. In [1], the author proposed a model that can automatically generate transformation sequences for a variety of programs. They also attempted to optimize a targeted program in terms of its size where they aimed to minimize the number of lines of code (LoC) of a targeted program as much as possible. A clear approach toward solving this issue is to randomly traverse the search space for possible solutions for a fixed number of iterations keeping the best sequence found so far. This can be seen as an exhaustive search of the search space, which will be infeasible as the number of combinations of transformation rules and their nodes of applications is exponential. For this reason, evolutionary and meta-heuristic search algorithms were employed to guide the search for good transformation sequences. The research made in [1] reported promising results of initial experiments on small synthetic program transformation problems. Results show that GA performs significantly better than either hill climbing or random search.

5   Conclusion

In this survey paper, works that focused on utilizing the concept of program analysis and transformation to support the process of software engineering were reviewed. Several researches have focused on utilizing GAs to support the process of software testing. Other researches have focused on program optimization using GAs. GA outperformed other techniques like hill climber and random search as the source program size increase and was utilized to reduce program size in terms of lines of code. However, random search out performed hill climber for certain programs. The fitness measure worked in such a way that a better individual (transformation sequence) is the one which definitely decreases produces an output program with size less than the original one. This technique punishes individuals that may not make the program size smaller but may guide the cursor to a point where a potentially good optimization may be exploited. It was also shown that hybridization is a possible way to build a competent GA as it solves tough problems quickly and accurately without the need of any human intervention. Embedding a search method with GA improves the search performance only if their role cooperates to achieve the optimization goal. The ability of a genetic-local hybrid algorithm to solve a tough problem quickly depends on the way of utilizing local search information as well as balancing genetic and local search. Results from comparing random test data generation methods with a new approach using GAs search, in which it was possible to apply GA techniques to find the most critical paths to improve software testing efficiency [2]. It was also shown that software testing using GA becomes efficient even with increasing number of test cases while in random testing methods data points didn't have a dependency with time. The application of GA as an optimization technique to automatically generate test cases succeeded to detect more errors in programs. A gap encountered in this survey is the absence of research related to utilizing a hybrid GA to search for the optimal transformation sequence to achieve a certain program transformation goal. This leaves a room for future investigation of combining GAs with a local search method to search for an optimal transformation sequence that achieves a program transformation goal, while avoiding the problems from using each algorithm solely. Other research areas might include the application of GAs and particle swarm optimization together, in addition to the use of GAs in regression testing and test case prioritization.

References

1. Fatiregun, D., et al.: Evolving transformation sequences using genetic algorithms. In: Fourth IEEE International Workshop on Source Code Analysis and Manipulation, 8 June 2009. https://doi.org/10.1109/scam.2004.11
2. Srivastava, P.R., Kim, T.: Application of genetic algorithm in software testing. Int. J. Softw. Eng. Appl. 3(4), 87–96 (2009)
3. Sharma, A., et al.: Software testing using genetic algorithms. Int. J. Comput. Sci. Eng. Surv. 7(2), 21–33 (2016). https://doi.org/10.5121/ijcses.2016.7203
4. El-Mihoub, T.A., et al.: Hybrid genetic algorithms: a review. Eng. Lett. 13, 124–137 (2006)
5. Pandian, S., Modrák, V.: Possibilities, obstacles and challenges of genetic algorithm in manufacturing cell formation. Adv. Logistic Syst. 3, 63–70 (2009)
6. Kudjo, P.K., Ocquaye, E., Ametepe, W.: Review of genetic algorithm and application in software testing. Int. J. Comput. Appl. 160(2), 1–6 (2017). https://doi.org/10.5120/ijca2017912965
7. Goyal, S., et al.: Software test case optimization using genetic algorithm. Int. J. Sci. Eng. Sci. 1(12), 69–73 (2018)
8. Rajkumari, R., Geetha, B.G.: Automated test data generation and optimization scheme using genetic algorithm. In: International Conference on Software and Computer Applications, vol. 9, pp. 52–56 (2011)
9. Cooper, K.D., et al.: Optimizing for reduced code space using genetic algorithms. In: Proceedings of the ACM SIGPLAN 1999 Workshop on Languages, Compilers, and Tools for Embedded Systems - LCTES 99 (1999). https://doi.org/10.1145/314403.314414
10. Baxter, I.D.: Transformation systems: domain-oriented component and implementation knowledge. In: Proceedings of the Ninth Workshop on Institutionalizing Software Reuse, Austin, TX, USA, January 1999
11. Bennett, K.H.: Do program transformations help reverse engineering? In: IEEE International Conference on Software Maintenance (ICSM 1998), Bethesda, Maryland, USA, pp. 247–254. IEEE Computer Society Press, Los Alamitos, November 1998
12. Darlington, J., Burstall, R.M.: A transformation system for developing recursive programs. J. ACM 24(1), 44–67 (1977)
13. Harman, M., Fox, C., Hierons, R.M., Hu, L., Danicic, S., Wegener, J.: Vada: a transformation-based system for variable dependence analysis. In: IEEE International Workshop on Source Code Analysis and Manipulation (SCAM 2002), Montreal, Canada, pp. 55–64. IEEE Computer Society Press, Los Alamitos (2002)

14. Harman, M., Hu, L., Hierons, R., Baresel, A., Sthamer, H.: Improving evolutionary testing by flag removal. In: Proceedings of the Genetic and Evolutionary Computation Conference, GECCO 2002, pp. 1359–1366. Morgan Kaufmann Publishers, New York, 9–13 July 2002
15. Somerville, I.: Software Engineering, 7th edn. Addison-Wesley, Boston
16. Mathur, A.P.: Foundation of Software Testing, 1st edn. Pearson Education, London (2008)
17. Alander, J.T., Mantere, T., Turunen, P.: Genetic Algorithm Based Software Testing (1997). http://citeseer.ist.psu.edu/40769.html
18. Mansour, N., Salame, M.: Data generation for path testing. Softw. Qual. J. **12**, 121–136 (2004)
19. Srivastava, P.R., et al.: Generation of test data using Meta heuristic approach. In: IEEE TENCON, 19–21 November 2008. India available in IEEEXPLORE
20. Wegener, J., Baresel, A., Sthamer, H.: Suitability of evolutionary algorithms for evolutionary testing. In: Proceedings of the 26th Annual International Computer Software and Applications Conference, Oxford, England, 26–29 August 2002
21. Berndt, D.J., Watkins, A.: Investigating the performance of genetic algorithm-based software test case generation. In: Proceedings of the Eighth IEEE International Symposium on High Assurance Systems Engineering (HASE 2004), pp. 261–262. University of South Florida, 25–26 March 2004
22. Korel, B.: Automated software test data generation. IEEE Trans. Softw. Eng. **16**(8), 870–879 (1990)
23. Jones, B.F., Sthamer, H.-H., Eyres, D.E.: Automatic structural testing using genetic algorithms. Softw. Eng. J. **11**, 299–306 (1996)
24. Aho, A.V., Sethi, R., Ullman, J.D.: Compilers: Principles, Techniques and Tools. Addison Wesley, Boston (1986)
25. Visser, E., Benaissa, Z., Tolmach, A.: Building program optimizers with rewriting strategies. In: Proceedings of the International Conference on Functional Programming (ICFP 1998), Baltimore, USA, September 1998
26. Ward, M.: Reverse engineering through formal transformation. Comput. J. **37**(5), 795–813 (1994)
27. Cooper, M.K.D., Schielke, P.J., Subramanian, D.: Optimising for reduced code space using genetic algorithms. In: Proceedings of the 1999 Workshop on Languages, Compilers and Tools for Embedded Systems (LCTES), May 1999 (1994)
28. Ward, M.P.: Assembler to C migration using the FermaT transformation system. In: IEEE International Conference on Software Maintenance (ICSM 1999), Oxford, UK, August 1999. IEEE Computer Society Press, Los Alamitos (1999)
29. Ryan, C.: Automatic Re-engineering of Software Using Genetic Programming. Kluwer Academic Publishers, Dordrecht (2000)
30. Goldberg, D.E.: Genetic Algorithms: in Search, Optimization & Machine Learning. Addison Wesley, Boston (1989)
31. Berndt, D.J., Fisher, J., Johnson, L., Pinglikar, J., Watkins, A.: Breeding software test cases with genetic algorithms. In: Proceedings of the Thirty-Sixth Hawaii International Conference on System Sciences (HICSS-36), Hawaii, January 2003
32. Last, M., Eyal1, S., Kandel, A.: Effective black-box testing with genetic algorithms. In: IBM Conference (2005)
33. Lin, J.C., Yeh, P.L.: Using genetic algorithms for test case generation in path testing. In: Proceedings of the 9th Asian Test Symposium (ATS 2000), Taipei, Taiwan, 4–6 December 2000
34. Baresel, A., Sthamer, H., Schmidt, M.: Fitness function design to improve evolutionary structural testing. In: Proceedings of the Genetic and Evolutionary Computation Conference (2002)

35. Rajappa, V., Biradar, A., Panda, S.: Efficient software test case generation using genetic algorithm based graph theory. In: First International Conference on Emerging Trends in Engineering and Technology, ICETET 2008, pp. 298–303 (2008)
36. Holland, J.: Adaptation in Natural and Artificial Systems: The University of Michigan (1975)
37. De Jong, K.: An analysis of the behavior of a class of genetic adaptive systems. Doctoral dissertation. The University of Michigan, Ann Arbor (1975)
38. Goldberg, D.E.: Genetic Algorithms in Search, Optimization, and Machine Learning. Addison-Wesley, Boston (1989)
39. Girgis, M.R.: Automatic test data generation for data flow testing, using a genetic algorithm. J. Univ. Comput. Sci. **11**(6), 898–915 (2005)
40. Last, M., Eyal, S., Kandel, A.: Effective black-box testing with genetic algorithms. Department of Computer Science and Engineering, Ben-Gurion University of the Negev, BeerSheva, Israel (2005)
41. Beasley, D., Bull, D.R., Martin, R.R.: An overview of genetic algorithms: part 1, fundamentals. Univ. Comput. **15**, 58–69 (1993)
42. Itkonen, J., Mantyla, M.V., Lassenius, C.: How do testers do it? An exploratory study on manual testing practices. In: Proceedings of the 2009 3rd International Symposium on Empirical Software Engineering and Measurement, pp. 494–497 (2009)
43. Miller, C.M.: Automated testing system, ed: Google Patents (1995)
44. Michael, C.C., McGraw, G.E., Schatz, M.A., Walton, C.C.: Genetic algorithms for dynamic test data generation. In: Proceedings of 12th IEEE International Conference Automated Software Engineering, pp. 307–308 (1997)
45. Goldberg, D.E.: Optimal initial population size for binary-coded genetic algorithms: clearinghouse for genetic algorithms, Department of Engineering Mechanics, University of Alabama (1985)
46. Reeves, C.R.: Using genetic algorithms with small populations. In: ICGA, p. 92 (1993)
47. Ahn, C.W., Ramakrishna, R.S.: A genetic algorithm for shortest path routing problem and the sizing of populations. IEEE Trans. Evol. Comput. **6**, 566–579 (2002)
48. Horn, J., Nafpliotis, N., Goldberg, D.E.: A niched Pareto genetic algorithm for multiobjective optimization. In: Proceedings of the First IEEE Conference on Evolutionary Computation, IEEE World Congress on Computational Intelligence, pp. 82–87 (1994)
49. Berndt, D.J., Watkins, A.: Investigating the performance of genetic algorithm-based software test case generation. In: Proceedings of Eighth IEEE International Symposium on High Assurance Systems Engineering, pp. 261–262 (2004)
50. Kosindrdecha, N., Daengdej, J.: A test case generation process and technique. J. Softw. Eng. **4**, 265–287 (2010)
51. Last, M., et al.: Effective black-box testing with genetic algorithms. LNCS, pp. 134–148. Springer, Heidelberg (2006)
52. Giuseppe, A., et al.: Testing web –applications: the state of art and future trends. Inf. Softw. Technol. **48**, 1172–1186 (2006)
53. Michael, C.C., McGraw, G.E., Schatz, M.A., Walton, C.C.: Genetic algorithms for dynamic test data generation. In: Proceedings of the 1997 International Conference on Automated Software Engineering (ASE 1997) (formerly: KBSE) 0-8186-7961-1/97 © 1997 IEEE
54. Asoh, H., Mühlenbein, H.: On the mean convergence time of evolutionary algorithms without selection and mutation. In: Davidor, Y., Schwefel, H.-P., Manner, R. (eds.) Parallel Problem Solving from Nature, PPSN III, pp. 88–97. Springer, Berlin (1994)
55. Thierens, D., Goldberg, D., Guimaraes, P.: Domino convergence, drift, and the temporal-salience structure of problems. In: 1998 IEEE International Conference on Evolutionary Computation Anchorage, pp. 535–540. IEEE (1998)

56. Mahfoud, S.W.: Boltzmann selection. In: Handbook of Evolutionary Computation, Back, T., Fogel, D.B., Michalewicz, Z. (eds.) pp. C2.5:1-4. IOP Publishing Ltd and Oxford University Press (1997)
57. Reeves, C.: Genetic algorithms and neighbourhood search. In: Fogarty, T.C. (ed.) Evolutionary Computing, AISB Workshop, vol. 865. LNCS, pp. 115–130. Springer, Heidelberg (1994)
58. Hart, W.E.: Adaptive global optimization with local search. Doctoral dissertation. University of California, San Diego (1994)
59. Lobo, F.G., Goldberg, D.E.: Decision making in a hybrid genetic algorithm. In: IEEE International Conference on evolutionary Computation, pp. 122–125. IEEE Press, Piscataway (1997)
60. De Jong, K.: Genetic algorithms: a 30 year perspective. In: Booker, L., Forrest, S., Mitchell, M., Riolo, R. (eds.) Perspectives on Adaptation in Natural and Artificial Systems. Oxford University Press, Oxford (2005)

An Improved Location Model for the Collection of Sorted Solid Waste in Densely Populated Urban Centres

Olawale J. Adeleke[1(✉)], David O. Olukanni[2],
and Micheal O. Olusanya[3]

[1] Department of Mathematics, College of Science and Technology,
Covenant University, Ota, Nigeria
wale.adeleke@covenantuniversity.edu.ng
[2] Department of Civil Engineering, College of Engineering,
Covenant University, Ota, Nigeria
[3] Department of Information Technology, Faculty of Accounting
and Informatics, Durban University of Technology, Durban,
Republic of South Africa

Abstract. This paper presents a facility location model for improving the collection of solid waste materials. The model is especially suitable for densely populated regions with several housing units as well as encourages initial sorting of wastes. Each individual house in the collection area is designated a customer, with randomly selected customers comprising the set of candidate hubs. The fundamental feature of the model is to group the customers into clusters by assigning each customer (house) to the nearest hub. Each cluster is then assigned to exactly one waste collection site drawn from the set of potential collection locations. The objective is to minimize the total number of activated waste collection sites such that all the customers' requests are satisfied without violating the capacity limit of each site. A simple Lagrangian relaxation heuristic is developed for the problem and solved with the CPLEX solver on the AMPL platform to find a feasible solution. Results from the numerical implementation of model show the model is efficient and competitive with existing solid waste collection facility location models.

Keywords: Facility location problem · Solid waste collection ·
Lagrangian relaxation · Lagrangian heuristic · Subgradient optimization

1 Introduction

Solid waste collection is an essential aspect of solid waste management, often requiring a large financial support from the government and private participation to ensure its success. Most certainly, the financial aspect of waste collection is controlled by the many factors, one of which is the locations of collection facilities. The ad-hoc placement of collection containers and the inappropriate selection of the number of such facilities have resulted in poor management system in many developing countries where the challenge is even further worsened by the lack of required funds (some local

© Springer Nature Switzerland AG 2019
R. Silhavy et al. (Eds.): CoMeSySo 2019, AISC 1047, pp. 125–135, 2019.
https://doi.org/10.1007/978-3-030-31362-3_13

factors affecting solid waste collections were highlighted in [13]). Hence, in order to promote maximum collection in the face of little budget, it is very essential to develop efficient facility location model whose objective is to find the optimal number of required facilities, their locations, through carefully regulated customer assignment, as well as the allocation of waste containers to the selected facility locations.

The study reported in this paper provides an efficient approach for locating waste collection facilities that is capable of reducing the total cost of operation such that the daily collection requirement is satisfied. The problem encountered in this study falls under the family of the facility location problem (FLP). The FLP is a well-researched combinatorial problem having a wide range of industrial applications. In its simplest description, the problem finds the best alternative of locations to operate new facilities in a given distribution network. Many variants of the problem exist, the simplest being the single FLP which locates a new single facility in a plane with the objective of minimizing the sum of distances between the proposed new facility and existing (planar) facilities (see [1] for the details of other variants of FLP models).

The applications of FLP in locating solid waste collection facilities abound in the literature. For instance, a study conducted in the southern part of Italy described an integer programming model for locating collection facilities so that all customers' demands are satisfied every day and each facility capacity is not violated [10]. The proposed optimization model was solved directly with the CPLEX solver as well as a construction heuristic. The study was extended in [11] through the introduction of a zoning criteria which help determine the service territories. Facility location is not restricted to only collection sites. The model may also be applicable at other stages of waste management. In locating landfills, for example, many FLP models have been developed and implemented in different areas and regions (see [8] for a summary of some of these models). Chauhan and Singh [6] proposed a multi-criteria model for selecting sustainable disposal locations in India. The study which focused on healthcare waste used a fussy optimization approach to achieve its objective. Two sets of datasets including, hypothetical and real-life cases from the Schuan province of China, were used in [21], a study proposed for the minimization of the total cost and risk of locating waste facilities and finding the optimal transportation routes. Their proposed mixed integer linear programming was solved using both exact and approximate techniques. The former used the CPLEX solver for a single objective case, while the latter used a customized multi-objective optimization approach. Situations of uncertainties were introduced in [19] where a stochastic model was proposed for minimizing the number of facility locations. The study used an interval analysis technique to obtain an approximate solution using hypothetical dataset derived from a case study.

A recent study in [12] used a column-and-constraint generation method to solve a mixed integer programming (MIP) model of FLP with multiple objective functions, with special focus on waste-to-energy initiative in the Shanghai city of China. Another MIP model for minimizing the total cost in the supply chain of sellable waste was presented in [5] for post-disaster wastes. With the aid of theoretical datasets, the study proposed a hybrid heuristic algorithm comprising of the particle swarm optimization and differential evolution for solving the model. The minimization of the total cost and the final priority weight for locating infectious waste facilities was examined in [18] and applied in north-east Thailand. The fuzzy analytical hierarchy process

(FAHP), hybrid FAHP and goal programming models developed in the study were solve with the LINGO solver. Aydemir-Karadag in [3] proposed an MIP location model with the objective of maximizing the total annual profit from hazardous waste. The author considered the application of the model in Turkey using the CPLEX solver.

Rabbani et al. in [15] proposed an MIP model for the minimization of multi-objective functions comprising of the total cost, total transportation risk of hazardous waste related to population exposure, and site risk. Through the use of hypothetical data, the linearized version of the model was solved by the CPLEX solver while the nonlinearized version was solved using a nondominated sorting genetic algorithm and multi-objective particle swarm optimization. A later study with similar objectives [16] used the genetic algorithm and the Monte Carlo simulation, and randomly generated datasets. Rathore and Sarmah in [17] used an optimization solver and the ArcGIS to optimize the selection of transfer stations for both segregated and unsegregated wastes. The study was implemented on a case instance in Bilaspur, India. A variable neigh-borhood search algorithm was proposed in [2] for minimizing the total cost of selecting solid waste facility locations and the associated transportation. The implementation of the algorithm used case study data derived from Tehran, the capital city of Iran.

The description of some of the most recent literature relating to finding the location of waste facilities has been presented in the foregoing paragraphs. In this paper, the focus is on locating solid waste facilities for the initial collection of waste materials from the direct sources of generation. The remaining parts of this paper are organized as follows. The problem and its mathematical formulation are presented in Sect. 2, while in Sect. 3 the Lagrangian relaxation of the problem is described. The numerical implementation of the model on a few small-sized problems is presented in Sect. 4. The concluding remarks are stated in Sect. 5.

2   Model Definition and Formulation

This section deals with the definition of the model and its mathematical formulation. The model is illustrated in Fig. 1. The main objective is to minimize the total number of waste collection sites to be located in highly populated areas. In particular, an urban centre with its numerous houses is partitioned into clusters. Each cluster is realized by selecting randomly a set of hubs from the collection of houses in the study area, and then assigning the houses to each hub based on the limitations imposed by the distances between the houses and the hubs, and the quantities of wastes from each house. From now on each house is referred to as a customer. Once the clusters have been realized, they are assigned to the candidate collection sites. Container allocation is done based on the volume of waste from the assigned clusters.

To describe the model mathematically, consider the problem on a graph $G(N, E)$, where $N = C \cup H \cup S$ is the set of nodes comprising of the set of customers, C, the set of hubs, H, and the set of candidate facility locations, S. The set E consists of all the edges connecting the modes in the system. Since the model considers the collection of different type of wastes, the set of all waste types is denoted with E. The decision variables of the model include, $x_k = 1$ if a facility is located at the site $k \in S$, otherwise $x_k = 0$; $y_{ij} = 1$ if a customer $i \in C$ is assigned to candidate hub $j \in H$, otherwise

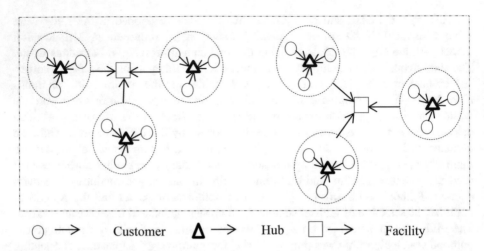

○ → Customer   △ → Hub   □ → Facility

Fig. 1. Graphical representation of the clustering and assignment procedures in the model formulation.

$y_{ij} = 0$; $z_{jk} = 1$ if a candidate hub with its assigned customer is assigned to a potential facility location $k \in S$, otherwise $z_{jk} = 0$; $g_i = 1$ if a customer $i \in C$ is selected as a hub, otherwise $g_i = 0$; $\gamma_{kt} = 1$ if a container of type $t \in T$ is allocated to facility location $k \in S$, otherwise $\gamma_{kt} = 0$. The parameters of the model are, n_t is the total number of containers of type $t \in T$ available for allocation, c_t is the capacity of each container of type $t \in T$, α_{it} is the quantity of waste of type $t \in T$ generated in at customer $i \in C$, β_{jt} is the volume of waste of type $t \in T$ shipped from each hub $j \in H$, d_{ij} is distance between a customer $i \in C$ and a candidate hub $j \in H$, e_{jk} is the distance between a candidate hub $j \in H$ and a collection site $k \in S$, μ is a threshold distance defined as the maximum distance allowed between a customer and a candidate hub, σ is a threshold distance defined for the maximum distance allowed between a candidate hub and a candidate collection site. Note that both μ and σ are constant parameters. The mathematical model for describing the customer-hub-site relationship discussed above is now presented as follows.

$$P{:}p = \min \sum_{k \in S} x_k$$

s.t.

$$\sum_{j \in H} y_{ij} = 1, \forall i \in C, \tag{1}$$

$$\sum_{k \in S} z_{jk} = 1, \forall j \in H, \tag{2}$$

$$\sum_{i \in C} y_{ij} \geq \sum_{i \in C} g_i, \forall j \in H, \tag{3}$$

$$\sum_{i \in C} \sum_{j \in H} \alpha_{it} y_{ij} \leq \sum_{j \in H} \beta_{jt}, \forall t \in T, \tag{4}$$

$$\sum_{j \in H} \sum_{t \in T} \beta_{jt} z_{jk} \leq \sum_{t \in T} c_t \gamma_{kt}, \forall k \in S, \tag{5}$$

$$\sum_{k \in S} \gamma_{kt} \leq n_t, \forall t \in T, \tag{6}$$

$$\delta x_k \leq \sum_{t \in T} \gamma_{kt}, \forall k \in S, \tag{7}$$

$$d_{ij} y_{ij} \leq \mu, \forall i \in C, j \in H, \tag{8}$$

$$e_{jk} z_{jk} \leq \sigma, \forall j \in H, k \in S, \tag{9}$$

$$x_k, y_{ij}, z_{jk}, g_i \in \{0, 1\}, \forall i \in C, j \in H, k \in S, \tag{10}$$

$$\gamma_{kt} \geq 0, \forall k \in S, t \in T, \tag{11}$$

Constraint Eq. (1) ensure that each customer $i \in C$ is assigned to one and only one candidate hub $j \in H$, while (2) ensure that each candidate hub $j \in H$ is assigned to exactly one candidate collection site $k \in S$. Equation (3) prevent the assignment of a customer $i \in C$ to a candidate hub $j \in H$ which has not been selected as a hub. For contrary cases like that, $y_{ij} = 0$. Constraints (4) ensure that the volume of waste of type $t \in T$ from a customer $i \in C$ assigned to a hub $j \in H$ does not exceed the quantity of waste of the same type that is generated at the cluster of hub $j \in H$. Similarly, constraints (5) ensure that the total quantity of waste of type $t \in T$ from a candidate hub $j \in H$ directed to site $k \in S$ does not exceed the total capacity of the containers of type $t \in T$ allocated to site $k \in S$. Constraints (6) ensure that the total number of containers of type $t \in T$ assigned to site $k \in S$ does not exceed the total available containers for the waste of type $t \in T$. Constraints (7) enforces that at least one container of waste type $t \in T$ is located at an activated site $k \in S$. δ is an arbitrary integer constant which is greater than or equal to the total number of clusters, that is, $\delta \geq |H|$. Constraints (8) and (9) are the distance coverage restrictions, while (10) and (11) are the bound constraints.

3 Lagrangian Relaxation Heuristic for Solving P

Lagrangian relaxation (LR) constitutes one of the efficient techniques for handling the complexity associated with FLP models. A wide range of location problems have been solved successfully with LR approaches. See [7, 9, 20] for detailed description of the method and implementations on FLP models. In this paper, three constraints are relaxed into the objective function namely, constraints (3), (5) and (7). The choice of these constraints is motivated by the separability of the problem into subproblems in the space of each decision variable. This is a possibility whenever constraints with linked variables are dualized into the objective function as will now be done. The result of this procedure is given in the following. Note that the multiplier vectors are each greater than or equal to zero, and $\lambda \in \mathbb{R}^{|S|}, \omega \in \mathbb{R}^{|S|}, \upsilon \in \mathbb{R}^{|H|}$.

$$P_{LR}(\lambda, \omega, \upsilon) = \min\left[\sum\nolimits_{k \in S} x_k + \sum\nolimits_{k \in S} \lambda_k \left(\sum\nolimits_{j \in H} \sum\nolimits_{t \in T} \beta_{jt} z_{jk} - \sum\nolimits_{t \in T} c_t \gamma_{kt}\right) \right.$$
$$+ \sum\nolimits_{k \in S} \omega_k \left(\delta x_k - \sum\nolimits_{t \in T} \gamma_{kt}\right) + \sum\nolimits_{j \in H} \upsilon_j \left(\sum\nolimits_{i \in C} g_i - \sum\nolimits_{i \in C} y_{ij}\right)\right]$$
$$= \min\left[\sum\nolimits_{k \in S} x_k(1 + \delta \omega_k) - \sum\nolimits_{i \in C} \sum\nolimits_{j \in H} \upsilon_j y_{ij} + \sum\nolimits_{j \in H} \sum\nolimits_{k \in S} \sum\nolimits_{t \in T} \lambda_k \beta_{jt} z_{jk} \right.$$
$$+ \sum\nolimits_{i \in C} \sum\nolimits_{j \in H} \upsilon_j g_i - \sum\nolimits_{k \in S} \sum\nolimits_{t \in T} (\lambda_k c_t + \omega_k) \gamma_{kt}\right]$$

s.t. $(1) - (2)$, (4), (6), $(8) - (11)$.

As mentioned above, P_{LR} can be decomposed into five different subproblems corresponding to the space of each of x, y, z, g and γ. The subproblem in the space of x is given by

$$P_{LR_x}(\omega) = \min \sum\nolimits_{k \in S} x_k(1 + \delta \omega_k),$$
$$\text{s.t.} x_k \in \{0, 1\}.$$

The solution to P_{LR_x} may be obtained by inspecting the signs of $(1 + \delta \omega_k)$ for each $k \in S$. The values of P_{LR_x} in this case are possible values of the lower bounds of the required numbers of activated collection locations. The subproblem in the space of y is given by

$$P_{LR_y}(\upsilon) = \min \sum\nolimits_{i \in C} \sum\nolimits_{j \in H} \upsilon_j y_{ij},$$
$$\text{s.t.} (1), (4), (8) \text{ and } y_{ij} \in \{0, 1\}.$$

P_{LR_y} is of the generalized assignment problem type with constraints (4) and (8) behaving as two separate capacity constraints. A simple approach for solving this problem comprises of dualizing (4) and (8) in P_{LR_y} so that a simple problem which can be easily evaluated is obtained. The subproblem in the space of z is given as

$$P_{LR_z}(\lambda) = \min \sum\nolimits_{j \in H} \sum\nolimits_{k \in S} \sum\nolimits_{t \in T} \lambda_k \beta_{jt} z_{jk},$$
$$\text{s.t.} (2), (9) \text{ and } z_{jk} \in \{0, 1\}.$$

The evaluation of P_{LR_z} is similar to that of P_{LR_y}. In this case, only (9) is considered a complicating constrained and therefore relaxed into the original P_{LR_z} problem. The subproblem in the space of g is given as

$$P_{LR_g}(\upsilon) = \min \sum\nolimits_{i \in C} \sum\nolimits_{j \in H} \upsilon_j g_i,$$
$$\text{s.t.} g_i \in \{0, 1\}.$$

The sign of υ_j at each $j \in H$ determines whether a customer will be selected as a hub location. As in the case of P_{LR_x}, the lower bound value to P_{LR_g} is the number of potential customer locations to be assigned the status of a hub location. Finally, the subproblem in the space of γ is given as

$$P_{LR_\gamma}(\lambda, \omega) = \min \sum_{k \in S} \sum_{t \in T} (\lambda_k c_t + \omega_k) \gamma_{kt},$$

s.t. (6) and (11)

The LR problem P_{LR_γ} is a typical unbounded knapsack problem and its optimal solution gives the lower bound value of the number of each type of containers to be allocated to each activated site. Solution approaches to this form of problem is broadly discussed in [14].

Using the information above requires maximizing a Lagrangian function written as

$$P_D(u) = \min_v L(v, u), \tag{12}$$

where $u = (\lambda, \omega, \upsilon)$ and $v = (x, y, z, g, \gamma)$. The fact that $P_D(u)$ provides a lower bound for the primal problem means an LR procedure is required to solve a maximization problem, called the Lagrangian dual function, of the form

$$P_D^*(u) = \max_u P_D(u) = \max_u \left[\min_v L(v, u) \right]. \tag{13}$$

The Lagrangian procedure begins by initializing the Lagrangian multiplier vectors and then iteratively updating them simultaneously as each of the subproblem is solved. This iterative process continues until a specified convergence criterion is satisfied. This criterion could be defined for a desired duality gap, that is, the difference between the primal and dual solutions, or the maximum number of iteration. A commonly used approach for updating the multipliers within the Lagrangian framework is the subgradient optimization. It has the advantage of providing very good lower bounds which are particularly very close to the optimal inter values [4].

In this work, the set of multiplier vector $u \in \mathbb{R}$, is updated using the subgradient optimization method. Subgradient optimization method is similar to the well-known gradient methods except that it replaces gradients with subgradients. The subgradient of $P_D(\cdot)$ at $\bar{u}$ is a vector ρ such that $P_D(u) \leq P(\bar{u}) + (u - \bar{u})^T \rho$. If ρ is unique at u, then it is the gradient at point u. The method works in the following way: Let u^0 be the initial vector of the decision variables, then the sequence of points $\{u^k\}$ is obtained by the iterative scheme

$$u_i^{k+1} = \max(0, u_i^k + \phi_k \rho_i^k), \tag{14}$$

where ρ_i^k is the ith component of the subgradient of $P_{LR}(u^k)$ at $u^k = u$ and ϕ_k is a positive scalar step-size similar to the step-size in the gradient methods. The value of ϕ_k is computed using

$$\phi_k = t_k \cdot \left(\left(P_{UP} - P_{LR}(u^k) \right) \Big/ \left\| \rho^k \right\|^2 \right), \tag{15}$$

where $t_k \in (0,2)$ is a user-defined parameter which, through common practice, is halved after a specific number of iterations have been performed without improvement to the lower bound. A known upper bound to the primal problem is denoted as P_{UP}. The following pseudo code describes the main steps required to solve the problem. The stopping criterion in Step 4 may either refer to the maximum number of iterations or the duality gap allowed in the process. The simple Lagrangian heuristic for obtaining a feasible solution to the primal problem comprises of tightening the solution for P_{LR_y} and P_{LR_z} by performing a linear relaxation with respect to their integral constraints. For these two subproblems, the process assumes. for LR_y and LR_z, the linear relaxation of P_{LR_y} and P_{LR_z}, respectively, that y_{ij} is redundant for all cases where $\sum_{i \in C} \alpha_{it} > \sum_{j \in H} \beta_{jt}$ $(\forall t \in T)$ and $d_{ij} > \mu$ $(\forall i \in C, j \in H)$, and z_{jk} is redundant for all cases where $e_{jk} > \sigma$ $(\forall j \in H, k \in S)$. All cases of redundancy are discarded.

Pseudo Code: Lagrangian Heuristic for P

1: identify the linking constraints and dualize in the primal problem;
2: initialize the subgradient parameters;
3: solve the subproblems $P_{LR}(u)$ to obtain the set of solution v;
4: **while** stopping criterion is unsatisfied **do**
5:     Update the vector v using subgradient optimization;
6:     Construct primal problem solution;
7:     Check for unsatisfied constrained and construct suitable
        heuristic to obtain a feasible solution;
8: **end**

---

4 Preliminary Numerical Implementation of the Model

The Lagrangian heuristic technique described in the preceding section was used to solve the proposed problem. The numerical implementation was based on hypothetical data derived from a near-real-world scenario in a West African city. The solution procedures were implemented on a PC with ADM C-70 APU with Radeon™ CPU Processor and an installed memory (RAM) of 2G. The algorithm was coded in AMPL and solved with the IBM ILOG CPLEX 12.6.3 solver. The data for the numerical experiments were generated using some online platforms. For instance, the study makes us of the Google Map to measure the locations of all the entities in the system. The locations are measured in the sense (a,b) where a is the longitude and b is the latitude. Using these values, the distances between the various locations are calculated on a Java-based software known as the Geographical Distance Matrix Calculator. The values of the quantities of individual waste were generated on the Minitab 17 package within appropriately specified intervals. These values correspond to the per capita rate of generation for each type of waste considered in this study. Each house is considered a potential hub. The experiment uses different value for e_{jk} ranging from 100 m to 200 m. A constant value of 50 m was assigned to d_{ij} throughout the experiment. Different values for the capacities of the containers were also tested. The preliminary test results are briefly summarized in Table 1 and Figs. 2 and 3 that follow. ITER denotes the number of iterations generated from the proposed algorithm. Clearly, as the number of customers increases, the numbers of iterations and the CPU times also increase. As the size of the e_{jk} increases, the numbers of activated open facilities reduce.

Table 1. Preliminary results for the number of open collection facilities ($c = 70$ kg)

$|C|$	e_{jk}	p^*	ITER	CPU Time
300	100	99	2610	1.1076
	125	66	4738	3.2448
	150	48	7721	5.7720
400	100	99	4067	2.0748
	125	67	7385	7.1604
	150	46	14977	21.8713
500	100	89	9087	7.8780
	125	64	130479	143.412
	150	47	754111	1346.0256

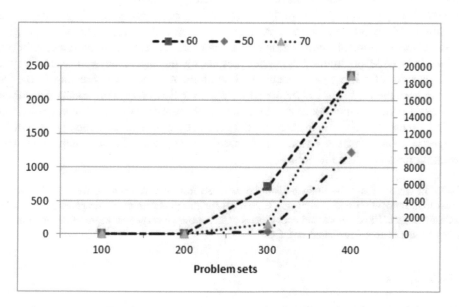

Fig. 2. The total CPU time for four problem sets

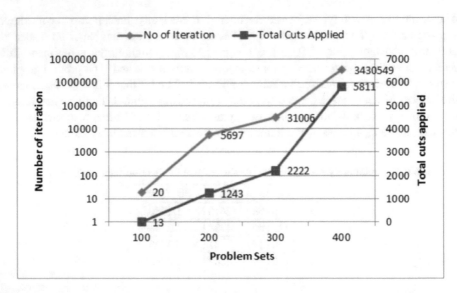

Fig. 3. Total cuts applied vs. number of iterations at $e_{jk} = 200$

5   Concluding Remarks

This study has described a new mathematical model for improving the collection of solid waste from densely populated urban areas. Distinctively, adequate measuring system is implemented in the study to describe the relationships existing between the locations of the various entities. A matheuristic method comprising of Lagrangian relaxation, subgradient optimization and a simple linear relaxation heuristic, is developed to provide preliminary numerical results on the model as well as testing the effectiveness of the proposed method of solution. As a way of directing interested researchers, the mathematical model requires the application of more robust algorithms to obtain better quality solutions. Practical implementations on real-world problems will be most desirable to further test the efficiency of the model. Finally, it will be important to compare the various results based on the proposed model with some other existing models in the literature.

Acknowledgements. The financial support received from the Covenant University Centre for Research, Innovation and Discovery (CUCRID) for the publication of this research is hereby acknowledged. The opinions and conclusions expressed in this paper are those of the authors and not necessarily to be attributed to CUCRID.

References

1. Adeleke, O.J.: Location-allocation-routing approach to solid waste collection and disposal. Ph.D. thesis, Covenant University, Ota (2017)
2. Aydemir-Karadag, A.: A profit-oriented mathematical model for hazardous waste locating-routing problem. J. Clean. Prod. **202**, 213–225 (2018)

3. Asefi, H., Lim, S., Maghrebi, M., Shahparvari, S.: Mathematical modelling and heuristic approaches to the location-routing problem of a cost-effective integrated solid waste management. Ann. Oper. Res. **273**(1–2), 75–110 (2019)
4. Beasley, J.E.: Lagrangean heuristics for location problems. Eur. J. Oper. Res. **65**(3), 383–399 (1993)
5. Boonmee, C., Arimura, M., Asada, T.: Location and allocation optimization for integrated decisions on post-disaster waste supply chain management: on-site and off-site separation for recyclable materials. Int. J. Disaster Risk Reduction **31**, 902–917 (2018)
6. Chauhan, A., Singh, A.: A hybrid multi-criteria decision making method approach for selecting a sustainable location of healthcare waste disposal facility. J. Clean. Prod. **139**, 1001–1010 (2016)
7. Chen, C.H., Ting, C.J.: Combining Lagrangian heuristic and ant colony system to solve the single source capacitated facility location problem. Transp. Res. Part E-Logistics Transp. Rev. **44**, 1099–1122 (2008)
8. Eiselt, H.A., Marianov, V.: Location modelling for municipal solid waste facilities. Comput. Oper. Res. **62**, 305–315 (2015)
9. Fisher, M.L.: The lagrangian relaxation method for solving integer programming problems. Manag. Sci. **50**(12), 1861–1871 (2004)
10. Ghiani, G., Laganà, D., Manni, E., Triki, C.: Capacitated location of collection sites in an urban waste management system. Waste Manag. **32**(7), 1291–1296 (2012)
11. Ghiani, G., Manni, A., Manni, E., Toraldo, M.: The impact of an efficient col- lection sites location on the zoning phase in municipal solid waste management. Waste Manag **34**(11), 1949–1956 (2014)
12. Hu, C., Liu, X., Lu, J.: A bi-objective two-stage robust location model for waste-to-energy facilities under uncertainty. Decis. Support Syst. **99**, 37–50 (2017)
13. Olukanni, D., Adeleke, J., Aremu, D.: A review of local factors affecting solid waste collection in Nigeria. Pollution **2**(3), 339–356 (2016)
14. Pissinger, D.: Algorithms for knapsack problems. Ph.D. dissertation, University of Copenhagen, Denmark (1995)
15. Rabbani, M., Heidari, R., Farrokhi-Asl, H., Rahimi, N.: Using metaheuristic algorithms to solve a multi-objective industrial hazardous waste location-routing problem considering incompatible waste types. J. Clean. Prod. **170**, 227–241 (2018)
16. Rabbani, M., Heidari, R., Yazdanparast, R.: A stochastic multi-period industrial hazardous waste location-routing problem: integrating NSGA-II and Monte Carlo simulation. Eur. J. Oper. Res. **272**(3), 945–961 (2019)
17. Rathore, P., Sarmah, S.P.: Modeling transfer station locations considering source separation of solid waste in urban centers: a case study of Bilaspur city, India. J. Clean. Prod. **211**, 44–60 (2019)
18. Wichapa, N., Khokhajaikiat, P.: Solving multi-objective facility location problem using the fuzzy analytical hierarchy process and goal programming: a case study on infectious waste disposal centers. Oper. Res. Perspect. **4**, 39–48 (2017)
19. Yadav, V., Bhurjee, A.K., Karmakar, S., Dikshit, A.K.: A facility location model for municipal solid waste management system under uncertain environment. Sci. Total Environ. **603**, 760–771 (2017)
20. Yang, Z., Chen, H., Chu, F.: A Lagrangian relaxation approach for a large scale new variant of capacitated clustering problem. Comput. Ind. Eng. **61**(2), 430–435 (2011)
21. Zhao, J., Huang, L., Lee, D.H., Peng, Q.: Improved approaches to the network design problem in regional hazardous waste management systems. Transp. Res. Part E: Logistics Transp. Rev. **88**, 52–75 (2016)

Atrial Fibrillation for Stroke Detection

Oi-Mean Foong[✉], Suziah Sulaiman, and Aimi Amirah Khairuddin

Department of Computer and Information Sciences,
Universiti Teknologi PETRONAS, 32610 Seri Iskandar, Perak, Malaysia
{foongoimean, suziah}@utp.edu.my,
aimiamirah02@gmail.com

Abstract. This paper presents Atrial Fibrillation Detection for Stroke Prevention. Stroke is the third leading cause of death in Malaysia after cancer. Furthermore, one of the risk factors for stroke is Atrial Fibrillation. The objective of this paper is to develop an IoT device that helps to perform early detection of irregular heart beat at affordable cost. However, the devices or gadgets are expensive and a doctor will not be always available to monitor patient's pulse regularly. Therefore, a real-time heart rate and rhythm monitoring device are presented in this paper. Experimental results show that the proposed device has achieved satisfactory performance in terms of user acceptance testing.

Keywords: Stroke · Atrial Fibrillation · Heart rate · Heart rhythm · Monitoring

1 Introduction

Atrial Fibrillation (AF) is a fast, irregular, twitching heartbeat caused by a breakdown in the hearts natural pacemaker. The condition can cause the blood flow to the heart to slow down or even to become stagnant particularly in the left atrium. For patients with atrial fibrillation, strokes are generally more severe and often have fatal outcomes due to the size of the clots formed in the left atrium becoming larger clots which eventually block the blood vessels in the brain (Friberg et al. 2014; Freedman and Lowres 2015).

According to Kooi et al. (2016), Stroke is a notable global health problem with increased morbidity and mortality in a country. Stroke is the third leading cause of death in Malaysia. The economic burden of stroke is going to increase substantially as the proportion of population increases (Lee et al. 2017). As atrial fibrillation is one of the risk factors for stroke, the AFD prototype is proposed to allow users to measure their heart rates and display the heart rhythms to identify the pattern of their heart beats whether they are regular or otherwise. The user can also store the records of their readings for data analytics.

There are many existing wearables and fitness wrist band that can passively measure the pulse/heart like smartwatch (Mela 2018; Turakhia et al. 2019; Hochstadt et al. 2019) using photoplethysmography (PPG). The Apple watch could detect the irregular pulse with AF effectively. However, the price of the smart watches is very costly. Therefore, it is crucial to have early detection and heart beat monitoring for patients.

© Springer Nature Switzerland AG 2019
R. Silhavy et al. (Eds.): CoMeSySo 2019, AISC 1047, pp. 136–143, 2019.
https://doi.org/10.1007/978-3-030-31362-3_14

2   Literature Review

According to the American Heart Association (2019), stroke risk factor can be divided into two, i.e. the risk that controllable and risks that not within the control. Controllable risk factor consists of high blood pressure, atrial fibrillation, diabetes, high cholesterol, physical inactivity and obesity, carotid artery disease, transient ischemic attacks (TIAs), certain blood disorders, tobacco use, excessive alcohol intake and illegal drug use. Whilst the uncontrollable risk factors include age, gender, family history and race. People should be aware of the warning signs of stroke because stroke could be fatal and might lead to permanent disability if left unattended immediately (NASAM 2018). The stroke warning signs including numbness in the left arm leg or left side of the face, trouble speaking and understanding other people speaking, complete or partial paralysis, trouble seeing on one or both eyes, hiccups, losing consciousness often, dizziness, breathing problems, hallucinations, epileptic attacks and personality changes.

According to Henzel et al. (2017), Atrial Fibrillation is one of the most common types of arrhythmia. The atria contracts quickly and irregularly at rates of 400 to 600 beats per minute in atrial fibrillation episodes. Their contraction is independent from ventricles, which themselves operate at much lower rate. Symptoms are not always evident but often include palpitations, irregular heartbeat, shortness of breath, chest pains and others. Detection of atrial fibrillation episodes and treatment is important in the absence of any symptoms. Untreated, severe complications can include stroke and heart attack. Atrial fibrillation is not a very common disease since it has an effect on around one in every hundred people, but the risk augments with age. For a population of people aged 75 years or more, one in ten people are affected by atrial fibrillation. In most cases, the detection of atrial fibrillation is based on heart rate analysis.

Atrial Fibrillation is a known risk factor for ischemic stroke, which can lead to death and disability and contribute to health system costs. The damage to the contractile activity of the atria, and thus the damage to blood removal from them, predisposes for clot formation, especially within the left atrial appendage. So, the risk of embolic complications, including ischemic stroke is increased. It should be pointing up that ischemic stroke occurring due to atrial fibrillation has a higher fatality rate and a higher level of potential disability for surviving patients, compared with patients with an ischemic stroke of other etiology. It is evaluated that the risk of ischemic stroke occurrence in patients with atrial fibrillation is six times higher than among the patients without the arrhythmia, regardless whether atrial fibrillation is of permanent or paroxysmal and symptomatic or asymptomatic nature. It was revealed that even a short-lasting (5–6 min), asymptomatic episode of atrial fibrillation may result in a 2.8 times higher risk of ischemic stroke with 2.5 times higher risk of a fatality (Roj et al. 2017). Schoonderwoerd et al. (2008) claimed that there are some new risk factors that lead to Atrial Fibrillation other than hypertension, cardiomyopathy, diabetes mellitus, thyroid disease and valvular disease. Other new risk factors such as obesity, sleep disorder, alcohol abuse, excessive sports practice and others are also susceptible to prevalence of stroke.

3   Methodology

The proposed Atrial Fibrillation model has a heart rate sensor, Arduino Uno and LCD. Figure 1 shows the system architecture. The heart rate sensor sends a digital signal to the Arduino Uno. The signal received is used to estimate the heart rate based on the code and display the output on the LCD, text file and serial plotter.

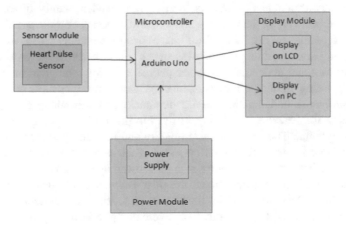

Fig. 1. Proposed Atrial Fibrillation model

3.1   Sensor Module

The sensor module can be connected to Arduino Uno to measure the heart rate in beat per minute (BPM). The heart rate (aka pulse) sensor is the input unit for this system. It sends digital signals to Arduino for further processing. The digital display of heartbeat is produced when your index finger is placed on the pulse sensor.

3.2   Microcontroller

The Microcontroller (Arduino Uno) takes the output from the sensor. The output of Sensor Module will first subject to A/D conversion. Through microcontroller programming, the digital signal is processed, the pulse rate (bpm) should be calculated and then sent to the output ports of the microcontroller.

3.3   Display Module

Display Module takes the output from the Microcontroller. It consists of three components. Firstly, it will display the pattern of heartbeat on the LCD. Secondly, the same heartbeat signals are displayed in real-time on a serial plotter from the serial port of a personal computer. This is also to confirm the device has received reasonable signals. Lastly, the third component is a recorded data that will be saved in a text file for user's data analysis.

4   Results and Discussion

This section describes the experiments conducted to evaluate the prototype device. Figure 2 shows the system set-up that involves AFil3, the proposed Atrial Fibrillation. The aim of the experiment is to test the accuracy and sensitivity of the proposed Atrial Fibrillation detection device (AFD). The accuracy refers to a system's ability to support user interaction to perform the tasks efficiently.

Fig. 2. Atrial Fibrillation Detection device (AFD)

4.1   System Performance Test

Thirty participants from different age groups and gender took part in the experiment. The heart rates of participants were captured using the conventional AF device. These measurement of heart rates were taken at regular time interval and served as the baseline for benchmarking. The heart rate reading using heart rate monitor and the proposed AF for 30 participants were recorded in Table 1. It is hypothesized that the AF device is susceptible to noise which causes data loss during the transmission from the IoT sensor to personal computer.

Table 1. Comparison between normal heart rates (baseline) and that of AF device.

No.	Baseline – heart rate monitor (bpm)	AF (bpm)	Accuracy
1	70	70	Accurate
2	83	83	Accurate
3	78	78	Accurate
4	73	73	Accurate
5	71	71	Accurate
6	74	74	Accurate
7	66	56	Inaccurate
8	63	63	Accurate
9	72	72	Accurate
10	76	76	Accurate
11	64	71	Inaccurate
12	79	79	Accurate

(continued)

Table 1. (*continued*)

No.	Baseline – heart rate monitor (bpm)	AF (bpm)	Accuracy
13	68	68	Accurate
14	66	66	Accurate
15	48	48	Accurate
16	71	98	Inaccurate
17	72	72	Accurate
18	58	58	Accurate
19	73	63	Inaccurate
20	64	64	Accurate
21	77	77	Accurate
22	72	72	Accurate
23	73	73	Accurate
24	74	68	Inaccurate
25	65	65	Accurate
26	73	73	Accurate
27	44	44	Accurate
28	66	71	Inaccurate
29	56	56	Accurate
30	63	63	Accurate

Next, the graph of participants' heart rate readings using AFD device was plotted as shown in Fig. 3.

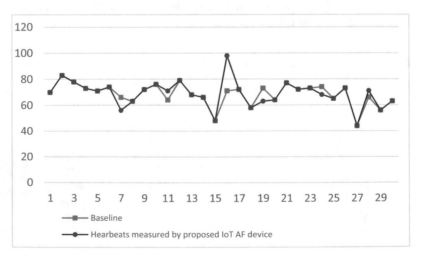

Fig. 3. Heart rate measurement for 30 respondents using AFD

From the graph, we can conclude that the normal average heart rate of the participants is between 70 to 80 bpm. After reading the heart rate from the AFD device, the participants heart rates were measured again using a simple heart rate monitor that people usually use to check their heart rates. The reason for doing this is to compare the readings taken from AFD device with the heart rate monitor as a baseline and determines the accuracy of proposed device.

Based on Table 1 that shows the comparison between heart rate reading using heart rate monitor (baseline) and AFD for 30 participants, the system prototype is evaluated and tested using the following Eq. 1:

$$\text{Accuracy} = \frac{\text{Number of correct measurement}}{\text{Total number of measurement}} \times 100\% \tag{1}$$

If the difference between the baseline device and AFD are ± 0.5, it is considered as accurate and vice versa. Therefore, the result of the accuracy of heart rate reading of 30 participants by using AFD has been recorded in the Table 2.

Table 2. Accuracy of the prototype device AFD.

Dataset (testing)	Number of participants	Accurate result	Inaccurate result	Result of accuracy (%)
30	24	6		80

Table 2 shows the accuracy percentage of the overall model. It showed that AFD device is 80% accurate.

4.2   User Acceptance Test

A set of questionnaires was distributed to the 30 participants after they completed the first part of the experiment. The survey uses five points Likert scale (rating from 1 to 5) which are strongly agree to strongly disagree. With rating 1 indicates strongly disagree, 2 for disagree, 3 for neutral, 4 for agree and 5 for strongly agree. There were five main questions involved ranging from the usefulness of AFD up to the device becoming a potential persuasive gadget.

Q1: Usefulness (AFD is useful in measuring heart rate to prevent stroke).
Q2: Usability (AFD is easy to use).
Q3: Behavioural intention (I intend to use AFD frequently).
Q4: Quality (Overall, I'm satisfied with AFD).
Q5: Use Persuasive technology.

Table 3. Participants' feedback on AFD

Questionnaires	Likert scale					Average score
	Strongly disagree (1)	Disagree (2)	Neutral (3)	Agree (4)	Strongly agree (5)	
Q1	0	0	0	5	25	4.83
Q2	0	3	0	3	24	4.4
Q3	0	1	2	3	24	4.67
Q4	0	2	1	0	27	4.73
Q5	0	0	0	4	26	4.87

Based on the average scores shown in Table 3, the highest value (i.e. 4.87) which is on Q5 regarding AFD as a persuasive technology could indicate participants' acceptance on the prototype device to be used in the future. The results also show that all 30 participants either agree or strongly agree with the potential of AFD. The second highest value (i.e. 4.83) is on the usefulness of AFD that is on questionnaire Q1. Similar to the result in Q5, all participants agree on the claim that AFD is being useful for measuring heart rate to prevent stroke.

Results for Q4, Q3, and Q2 which scored 4.73, 4.67, and 4.4, respectively on the average score have a mixture of participants' agreement and disagreement on the questions asked pertaining to AFD. Despite some negative feedback received, on the average majority of the participants agreed that the quality of AFD (Q4), using AFD frequently (Q3), and usability of AFD (Q2). Those who were not happy with AFD may find the presentation of the prototype is a little unappealing, and cumbersome as they have to click on many items in order to check for the heart pattern and the recorded data. In general, the participants were satisfied with the device. Most of them agreed that the device will ease them in monitoring their heart rate regularly.

5   Conclusion and Recommendations

Healthcare is one of the most delicate and important field that should be focused on for developing and enhancing smart systems or devices that are simple, using low energy consumption and having real-time feedback. The proposed AFD prototype/device has the potential in improving the quality of health services and reduce the total cost of healthcare by avoiding unnecessary hospitalizations and ensuring that those who need urgent care get it sooner. The proposed IoT AFD device has proven to help perform early detection of irregular heart beat by providing the user with the heart rate readings, heart rate pattern and recorded data for future use. Thus, this device can be an advantage to the elderly society by assisting them in getting quality assistance at their own houses.

For the recommendation, AFD device can be upgraded to measure body temperature, blood pressure, and sugar. The appearance of the prototype as well as the operations needed to use the device need improvement. Besides, the developer may use a smartphone instead of a laptop to enhance user's satisfaction and experience when using the device.

Acknowledgment. We would also like to thank UTP and all staffs who provide support either directly or indirectly in the research. Special gratitude goes to visiting Professor Dr Merienne for giving valuable feedback on the paper.

References

Hochstadt, A., Chorin, E., Viskin, S., Schwartz, A.L., Lubman, N., Rosso, R.: Continuous heart rate monitoring for automatic detection of atrial fibrillation with novel bio-sensing technology. J. Electro cardiology **52**, 23–27 (2019)

American National Stroke Association, January 2019. http://www.strokeassociation.org/strokeorg/

Camm, A.J., Kirchhoff, P., Lip, G.Y., Schotten, U., Savelieva, I., Ernst, S., Van Gelder, I.C., Al-Attar, N., Heidbuchel, H.: Guidelines for the management of atrial fibrillation: the Task Force for the Management of Atrial Fibrillation of the European Society of Cardiology (ESC). Eur. Heart J. **31**(19), 2369–2429 (2010)

Kooi, C.W., Peng, H.C., Aziz, Z.A., Looi, I.: A review of stroke research in Malaysia from 2000–2014. Med. J Malays. **71**(1), 58–68 (2016)

Freedman, B., Lowres, N.: Asymptomatic atrial fibrillation: the case for screening to prevent stroke. JAMA **314**, 1911–1913 (2015)

Friberg, L., Rosenqvist, M., Lindgren, A., Terent, A., Norrving, B., Asplund, K.: High prevalence of atrial fibrillation among patients with ischemic stroke. In: Stroke 2014, vol. 45, no. 9, pp. 2599–2605 (2014)

Henzel, N., Wróbel, J., Horoba, K.: Atrial fibrillation episodes detection based on classification of heart rate derived features. In: 24th IEEE International Conference in Mixed Design of Integrated Circuits and Systems, pp. 571–576 (2017)

Lee, Y.Y., Shafieb, A.A., Sidekc, N.N., Azizd, Z.A.: Economic burden of stroke in Malaysia: results from national neurology registry. J. Neurol. Sci. **381**(Supplement), 167–168 (2017)

Mela, T.: Smartwatches in the fight against atrial fibrillation: the little watch that could. J. Am. Coll. Cardiol. **71**(21), 2389–2391 (2018)

Ministry of Health Malaysia. Clinical Practice Guidelines (CPG) (2019). http://www.moh.gov.my

Turakhia, M.P., Desai, M., Hedlin, H., Rajmane, A., Talati, N., Ferris, T., Desai, S., Nag, D., Patal, M., Kowey, P., Rumsfeld, J.S., Russo, A.M., Hills, M.T., Granger, C.B., Mahaffey, K. W., Perez, M.V.: Rationale and design of a large-scale app-based study to identify cardiac arrhythmias using a smartwatch: The Apple Heart Study. Am. Heart J. **27**, 66–75 (2019)

National Stroke Association Malaysia (NASAM). Stroke is fatal. http://www.nasam.org

Roj, D., Wrobel, J., Matonia, A., Sobotnicka, E.: Hardware design issues and functional requirements for smart wristband monitor of silent atrial fibrillation. In: 24th International Conference IEEE Mixed Design of Integrated Circuits and Systems, MIXDES, pp. 596–600 (2017)

Schoonderwoerd, B.A., Smit, M.D., Pen, L., Van Gelder, I.C.: New risk factors for atrial fibrillation: causes of 'not-so-lone atrial fibrillation'. Europace **10**(6), 668–673 (2008)

Stroke Patients' at Hospital Universiti Kebangsaan Malaysia (HUKM). Med. J. Malays. **58**(4), 499–505

A Self-adapting Immigrational Genetic Algorithm for Solving a Real-Life Application of Vehicle Routing Problem

Adrian Petrovan[✉], Rudolf Erdei, Petrica Pop-Sitar, and Oliviu Matei

Technical University of Cluj-Napoca, North University Center of Baia Mare,
Baia Mare, Romania
adrian.petrovan@cunbm.utcluj.ro

Abstract. The present article presents the research for optimizing a real-life instance of heterogeneous vehicle routing problem, used intensively by shipping companies. The experiments have been carried out on the data provided by real companies, with constrains on the number and capacity of the vehicles, minimum and maximum number of stops for each route, along with the margins which can be take into account when optimizing the load of each truck. The optimization is performed using genetic algorithms hybridized with techniques for avoiding local optima, such as self adaptation and immigration. It turns out that more sophisticate approaches perform better, with very little compromise on execution time. It is a new proof of the importance of immigration techniques in bringing diversity in the genetic population.1.

1 Introduction

Combinatorial optimization is a branch of mathematical optimization with a large number of applications in everyday life. Whether it's the car navigation system, the software used to create high school calendars, or the decision support systems used in production and logistics, you can be sure that combinatorial optimization techniques are used.

1.1 Related Work

The vehicle routing problem (VRP) is a generalization of the traveler salesman problem and is often used when it is desired to optimize the customer's route to minimize the distance traveled. For this kind of problem, the computation cannot be done by using simple methods, especially if the distance is not Euclidean, nevertheless time or cost serves as a basis for calculating the route.

The aim of any variant of a VRP is to find a set of routes for a fleet of vehicles and to serve a set of customers to a minimum total route cost. There are three main constraints which are subject to the objective function: (i) each route begins and ends at the depot; (ii) the customers are visited exactly once and (iii) the total demands for each route does not exceed the vehicle capacity.

© Springer Nature Switzerland AG 2019
R. Silhavy et al. (Eds.): CoMeSySo 2019, AISC 1047, pp. 144–156, 2019.
https://doi.org/10.1007/978-3-030-31362-3_15

The mathematical formulation for this problem was first proposed by Dantzig and Ramser [2] and few years later first heuristic algorithms have been presented by Clarke and Wright [1]. A much more exhaustive survey was done by Toth and Vigo [16] and Laporte [8]. The VRP belongs to the category of NP hard problems and can be solved exactly only for instances with reduced number of customers. Heuristics have the property to cover more complex configurations and researchers have concentrated attention on developing algorithms to solve this problem [5,9].

An important work has been reported by Matei et al. [11], which embeds immigration in genetic algorithms for solving heterogeneous fixed fleet vehicle routing problem.

1.2   The Business Case

This problem was brought to our attention by a company from Oradea, Romania that manufactures furniture for export. Every week they have to ship a few hundred packages, ranging from 0.1 m^3 to 100 m^3. They provided a dataset for us to test that has around 700 packages. Each package is for a single client and has GPS coordinates assigned.

The company does not have their own fleet of trucks, so for each week they search for available trucks from local transport companies. This means that different truck capacities are available that have to be taken into account.

Also, each truck has to make a maximum of 15 unloads. More would only slow the deliveries to unacceptable values.

Taking into considerations these real practical business assertions our case study is a heterogenous (or mixed) fleet capacity vehicle routing problem which was firstly defined by Golden et al. [6]. More case studies related to heterogenous vehicle routing problem can be found in [14,15].

1.3   Notations and Problem Definition

The problem referred as heterogeneous capacity vehicle routing problem (HCVRP) may be stated as follows. A directed graph $G = (V, A)$ is given where $V = \{0, 1, 2, ..., n\}$ is a set of nodes and A is the set of arcs between them, Node 0 represent the depot for the fleet of p vehicles of different capacity $C_k, k \in M = \{1, 2, ...p\}$ and the remaining node set $V' = V \backslash \{0\}$ corresponds to the n customers. For each arc $(i, j) \in A$ and for each vehicle there is a non-negative routing cost c_{ij}^k. In the depot there are a set of N type of goods each one of a volume v. Each customer $i \in V'$ orders a quantity q_i^l of each good $l \in N$. A route R defined as a pair (R, k) is a circuit in G and k is the vehicle associated with the route. A route is feasible only if total demands carried by vehicle k for the customers visited by the route does not exceed the vehicle capacity C_k.

The general solution for HCVRP consists of finding a set of feasible routes with the following constraints: (i) the demand of a customer is serviced by only one vehicle, (ii) the total demands within one route does not exceed the vehicle capacity, (iii) each vehicle does not exceed a maximum number of customers per each route and (iv) minimizes the cost function of the total distance traveled by all vehicles.

The formulation for HCVRP for the most general variant is as follows:

$$Min \sum_{k \in M} \sum_{i,j \in V} c_{ij}^k x_{ij}^k \tag{1}$$

$$s.t. \sum_{k \in M} \sum_{i \in V} x_{ij}^k = 1, \forall j \in V' \tag{2}$$

$$\sum_{i \in V} x_{ip}^k = \sum_{j \in V} x_{pj}^k, \forall p \in V', \forall k \in M \tag{3}$$

$$\sum_{l \in N} \sum_{i \in V} \sum_{j \in V'} q_i^l v^l x_{ij}^k \leq C_k, \forall k \in M \tag{4}$$

$$x_{ij}^k \in \{0,1\}, \forall i, j \in V, i \neq j, \forall k \in M \tag{5}$$

In the above problem formulation the objective function (1) minimizes the total travel cost, constraints (2) ensure that a customer is visited by exactly one vehicle, constraint (3) guarantee that if a customer is visited by a vehicle it must also depart from it. In constraint (4) the sum of the demands of customers into one route is less than or equal the capacity of the vehicle on that route, in other words, the capacity is never exceeded. Constraint (5) states a condition if there exist a route for a vehicle between nodes i and j.

2   Vehicle Routing Problem

The vehicle routing problem (VRP) is a combinatorial optimization and integer programming problem seeking to service a number of customers with a fleet of vehicles. Proposed by Dantzig and Ramser in 1959 [2], VRP is an important problem in the fields of transportation, distribution and logistics [12]. The common application is that of delivering goods located at a central depot to customers, with the goal of minimizing the cost of the distribution (Fig. 1).

Several variations and specializations of the vehicle routing problem exist, such as Vehicle Routing Problem with Pickup and Delivery (VRPPD) [17], Vehicle Routing Problem with LIFO [10], Vehicle Routing Problem with Time Windows (VRPTW) [3] and Capacitated Vehicle Routing Problem (with or without Time Windows) [7].

3   Genetic Algorithms for VRP

The Genetic Algorithms (GA), introduced for the first time back in 1979s, mimic the natural evolution, as defined by Darwin [12]. The GA's manage a pool of potential solutions, called *individuals* or *chromosome*, which undergo various

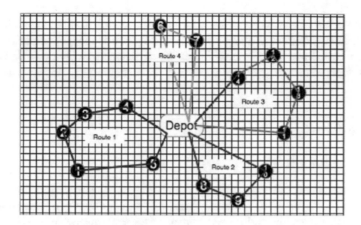

Fig. 1. Example of VRP solution

evolution operators. The most important ones are **crossover**, which combines two individuals for forming two new ones, **mutation**, which performs random changes in the solutions and **selection operator**, which simulates the natural selection, by forming a new population out of the best chromosomes.

At the end of each iteration the offspring together with the solutions from the previous generation form a new generation, after undergoing a selection process to keep a constant population size. The solutions are evaluated in terms of their fitness values identical to the fitness of individuals. An outline of a GA is shown in Fig. 2:

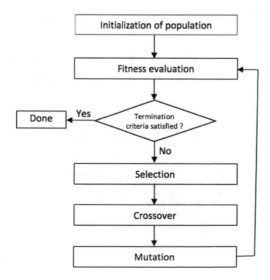

Fig. 2. Workflow of a genetic algorithm

Genetic algorithms have proven to be a powerful and successful problem-solving strategy, demonstrating the power of evolutionary principles. Genetic algorithms have been used in a wide variety of fields to evolve solutions to combinatorial optimization problems, see e.g. Fleurent et al. [4], Pop et al. [13], etc.

4   Case Study

4.1   Experimental Setup

In our experimental dataset, sent by the client there were around 600 packages that need to be delivered in multiple locations (each package to its own location).

The locations are spread across Europe, the vast majority of them being in France, as suggested in Fig. 3:

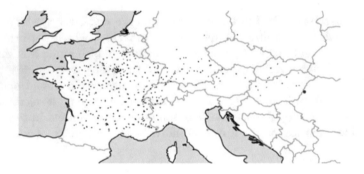

Fig. 3. Map and GPS coordinates of customers

As the company did not have their own fleet of trucks, the truck capacity varies with each order. For our experimental data set, the fleet looked like this (Table 1):

Table 1. The fleet of vehicles

#	Type	No. of trucks
1	100 m^3 truck	10
2	90 m^3 truck	25
3	80 m^3 truck	Unlimited

4.2 Genetic Algorithm Parameters Setup

The setup involves selecting a way to represent the individuals (the chromosome). There are multiple ways that this can be done, so the first step was to select which representation gave the best results (in ease of use and speed).

The **first representation** that was tested was the classic VRP representation in which each of the locations is given a number (zero being the headquarters). The order of the locations between two consecutive zero's represents the order in which the truck has to visit the locations. In this case, a chromosome will be an array of integers like:

$$I_i = (p_{i1}, p_{i2}, \ldots p_{in}) \tag{6}$$

where p_{ik} is a k package. A practical example of a chromosome would be (Fig. 4):

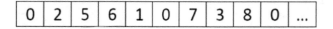

Fig. 4. Representation of a classical VRP chromosome

The **second one** is very different and a little counter intuitive. Each position in the chromosome (the index of the array) represents a package. The value is the truck it is loaded on. This representation has the advantage that a package cannot be loaded simultaneously on two trucks, so we don't have to check for this constraint, speeding up the code a little bit.

$$I_i = (t_{i1}, t_{i2}, \ldots t_{in}) \tag{7}$$

where t_{ik} is the truck delivering the package p_{ik}. In this case, a practical example of a chromosome would be (Fig. 5):

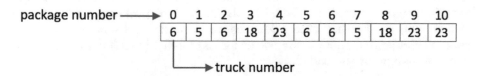

Fig. 5. Representation of a modified VRP chromosome

As the code was cleaner, quicker, with less bugs and also the first generation was produced easier, we used the second representation.

The fitness function is the total distance travelled by the trucks for delivering all the packages to all locations.

The second step of designing the algorithm was to decide how the selection will be done. We tested a few ways, including tournament selection, elite, and agreed that the elite gave us the quickest growth. But, in the long run, this meant that the algorithm has a higher chance to get stuck in a local optimum. So, our algorithm selects 2 individuals pseudo-randomly (Gaussian distributed with 1.0 standard deviation).

Crossover, in our case is possible with single or multiple cutting points. The algorithm randomly selects the number of cutting points between 10 and 200. The new individual has alternative parts of chromosomes from both parents. Doing this we get two different individuals that both go in the new population pool.

Mutation means randomly swapping 2 packages from different trucks (in the same individual). For our algorithm we decided to test a self-adapting genetic algorithm, so we designed a function that gave a number of mutations to be done on the individual, inversely proportional with the fitness of the individual, as seen in Fig. 6:

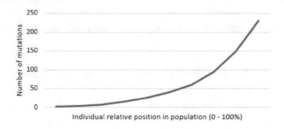

Fig. 6. Representation of the number of mutations

This means that better solutions will be less mutated, but worse ones will be more mutated, in an effort to produce better individuals.

The classic approach to mutation is to check every gene of the chromosome against a mutation probability, but this direction did not work in this case. The very large number of genes (equal to the number of packages, so somewhere around 600) means that even a low chance of mutation (something like 5%) resulted in a large number of mutations for good individuals (around 30 swaps), and not enough for the bad ones.

Self Adapting Genetic Algorithm. One of the biggest problems of the genetic approach is that the algorithm sometimes tends to get stuck in local optimums. Regarding this, for our case, there are 2 ways to optimize a GA:

1. A fast pacing, fast growth algorithm that has a higher chance to get stuck in a local optimum (small population, only the very best are selected for breeding). This version ensures very fast growth both in time and epochs;

2. A slow pace, slow growth algorithm that has a smaller chance of getting stuck in a local optimum (very large population, almost random selection for breeding). This version has small growth in successive epochs and also the code runs for much more time, but has the advantage to be able to exit local optimums.

A self adapting GA has to be able to detect that it's stuck in a local optimum and try to exit this situation. The implementation means writing both optimized variants and switching between them when a local optimum is detected. This way, the second variant of the algorithm will eventually succeed to exit the local optimum, spreading the exploration of the solutions area in a more random manner.

A local optimum can be detected by analyzing each epoch (or 10 epochs) and comparing the results with the previous. Useful metrics are the fitness of the best individual ever generated, the best in this epoch, the difference between the best and the worst (the spread), the growth in the last 10 epochs and so on.

If the algorithm runs the fast version and detects a local optimum, for example the growth in the last 10 epochs was less than a certain amount and/or the spread of the fitness's of each individual is not very good, it automatically switches to the slow version. This means using a larger population and a random selection for breeding, resulting in a better spread.

After successfully exiting the local optimum, it can switch back to the fast version. Doing this, the run time of the algorithm can be improved by a very large amount.

Also, the population size is subject to self-adaptation. In the beginning, we start with a small population and each 100 epochs the number is increased by an amount (10–30%). This has the benefit of reaching good individuals very fast in the beginning and using larger populations only when they are needed, later in the algorithm.

Immigrating Individuals Model. Another problem is that in time the individuals tend to be very similar, as a very fit individual has a higher chance to have multiple offspring and a very low chance of mutation. But raising the mutation probability results in a slower growth over time, so a better solution was an immigration approach.

In a separate thread, every 10 epochs an individual was put through a process similar to simulated annealing: in an infinite loop, random mutations were applied to it and if the results were better the individual is saved and used in the next loops.

At the end of the 10 epochs, the resulting individual is very different from the ones existing already in the population pool of the GA. If the fitness of this individual is comparable (equal or better) to the ones in the pool, it is inserted in the pool as an immigrant.

In the first stages, this approach gave an extremely large growth, as the immigrants were much better than the ones in the pool (around 30–60%). But, after the first 1000 epochs, the immigrants' importance kept getting lower and lower, to the point that it was not feasible to keep a separate thread running for it, so after this point, this thread is destroyed.

The last thing that could be optimized is the algorithm's behavior when it reaches the end. At around 5–10% epochs remaining (this means around 100 epochs) if it detects fast growth, it can add more epochs, so it can try to get an even better result. Of course this behavior will be disabled in testing and only be activated once the algorithm reaches production.

Final Adjustments. The last step of the design was to decide if we allowed the old generation individuals to "live" in the new generation. The compromise was to put both in a list and select only the best.

After the design was done, we went on to fine-tune the parameters in order to achieve best results. We tested different values for the number of individuals in a generation, number of cutting points, mutation probability and even the "life" of the individuals (how many generations is it allowed to live).

On average, the best possible solution was reached after around 5000 epochs. After experimenting with population sizes, trying different numbers between 50 and 10.000 individuals per generation, the fastest growth over time was with a population of just 100. This enabled the algorithm to run very fast, around 10 epochs per second on our machine.

Naturally, in some circumstances, the algorithm reached a local optimum, but the auto-adaptive part of the algorithm took care of this, managing to eliminate this situation in about 99.9% of times.

4.3   Experimental Results

For the experiment, the genetic algorithm was run 10 times for each benchmark set, a maximum of 3000 epochs. The tests have been performed on a machine with i7 4790K processor (4 cores, 8 threads). Implementation was done in Java JDK 1.8, using the `parallel` function whenever possible.

For each set of results, the best value, the average and the standard deviation are computed. The results, in term of distance and execution time are concluded in Table 2 The columns *Best*, *Average* and *Stdev* summarize the best, average, respectively standard deviation of the results for the genetic algorithm.

Table 2. Quality of the results

Indicator	Genetic algorithm		
	Best	Average	Stdev
Time (minutes)	7:15	10:22	1:47
Minimum distance (km)	182306	185554	880

The time, epoch and best individual evolution table (for one of the runs) is graphed bellow:

Table 3. Values evolution for one of the runs

Elapsed time	Epoch	Best solution
00:00	1	230825
01:00	458	202517
02:00	1269	191343
03:00	1781	188301
04:00	2138	186493
05:00	2272	185732
06:00	2508	184929
07:00	2669	184430
08:00	2823	183803
09:00	2954	183342
09:10	3000	183127

Table 4. Table of running times

Run	GA	Self-adaptation	Self-adaptation + migration	Migration
1	05:34.0	07:51.0	12:18.0	06:06.0
2	05:32.0	**06:25.0**	**07:15.0**	06:11.0
3	05:41.0	11:40.0	08:42.0	06:21.0
4	05:45.0	07:48.0	11:37.0	06:11.0
5	05:34.0	10:19.0	10:50.0	06:02.0
6	**05:27.0**	07:01.0	09:53.0	**05:46.0**
7	05:58.0	06:44.0	09:10.0	06:03.0
8	05:33.0	08:16.0	13:18.0	06:05.0
9	05:46.0	10:37.0	09:43.0	06:04.0
10	05:44.0	08:25.0	10:57.0	05:53.0
mean	05:39.4	08:30.6	10:22.3	06:04.2
stddev	00:09.1	01:46.7	01:47.8	00:09.7
min	05:27.0	06:25.0	07:15.0	05:46.0
max	05:58.0	11:40.0	13:18.0	06:21.0

As it can be noticed, the largest growth is in the first three minutes of the running time, in the rest of the time the improvement is rather small (10%).

In the first minute, the speed of the GA is slower due to the fact that Java's JIT Compiler did not yet fully integrate all the necessary code. After the 4th minute, the slowness is due to the local optimums that force the algorithm to switch to the slow version (Fig. 7 and Tables 4 and 5).

Table 5. Table of calculated distances

Run	GA	Self-adaptation	Self-adaptation + migration	Migration
1	183833	**182600**	**182306**	184260
2	184655	185310	184101	**182785**
3	185393	182962	184021	185031
4	185550	183830	183303	183745
5	183701	183126	183837	183644
6	183935	186108	183921	184256
7	183695	186531	183127	185587
8	186097	184032	182881	186257
9	**183064**	184205	183325	184656
10	183387	184292	185554	184123
mean	184331	184300	183638	184434
stddev	1030	1318	880	1002
min	183064	182600	182306	182785
max	186097	186531	185554	186257

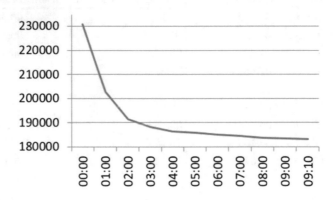

Fig. 7. Graph of the values for Table 3

On average, the fast version runs at a speed of 1 s for 10 epochs. The slow version takes from 11 to 36 s on 10 epochs (so 11 up to 36 times slower). This means that in a set amount of time, a self-adapting algorithm can reach a much better solution, also being able to exit local optimums.

In the end, the algorithm tends to remain on the slow version, due to the small growths and many local optimums, but this behavior is expected and acceptable. If it would have run the slow version from the beginning, the run time would be at least 10 times longer (so around 2 h instead of just 12 min) for the same solution. It's up to the client's choice to continue to run the algorithm after the first 3000 epochs, so an infinite running mode was implemented.

5   Conclusions

The article hybridizes a genetic algorithm for improving its performance. The hybridization techniques are:

- a self-adaptation method for getting out of local optima;
- use of immigrants for increasing the population diversity;
- a combination of the two.

Of course, the execution time increases along with the complexity of the hybridization, yet in reasonable range. However, the quality of the solution is better after embedding these techniques. More, the standard deviation of the solutions decreases, which means that the results are more reproducible on long term.

References

1. Clarke, G., Wright, J.W.: Scheduling of vehicles from a central depot to a number of delivery points. Oper. Res. **12**(4), 568–581 (1964)
2. Dantzig, G.B., Ramser, J.H.: The truck dispatching problem. Manag. Sci. **6**(1), 80–91 (1959)
3. Dror, M.: Note on the complexity of the shortest path models for column generation in VRPTW. Oper. Res. **42**(5), 977–978 (1994)
4. Fleurent, C., Ferland, J.A.: Genetic and hybrid algorithms for graph coloring. Ann. Oper. Res. **63**(3), 437–461 (1996)
5. Gendreau, M., Potvin, J.Y., et al.: Handbook of Metaheuristics, vol. 2. Springer, Heidelberg (2010)
6. Golden, B., Assad, A., Levy, L., Gheysens, F.: The fleet size and mix vehicle routing problem. Comput. Oper. Res. **11**(1), 49–66 (1984)
7. Golden, B.L., Raghavan, S., Wasil, E.A.: The vehicle routing problem: latest advances and new challenges, vol. 43. Springer, Heidelberg (2008)
8. Laporte, G.: Fifty years of vehicle routing. Transp. Sci. **43**(4), 408–416 (2009)
9. Laporte, G., Ropke, S., Vidal, T.: Heuristics for the vehicle routing problem. In: Vehicle Routing: Problems, Methods, and Applications, Second Edition, pp. 87–116. SIAM (2014)
10. Lin, C., Choy, K.L., Ho, G.T., Chung, S.H., Lam, H.: Survey of green vehicle routing problem: past and future trends. Expert Syst. Appl. **41**(4), 1118–1138 (2014)
11. Matei, O., Pop, P.C., Sas, J.L., Chira, C.: An improved immigration memetic algorithm for solving the heterogeneous fixed fleet vehicle routing problem. Neurocomputing **150**, 58–66 (2015)
12. Oliviu, M.: Theoretical and practical applications of evolutionary computation in solving combinatorial optimization problems. Ph.D. thesis, Technical University of Cluj-Napoca (2012)
13. Pop, P.C., Matei, O., Sitar, C.P.: An improved hybrid algorithm for solving the generalized vehicle routing problem. Neurocomputing **109**, 76–83 (2013)
14. Prins, C.: Efficient heuristics for the heterogeneous fleet multitrip VRP with application to a large-scale real case. J. Math. Model. Algorithms **1**(2), 135–150 (2002)

15. Semet, F., Taillard, E.: Solving real-life vehicle routing problems efficiently using tabu search. Ann. Oper. Res. **41**(4), 469–488 (1993)
16. Toth, P., Vigo, D.: Vehicle Routing: Problems, Methods, and Applications. SIAM (2014)
17. Yanik, S., Bozkaya, B., deKervenoael, R.: A new VRPPD model and a hybrid heuristic solution approach for e-tailing. Eur. J. Oper. Res. **236**(3), 879–890 (2014)

Scalable Hypothesis Tests for Detection of Epileptic Seizures

Dorin Moldovan[✉]

Faculty of Automation and Computer Science, Technical University of Cluj-Napoca,
Cluj, Romania
dorin.moldovan@cs.utcluj.ro

Abstract. Epilepsy is a disease that affects a large part of the population and the monitoring of the cerebral electrical activity using Electroencephalogram sensors provides a real time graphical representation of the brain function that can be used to detect the epileptic seizures. The detection of epileptic seizures is challenging because the brain activity is very complex, unique for each person and can not be completely captured even though the technology advanced very much in the last years and the sensors that capture this data are very complex in the present. In this research article the detection of epileptic seizures is approached using a machine learning methodology that is based on a combination of some of the latest technological advancements in the fields of big data and machine learning in which the features are extracted using a method based on scalable hypothesis tests and then the data is predicted using three machine learning classifiers namely, AdaBoost, Gradient Boosted Trees and Random Forest.

Keywords: Time series · Machine learning · Epileptic seizures ·
Features extraction · Classification

1 Introduction

The survey results of National Society for Epilepsy indicate that globally around 10 million people suffer from epilepsy [1] and this chronic neurological disorder is defined by recurrent and unprovoked seizures. Epilepsy is a symptom with a strong genetic predisposition [2] and it is defined as two unprovoked seizures that occur more than 24 h apart [3]. The percentage of the world population that suffers from epileptic seizures is around 2% and the occurrence of epileptic seizures is often uncertain [4]. The percentage of the world population that experienced at least one seizure in their life is approximately 5% [5] and epilepsy results in many types of injuries such as accidents, burns and fractures [6]. The monitoring of cerebral electrical activities using Electroencephalogram (EEG) [7] provides a graphical representation in real time of the brain function and this data can be used to detect epileptic seizures. In [8] the epileptic seizures are defined as a transient occurrence of symptoms and signs due to synchronous or

© Springer Nature Switzerland AG 2019
R. Silhavy et al. (Eds.): CoMeSySo 2019, AISC 1047, pp. 157–166, 2019.
https://doi.org/10.1007/978-3-030-31362-3_16

excessive neuronal activity in brain [9] and they may consist of body shaking, lack of awareness or difficulty in responding. Even though brain imaging and EEG are the main techniques for diagnostic testing of seizures, the EEG after the first seizure indicates abnormalities up to 73% of the time [10].

The recognition of epileptic seizures was approached in literature using various machine learning classification algorithms such as Support Vector Machine (SVM) [11–13], Bayesian Linear Discriminant Analysis (BLDA) [14], Artificial Neural Network (ANN) [15,16] and fuzzy rule-based systems [17]. In [18] is presented a methodology for the prediction of epileptic seizures that exploits the spatiotemporal relationship of the EEG signals which is based on phase correlation and the accuracy of the method described in that article is approximately 91.95%. The proposed methodology consists of four major steps: (a) preprocessing, (b) feature extraction, (c) classification and (d) post-processing. The extraction of features in the case of epileptic seizures datasets was approached in literature using a variety of methods such as univariate features [19], bivariate features [20] and spike rate [21]. In [22] the authors propose a novel algorithm based on an optimal allocation technique (OAT) in order to classify EEG signals and they validate their approach using the Multiclass Least Square Support Vector Machine (MLSSVM) algorithm. Their results are much better than the ones presented in [23] where spectral analysis techniques are applied. The OAT technique is also used in [24] where the authors compare the performance of three classification models namely, Logistic Model Trees (LMT) [25], Support Vector Machine (SVM) [26] and Multinomial Logistic Regression (MLR) [27].

The main contributions of this research article are:

- a critical study of the state of the art approaches that are used for the prediction of epileptic seizures;
- the development of a machine learning methodology for detection of epileptic seizures based on scalable hypothesis tests;
- the evaluation of the performance of the proposed methodology using three machine learning classifiers: AdaBoost (AB) [28], Gradient Boosted Trees (GBT) [29] and Random Forest (RF) [30];

The rest of the paper is structured as follows: Sect. 2 presents the machine learning methodology, Sect. 3 presents the experimental results and Sect. 4 presents the discussions.

2   Methods

The methodology for prediction of epileptic seizures is presented in Fig. 1 and it contains the following phases: the collection of EEG sensors data, the processing of EEG sensors data, the extraction of features from time series data using scalable hypothesis tests and the prediction of epileptic seizures using AB, GBT and RF.

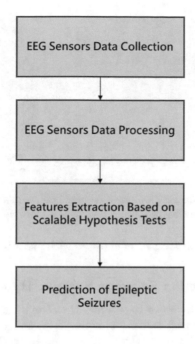

Fig. 1. Machine learning methodology for prediction of epileptic seizures

2.1 EEG Sensors Data Collection

The dataset that is used in this article is taken from University of California at Irvine (UCI) Machine Learning Repository [31,32]. The original dataset consists of information from 500 individuals and for each individual the brain activity is recorded for 23.6 s. The dataset derived from the original dataset contains 11500 samples and each sample has 178 features that correspond to 178 data points recorded for one second and a label. The values of the features are both positive and negative and the labels take values from the set $\{1, 2, 3, 4, 5\}$. Although the labels can take 5 different values, only the subjects that fall in the class 5 have epileptic seizures.

2.2 EEG Sensors Data Processing

Prior to the application of the features extraction step the data is normalized so that the features take values from the interval $[0, 1]$, the labels $\{1, 2, 3, 4\}$ are replaced with 0 and the label 5 is replaced with 1. Label 0 describes the cases when the subjects do not have epileptic seizures and label 1 describes the cases when the subjects have epileptic seizures. Figure 2 presents two data samples after data processing. One of the samples corresponds to a subject that has epileptic seizures and the other one corresponds to a subject that does not have epileptic seizures.

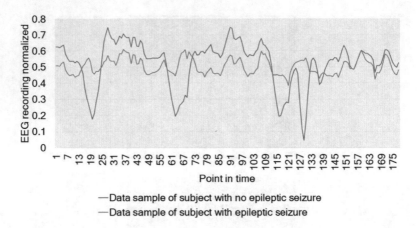

Fig. 2. Illustrative data samples of subjects with and without epileptic seizure

2.3 Features Extraction Based on Scalable Hypothesis Tests

In literature there are two major techniques that can be used for the classification of time series data. The first technique is based on using a specific method such as a Recurrent Neural Network (RNN) or a Long Short-Term Memory (LSTM) artificial neural network. The other technique is based on extraction of features such as mean, median, maximum, minimum and so on for each time series and then to use a supervised learning approach for classification. In this article the second technique is adopted and the features are extracted using a method that is based on scalable hypothesis tests using the tsfresh package from Python. This method is described in more details in [33] and its main three steps are (1) features extraction, (2) features relevance and (3) multiple testing. These three steps are illustrated in Fig. 3.

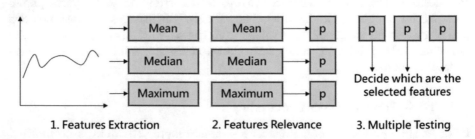

Fig. 3. Scalable extraction of features based on hypothesis tests

2.4 Prediction of Epileptic Seizures

The prediction of epileptic seizures is realized using three classification algorithms: AB, GBT and RF. These algorithms were chosen after a number of experiments using different types of classifiers ranging from Logistic Regression to Recurrent Neural Networks. One of the main characteristics of these algorithms that make them the best choice for the classification problem described in this article is the fact that all of them are ensemble classifiers. The parameters that are used for tuning these algorithms are the default ones from sklearn package and Table 1 presents illustrative parameters that are tuned for each algorithm.

Table 1. Parameters description for the classification algorithms used for predicting epileptic seizures

Algorithm	Parameter
AdaBoost	Number of estimators
	Learning rate
	Random state
Gradient Boosted Trees	Number of estimators
	Learning rate
	Maximum depth
Random Forest	Number of estimators
	Bootstrap
	Maximum depth

3   Results

This section presents the results obtained after the application of the machine methodology proposed in this article. The metrics that are used for the evaluations of the three classifiers that are compared are:

$$Precision = \frac{TP}{TP + FP} \tag{1}$$

$$Recall = \frac{TP}{TP + FN} \tag{2}$$

$$Accuracy = \frac{TP + TN}{TP + TN + FP + FN} \tag{3}$$

$$F1Score = 2 \times \frac{Precision \times Recall}{Precision + Recall} \tag{4}$$

where:

- TP (True Positive) - is the number of samples for which the actual value is positive and the predicted value is positive;

- FN (False Negative) - is the number of samples for which the actual value is positive and the predicted value is negative;
- FP (False Positive) - is the number of samples for which the actual value is negative and the predicted value is positive;
- TN (True Negative) - is the number of samples for which the actual value is negative and the predicted value is positive;

3.1   Features Extraction Results

After the application of features extraction, the number of features extracted for each time series becomes 285. Some example of features that are extracted are: (1) Friedrich coefficients, (2) autocorrelation, (3) variance, (4) kurtosis, (5) sum of changes, (6) maximum, (7) minimum, (8) mean, (9) quantile and (10) partial autocorrelation. In Fig. 4 is presented the matrix of correlation of the original dataset used in experiments before the application of the features extraction technique that is based on scalable hypothesis tests and in Fig. 5 is presented the matrix of correlation after the application of the features extraction technique described in this article.

Fig. 4. Matrix of correlation before application of features extraction

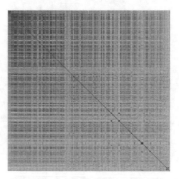

Fig. 5. Matrix of correlation after application of features extraction

3.2   Classification Results

In Fig. 6 are presented the classification results for training data. The training data represents 80% from the initial data and the testing data represents 20% from the initial data.

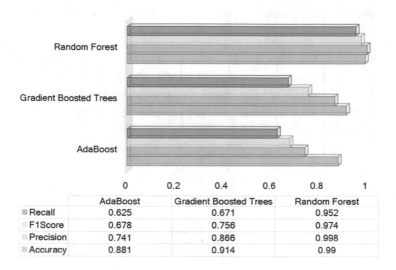

	AdaBoost	Gradient Boosted Trees	Random Forest
▪ Recall	0.625	0.671	0.952
F1Score	0.678	0.756	0.974
Precision	0.741	0.866	0.998
▪ Accuracy	0.881	0.914	0.99

Fig. 6. Classification results for training data

In Fig. 7 are presented the classification results for testing data.

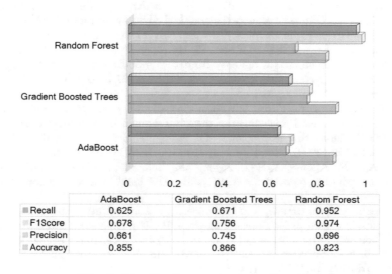

	AdaBoost	Gradient Boosted Trees	Random Forest
▪ Recall	0.625	0.671	0.952
F1Score	0.678	0.756	0.974
Precision	0.661	0.745	0.696
▪ Accuracy	0.855	0.866	0.823

Fig. 7. Classification results for testing data

4   Discussions

In this article was proposed an approach for detection of epileptic seizures that is based on scalable hypothesis tests. Compared to some approaches from literature that use the same experimental dataset as support, in this article each time series is considered as a singular point and for each time series a set of features is extracted. Even though the prediction results are not the best ones compared to the best results obtained by other authors that studied the same dataset, the approach based on Random Forest gave very good results for the training data: an accuracy of 99% and a precision of 99.8%. Future research directions are: (1) the tuning of the initial parameters of the algorithms using either literature approaches or approaches that are adapted for this specific type of data, (2) the combination of the methodology presented in this article with methods used by other authors in literature in order to obtain better precision and classification results and (3) the generalization of the methodology.

References

1. Raghu, S., Sriraam, N., Kumar, G.P., Hegde, A.S.: A novel approach for real-time recognition of epileptic seizures using minimum variance modified fuzzy entropy. IEEE Trans. Biomed. Eng. **65**(11), 2612–2621 (2018). https://doi.org/10.1109/TBME.2018.2810942
2. Thijs, R.D., Surges, R., O'Brien, T.J., Sander, J.W.: Epilepsy in adults. Lancet **393**(10172), 689–701 (2019). https://doi.org/10.1016/S0140-6736(18)32596-0
3. Fisher, R.S., Acevedo, C., Arzimanoglou, A., Bogacz, A., Cross, J.H., Elger, C.E., Engel Jr., J., Forsgren, L., French, J.A., Glynn, M., Hesdorffer, D.C., Lee, B.I., Mathern, G.W., Moshe, S.L., Perucca, E., Scheffer, I.E., Tomson, T., Watanabe, M., Wiebe, S.: ILAE official report: a practical clinical definition of epilepsy. Epilepsia **55**(4), 475–482 (2014). https://doi.org/10.1111/epi.12550
4. Fisher, R.S., Vickrey, B.G., Gibson, P., Hermann, B., Penovich, P., Scherer, A., Walker, S.: The impact of epilepsy from the patients perspective I. Descriptive and subjective perceptions. Epilepsy Res. **41**(1), 39–51 (2000). https://doi.org/10.1016/S0920-1211(00)00126-1
5. Netoff, T., Park, Y., Parhi, K.: Seizure prediction using cost-sensitive support vector machine. In: Proceedings of the 2009 Annual International Conference of the IEEE Engineering in Medicine and Biology Society, Minneapolis, MN, USA, pp. 3322–3325 (2009). https://doi.org/10.1109/IEMBS.2009.5333711
6. Mormann, F., Andrzejak, R.G., Elger, C.E., Lehnertz, K.: Seizure prediction: the long and winding road. Brain **130**(2), 314–333 (2007). https://doi.org/10.1093/brain/awl241
7. Bhattacharyya, S., Biswas, A., Mukherjee, J.: Feature selection for automatic burst detection in neonatal electroencephalogram. IEEE J. Emerg. Sel. Topics Circ. Syst. **1**(4), 469–479 (2011). https://doi.org/10.1109/JETCAS.2011.2180834
8. Johnson, E.L.: Seizures and epilepsy. Med. Clin. N. Am. **103**(2), 309–324 (2019). https://doi.org/10.1016/j.mcna.2018.10.002

9. Fisher, R.S., van Emde Boas, W., Blume, W., Elger, C., Genton, P., Lee, P., Engel Jr., J.: Epileptic seizures and epilepsy: definitions proposed by the international league against epilepsy (ILAE) and the international bureau for epilepsy (IBE). Epilepsia **46**(4), 470–472 (2005). https://doi.org/10.1111/j.0013-9580.2005.66104.x

10. Jackson, A., Teo, L., Seneviratne, U.: Challenges in the first seizure clinic for adult patients with epilepsy. Epileptic Disord. **18**(3), 305–314 (2016). https://doi.org/10.1684/epd.2016.0853

11. Islam, M.K., Rastegarnia, A., Yang, Z.: A wavelet-based artifact reduction from scalp EEG for epileptic seizure detection. IEEE J. Biomed. Health Inform. **20**(5), 1321–1332 (2016). https://doi.org/10.1109/JBHI.2015.2457093

12. Esbroeck, A.V., Smith, L., Syed, Z., Singh, S., Karam, Z.: Multi-task seizure detection: addressing intra-patient variation in seizure morphologies. Mach. Learn. **102**(3), 309–321 (2016). https://doi.org/10.1007/s10994-015-5519-7

13. Sriraam, N., Raghu, S.: Classification of focal and non focal epileptic seizures using multi-features and SVM classifier. J. Med. Syst. **41**(10), 160 (2017). https://doi.org/10.1007/s10916-017-0800-x

14. Zhou, W., Liu, Y., Yuan, Q., Li, X.: Epileptic seizure detection using lacunarity and Bayesian linear discriminant analysis in intracranial EEG. IEEE Trans. Biomed. Eng. **60**(12), 3375–3381 (2013). https://doi.org/10.1109/TBME.2013.2254486

15. Srinivasan, V., Eswaran, C., Sriraam, N.: Artificial neural network based epileptic detection using time-domain and frequency-domain features. J. Med. Syst. **29**(6), 647–660 (2005). https://doi.org/10.1007/s10916-005-6133-1

16. Srinivasan, V., Eswaran, C., Sriraam, N.: Approximate entropy-based epileptic EEG detection using artificial neural networks. IEEE Trans. Inf. Technol. Biomed. **11**(3), 288–295 (2007). https://doi.org/10.1109/TITB.2006.884369

17. Aarabi, A., Fazel-Rezai, R., Aghakhani, Y.: A fuzzy rule-based system for epileptic seizure detection in intracranial EEG. Clin. Neurophysiol. **120**(9), 1648–1657 (2009). https://doi.org/10.1016/j.clinph.2009.07.002

18. Parvez, M.Z., Paul, M.: Epileptic seizure prediction by exploiting spatiotemporal relationship of EEG signals using phase correlation. IEEE Trans. Neural Syst. Rehabil. Eng. **24**(1), 158–168 (2016). https://doi.org/10.1109/TNSRE.2015.2458982

19. Rasekhi, J., Mollaei, M.R.K., Bandarabadi, M., Teixeira, C.A., Dourado, A.: Preprocessing effects of 22 linear univariate features on the performance of seizure prediction methods. J. Neurosci. Methods **217**(1–2), 9–16 (2013). https://doi.org/10.1016/j.jneumeth.2013.03.019

20. Mirowski, P., Madhavan, D., LuCun, Y., Kuzniecky, R.: Classification of patterns of EEG synchronization for seizure prediction. Clin. Neurophysiol. **120**(11), 1927–1940 (2009). https://doi.org/10.1016/j.clinph.2009.09.002

21. Li, S., Zhou, W., Yuan, Q., Liu, Y.: Seizure prediction using spike rate of intracranial EEG. IEEE Trans. Neural Syst. Rehabil. Eng. **21**(6), 880–886 (2013). https://doi.org/10.1109/TNSRE.2013.2282153

22. Siuly, Li, Y.: A novel statistical algorithm for multiclass EEG signal classification. Eng. Appl. Artif. Intell. **34**(1), 154–167 (2014). https://doi.org/10.1016/j.engappai.2014.05.011

23. Ubeyli, E.D.: Decision support systems for time-varying biomedical signals: EEG signals classification. Expert Syst. Appl. **36**(2), 2275–2284 (2009). https://doi.org/10.1016/j.eswa.2007.12.025

24. Kabir, E., Siuly, Zhang, Y.: Epileptic seizure detection from EEG signals using logistic model trees. Brain Inform. **3**(2), 93–100 (2016). https://doi.org/10.1007/s40708-015-0030-2

25. Landwehr, N., Hall, M., Frank, E.: Logistic model trees. Mach. Learn. **59**(1–2), 161–205 (2005). https://doi.org/10.1007/s10994-005-0466-3

26. Yin, X., Ng, B.W.-H., Fischer, B.M., Ferguson, B., Abbott, D.: Support vector machine applications in terahertz pulsed signals feature sets. IEEE Sens. J. **7**(12), 1597–1608 (2007). https://doi.org/10.1109/JSEN.2007.908243

27. Le Cessie, S., Van Houwelingen, J. C.: Ridge estimators in logistic regression. J. Roy. Stat. Soc. Ser. C (Appl. Stat.) **41**(1), 191–201 (1992). https://doi.org/10.2307/2347628

28. Chengsheng, T., Bing, X., Huacheng, L.: The application of the adaboost algorithm in the text classification. In: Proceedings of the 2018 2nd IEEE Advanced Information Management, Communicates, Electronic and Automation Control Conference (IMCEC), Xi'an, China, pp. 1792–1796 (2018). https://doi.org/10.1109/IMCEC.2018.8469497

29. Sakata, R., Ohama, I., Taniguchi, T.: An extension of gradient boosted decision tree incorporating statistical tests. In: Proceedings of the 2018 IEEE International Conference on Data Mining Workshops (ICDMW), Singapore, Singapore, pp. 964–969 (2018). https://doi.org/10.1109/ICDMW.2018.00139

30. Feng, Z., Mo, L., Li, M.: A random forest-based ensemble method for activity recognition. In: Proceedings of the 2015 37th Annual International Conference of the IEEE Engineering in Medicine and Biology Society (EMBC), Milan, Italy, pp. 5074–5077 (2015). https://doi.org/10.1109/EMBC.2015.7319532

31. Dua, D., Karra Taniskidou, E.: UCI Machine Learning Repository. http://archive.ics.uci.edu/ml/index.php

32. Andrzejak, R.G., Lehnertz, K., Mormann, F., Rieke, C., David, P., Elger, C.E.: Indications of nonlinear deterministic and finite-dimensional structures in time series of brain electrical activity: dependence on recording region and brain state. Phys. Rev. E **64**(1), 1–8 (2001). https://doi.org/10.1103/PhysRevE.64.061907

33. Christ, M., Braun, N., Neuffer, J., Kempa-Liehr, A.W.: Time series feature extraction on basis of scalable hypothesis tests (tsfresh - a Python package). Neurocomputing. **307**(1), 72–77 (2018). https://doi.org/10.1016/j.neucom.2018.03.067

The Exact Frequency Domain Solution for the Periodic Synchronous Averaging Performed in Discrete-Time

Oksana Guschina[1,3] 📵, Timofey Shevgunov[1,2(✉)] 📵,
Evgeniy Efimov[1] 📵, and Vladimir Kirdyashkin[1] 📵

[1] Moscow Aviation Institute (National Research University),
Volokolamskoe shosse 4, 125993 Moscow, Russia
busya_03@mail.ru, v_kirdyashkin@mail.ru,
shevgunov@gmail.com, omegatype@gmail.com
[2] National Research University Higher School of Economics,
Myasnitskaya Ulitsa 20, 101000 Moscow, Russia
[3] Microwave Systems JSC, Nizhnaya Syromyatnicheskaya street 11,
105120 Moscow, Russia

Abstract. This paper deals with the technique known as the periodic synchronous averaging. The exact analytical expression for the fast Fourier transform (FFT) representing the digital spectrum of the signal undergoing periodic synchronous averaging is derived using the general signal and spectral framework. This formula connects the coefficient of Fourier series of the original continuous-time signal with the FFT of the averaged sampled version taking into consideration all the effects such as difference between the true and averaging periods, the attenuation and the leakage. The results of the numerical simulation are presented for the case of periodic signal, which was chosen a train of triangle pulses, the spectrum of which possesses a closed form and whose Fourier series coefficients rapidly decrease with the index. The chosen example allows the authors to illustrate that the waveform of the recovered signal can vary significantly, despite a rather slight difference in values between the true and averaging periods. Another important effect emphasized in the presented paper is that overall distinction between the original and averaged signals measured by means of relative mean square error raises if the total observation length increases while the other parameters remains fixed.

Keywords: Synchronous periodic averaging · Shuster's method ·
Buys Ballot scheme · Spectral density · FFT · RMSE

1 Introduction

The problem of detecting periodic signal against the background noise is often encountered in modern radio-electronic systems performing signal processing algorithms. Although there are many ways to identify a periodic signal with known parameters, there are cases where almost nothing is known about the signal, except its period with huge uncertainty. In this situation one may give a try to the well-known

© Springer Nature Switzerland AG 2019
R. Silhavy et al. (Eds.): CoMeSySo 2019, AISC 1047, pp. 167–175, 2019.
https://doi.org/10.1007/978-3-030-31362-3_17

method of synchronous periodic averaging proposed by Shuster in 1898 [1] to search for so-called hidden periodicities in time series describing meteorological and geological phenomena. Since the first publication on this method [2] seemed to be out fifty years before, it sometimes referred as Buys Ballot scheme.

Among others, Javorskyj made a significant contribution to the statistical analysis of the periodic averaging method, investigating the estimators for the coherent covariance of a periodically correlated random process with an unknown period of nonstationarity [3–6] in the identification problems related to mechanical systems. Some recent examples of the successful use of this method are given in [7] for analysis of the seasonality in time series describing flow of airline passengers and in [8] for estimating the trend and seasonality in time series of asset prices in a stock exchange.

The synchronous periodic averaging method plays a significant role in the analysis of signals that are considered to be realizations of cyclostationary [9, 10] or periodically correlated [11] random processes. Moreover, this method can be a starting point for the understanding of the cyclostationary concept from the time-domain perspective [12]. Besides, some algorithms for spectral correlation function estimation [13, 14] also involve the synchronous averaging. Even though the averaging can be carried out implicitly for a sample sequence of the finite length, it will lead to software implementation that can run in the real-time mode [15] due to its higher performance.

Despite the fact the periodic averaging originates from simple yet outstanding concept developed in the 19th century, its exhaustive analytical description in frequency domain has not been obtained yet for the case where digital processing of the continuous waveform is performed. The current paper aims at filling this gap and is organized as follows. The description of spectral transformation accompanying the synchronous periodical averaging is given in Sect. 2. The numerical examples of the averaging for different period ratios and observation lengths are presented in Sect. 3. Section 4 provides the way to measure the overall distinction between the original and averaged signals. The main results are summarized in the conclusion.

2   Spectral Description of Periodic Averaging

The starting point for further transformations is an analog signal $s(t)$ with period T:

$$s(t) = s(t+T). \tag{1}$$

Its representation as the Fourier series can be written as:

$$s(t) = \sum_{k \in \mathbb{Z}} S_k e^{+j2\pi \frac{k}{T} t}, \tag{2}$$

where the complex numbers S_k are the coefficients of the Fourier series:

$$S_k = \left\langle s(t) e^{-j2\pi \frac{k}{T} t} \right\rangle_T = \int_0^T s(t) e^{-j2\pi \frac{k}{T} t} dt. \tag{3}$$

The spectral density of the signal $s(t)$ can be written as follows:

$$S(f) = \sum_{k \in \mathbb{Z}} S_k \delta\left(f - \frac{k}{T}\right), \tag{4}$$

where $\delta(f)$ stands for a Dirac delta function in the frequency domain.

The first step of synchronous periodic averaging is the division of the signal into L non-overlapping in time segments of length T_a, which is chosen equal to the averaging signal period. The segments are then summed up, and the sum is divided by the total number L, provided $T_{obs} = LT_a$ is the total length of the signal realization. Therefore, the averaged signal can be represented as a sum of the segments of the signal $s(t)$ shifted by lT_a ($l = 0,\ldots, L{-}1$):

$$x_a(t) = \frac{1}{L} \sum_{l=0}^{L-1} s(t + lT_a). \tag{5}$$

It is crucial to note that the signal obtained by means of the averaging (5) is also a periodic signal of the same period T, regardless of T_a. That means it can be expressed as its Fourier series:

$$x_a(t) = \sum_{k \in \mathbb{Z}} X_{ak} e^{+j2\pi\frac{k}{T}t}, \tag{6}$$

where the coefficients X_{ak} relate to the coefficients (3):

$$X_{ak} = S_k D_L\left(\pi\frac{k}{T}T_a\right) e^{+j\pi\frac{k}{T}(L-1)T_a}. \tag{7}$$

where $D_L(v)$ stands for the Dirichlet kernel:

$$D_L(v) = \frac{\sin(Lv)}{L\sin(v)}. \tag{8}$$

This allows one to write down the spectral density of the signal $x_a(t)$:

$$X_a(f) = \sum_{k \in \mathbb{Z}} X_{ak} \delta\left(f - \frac{k}{T}\right). \tag{9}$$

The second step of synchronous averaging is the extraction of the signal related to the period T_a. This can be carried out by the limitation of the signal $x_a(t)$ with the rectangular window $w(t)$:

$$w(t) = rect\left(\frac{t - 0,5T_a}{T_a}\right). \tag{10}$$

The observation of one-period signal $x(t)$ can be simply expressed as the product:

$$x(t) = x_a(t)w(t). \tag{11}$$

However, if the signal $s(t)$ is processed in the digital domain, one deals with a sampled signal, the series of numbers taken with the sampling period T_s, rather than a continuous waveform:

$$s[n] = s(nT_s). \tag{12}$$

In this case, the averaging period T_a must be chosen as a multiple of T_s unless some resampling scheme is considered to be involved. One can assume that $T_a = NT_s$ and, it will be reasonable in turn to consider the whole observation as a multiple of the sampling period: $T_{obs} = LNT_s$. In fact, any observation can be truncated to the closest multiple of LT_s if it appears to be too long.

Thus, using a Dirac delta function in time domain $\delta(t)$, the sampled version of the continuous signal $x(t)$ (11) can be rewritten as a finite discrete-time signals:

$$x_\Delta(t) = \sum_{n=0}^{N-1} x[n]T_s\delta(t - nT_s), \tag{13}$$

where N is the number of samples in the period T_a, and $x[n] = x(nT_s)$ are the digital samples of the signal.

With the necessary transformations, one can obtain the following expression for the spectrum of the discrete signal $x_\Delta(t)$:

$$X_\Delta(f) = \sum_{k\in\mathbb{Z}} S_k D_L\left(\pi\frac{k}{T}T_a\right)e^{+j\pi\frac{k}{T}(L-1)T_a}T_a D_N\left(\pi\left(f - \frac{k}{T}\right)T_s\right)e^{-j\pi\left(f-\frac{k}{T}\right)(N-1)T_s}. \tag{14}$$

The formula (14) corresponds to a discrete-time Fourier transform (DTFT) [16] of the digital signal $x[n]$ consisting of a finite number N of samples taken with the sampling period T_s. However, in digital signal processing, a researcher usually deals with a discrete Fourier transform (DFT), that is exactly what is evaluated after applying a fast Fourier transform (FFT) to the result of averaging $x[n]$ as if it is done algorithmically on a computer. The FFT spectrum in the form of spectral density can be described by the formula:

$$X_{DFT}(f) = \sum_{m=0}^{N-1} X[m]\delta\left(f - \frac{m}{T_a}\right). \tag{15}$$

Nevertheless, the samples of FFT $X[m]$ relate to DTFT (14) in the similar manner [17] to how the samples $x[n]$ relate to the continuous signal $x(t)$:

$$X[m] = \frac{1}{T_a} X_\Delta \left(\frac{m}{T_a} \right). \tag{16}$$

Having substituted (14) into (16) and carried out the necessary transformations, one can obtain to following expression for $X[m]$:

$$X[m] = \left\{ \sum_{k \in Z} S_k D_L \left(\pi \frac{k}{T} T_a \right) D_N \left(\frac{\pi}{N} \left[m - \frac{k}{T} T_a \right] \right) e^{+j\pi \frac{k}{T} T_a \frac{NL-1}{N}} \right\} e^{-j\frac{\pi}{N} m(N-1)}. \tag{17}$$

It is important to highlight that the part in braces, involving Fourier series coefficients (3), depends on the ratio T_a/T of two periods as well as the term NL which is the total length of the observation in samples.

3 Simulation

A numerical simulation of synchronous periodic averaging was performed. A periodic train of triangle pulses was chosen as a good example of signal $s(t)$. Fig. from 1 to 4 show one period of the original signal $s(t)$ and samples $x[n]$ of the recovered signal $x(t)$, as well as their amplitude spectra $|S(f)|$ and $|X_{FFT}([m/T_a])|$ respectively. We consider the following cases in more details below.

Case 1. The results for $T/T_a = 1$, the total length of the realization $NL = 1000$ are shown in Fig. 1. The recovered signal and a period of the original signal match each other completely as well as their spectra; therefore, the signal can be recovered with no loss.

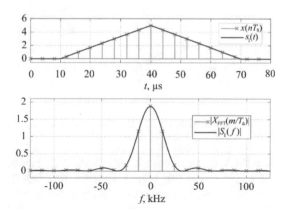

Fig. 1. Comparison of one period $s(t)$ и $x(t)$ and their spectra. $T/T_a = 1$, $NL = 1000$.

Case 2. Figure 2 illustrates the case where $T/T_a = 1.0005$, $NL = 1000$. One can easily see that with slight difference between the periods T_a and T there is insignificant distortion when the signals are compared.

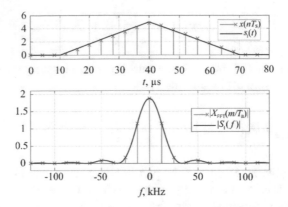

Fig. 2. Comparison of one period $s(t)$ и $x(t)$ and their spectra. $T/T_a = 1.0005$, $NL = 1000$.

Case 3. When the difference in periods reaches 5 percent: $T/T_a = 1.05$, $NL = 1000$, one can inevitably obtain the results shown in Fig. 3 where the distinction has become so significant that the shape of the original signal cannot be visually observed, and there is a significant attenuation of the spectral components with respect to the component at zero frequency.

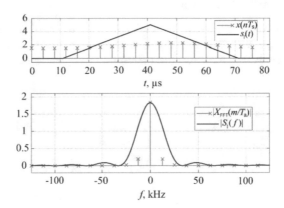

Fig. 3. Comparison of one period $s(t)$ и $x(t)$ and their spectra. $T/T_a = 1.05$, $M = 1000$.

Case 4. Let us choose the ratio from case 2: $T/T_a = 1.0005$, but the total length will increase 10 times up to $NL = 10\ 000$. Figure 4 shows that this increase leads to a noticeable distortion in the recovered signal. This is due to the fact that if the periods do not coincide with an increase in NL the number of averaging periods L increases and the multiplier D_L in (17) decreases.

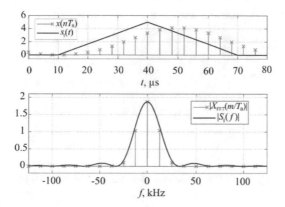

Fig. 4. Comparison of one period $s(t)$ и $x(t)$ and their spectra. $T/T_a = 1.0005$, $NL = 10\,000$.

4 Distinction Estimation

As far as sampled signals are concerned, the comparison have to be made between the original signal $s[n]$ and the signal $x_T[n]$ obtained as a repetition of the signal $x[n]$ evaluated by the synchronous averaging. The relative mean square error (RMSE) ε^2 can be evaluated by formula:

$$\varepsilon^2 = \frac{\sum\limits_{n=1}^{M} (s[n] - x_T[n])^2}{\sum\limits_{n=1}^{M} s[n]^2}. \tag{18}$$

The plots of the dependence of the RSME on the parameter γ, which measures the deflection of the periods T and T_a as the difference of its ratio from one:

$$\frac{T}{T_a} = 1 + \gamma \tag{19}$$

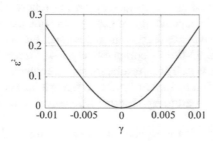

Fig. 5. RMSE of determining the period, $M = 1000$.

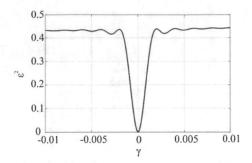

Fig. 6. RMSE of determining the period, $M = 10\,000$.

are shown for two values of the sample length of the signal realization $NL = 1000$ and $10\,000$ in Figs. 5 and 6 respectively. Those plots allow one to conclude that the RMSE goes up for the same values of γ when the number of samples involved in the averaging increases.

5   Conclusion

The original theoretical result obtained in the present research is (17), which provides the exact analytical expression for evaluating the samples of DFT in case of the Fourier series coefficients describing the original continuous waveform are known. This formula takes under consideration all underlying effects such as averaging attenuation and spectral leakage.

By using the example of a periodic train of triangle pulses, it was shown how the difference between the averaging period and the true period affects the signal being obtained as a result of the synchronous averaging. As it could have been anticipated, the signal is recovered without any distortion in the case of complete coincidence of the true signal period and the averaging period. However, in the case of a period mismatch, the recovered signal differs from the true signal. Moreover, the difference will increase as long as the deflection between the periods goes up or the sample length of the signal realization increases.

The distinction of the original and recovered signals measured by relative mean square error was estimated as a dependency on the relative difference in the period lengths. According to this dependency, one may conclude that RMSE certainly rises if the value of the difference increases. On the other hand, the neighborhood of the period ratio around one, where RMSE is comparatively low, becomes narrower if the total number of samples increases. These results reveal a simple procedure that may well be used for a priori estimation of the maximum difference between periods providing the acceptable recovery for an arbitrary deterministic waveform.

Acknowledgement. The work was supported by state assignment of the Ministry of Education and Science of the Russian Federation (project 8.8502.2017/BP).

References

1. Shuster, A.: On the investigation of hidden periodicities with application to a supposed 26 day period of metrological phenomena. J. Geophys. Res. **3**(1), 3–41 (1898)
2. Buys-Ballot, Les Changements Periodiques de Temperature, Utrecht (1847)
3. Javorskyj, I., Yuzefovych, R., Matsko, I., Zakrzewski, Z., Majewski, J.: Coherent covariance analysis of periodically correlated random processes for unknown non-stationarity period. Digit. Sig. Process. **65**, 27–51 (2017). https://doi.org/10.1016/j.dsp.2017.02.013
4. Javorskyj, I., Yuzefovych, R., Matsko I., Zakrzewski, Z., Majewski, J.: Statistical analysis of periodically non-stationary oscillations for unknown period. In: Proceedings of 2017 MIXDES, Bydgoszcz, pp. 543–546 (2017). https://doi.org/10.23919/MIXDES.2017.8005271
5. Javorskyj, I., Leśkow, J., Kravets, I., Isayev, I., Gajecka, E.: Linear filtration methods for statistical analysis of periodically correlated random processes – part i: coherent and component methods and their generalization. Sig. Process. **92**(7), 1559–1566 (2012). https://doi.org/10.1016/j.sigpro.2011.09.030
6. Javorskyj, I., Leśkow, J., Kravets, I., Isayev, I., Gajecka, E.: Linear filtration methods for statistical analysis of periodically correlated random processes – part ii: harmonic series representation. Sig. Process. **91**(11), 2506–2519 (2011). https://doi.org/10.1016/j.sigpro.2011.04.031
7. Fomby, T.B.: Buys Ballot Plots: Graphical Methods for Detecting Seasonality in Time Series (2008). http://faculty.smu.edu/tfomby/eco5375/data/season/buys%20ballot%20plots.pdf
8. Iwueze, I.S., Nwogu, E.C., Johnson, O., Ajaraogu, J.C.: Uses of the Buys-Ballot table in time series analysis. Appl. Math. **2**(5), 633–645 (2011). https://doi.org/10.4236/am.2011.25084
9. Gardner, W.A.: Statistical Spectral Analysis A Nonprobabilistic Theory. Prentice Hall, Upper Saddle River (1988)
10. Gardner, W.A., Napolitano, A., Paura, L.: Cyclostationarity: half a century of research. Sig. Process. **86**(4), 639–697 (2006). https://doi.org/10.1016/j.sigpro.2005.06.016
11. Hurd, H.L., Miamee, A.: Periodically Correlated Random Sequences: Spectral Theory and Practice. Wiley-Interscience (2007)
12. Shevgunov, T.: A comparative example of cyclostationary description of a non-stationary random process. J. Phys.: Conf. Ser. **1163**, 012037 (2019). https://doi.org/10.1088/1742-6596/1163/1/012037
13. Efimov, E., Shevgunov, T., Kuznetsov, Y.: Time delay estimation of cyclostationary signals on PCB using spectral correlation function. In: Proceedings of Baltic URSI Symposium, Poznan, pp. 184–187 (2018). https://doi.org/10.23919/URSI.2018.8406726
14. Shevgunov, T., Efimov, E., Zhukov, D.: Averaged absolute spectral correlation density estimator. In: Proceedings of Moscow Workshop on Electronic and Networking Technologies (MWENT), pp. 1–4 (2018). https://doi.org/10.1109/MWENT.2018.8337271
15. Shevgunov, T., Efimov, E.: Software implementation of spectral correlation density analyzer with RTL2832U SDR and Qt framework. In: Advances in Intelligent Systems and Computing, vol. 764. Springer, Cham (2019). https://doi.org/10.1007/978-3-030-19813-8_18
16. Oppenheim, A.V., Schafer, R.W.: Digital Signal Processing, 2nd edn. Prentice Hall, Upper Saddle River (1999)
17. Marple Jr., S.L.: Digital Spectral Analysis: With Applications. Prentice Hall, Upper Saddle River (1987)

An Adaptive Cognitive Temporal-Causal Model for Extreme Emotion Extinction Using Psilocybin

Seyed Sahand Mohammadi Ziabari[✉] and Jan Treur

Social AI Group, Vrije Universiteit Amsterdam, De Boelelaan 1105, Amsterdam,
The Netherlands
sahandmohammadiziabari@gmail.com, j.treur@vu.nl

Abstract. In this paper, an adaptive cognitive temporal-causal model using psilocybin for a reduction in extreme emotion is presented. Extreme emotion has an effect on some brain components such as visual cortex, auditory cortex, gustatory cortex, and somatosensory cortex as well as motor cortex such as primary motor cortex, and premotor cortex. Neuroscientific literature reviews show that using psilocybin has a significant effect mostly on two brain components, cerebral cortex, and thalamus. Network-oriented modeling via temporal-causal network-oriented modeling is presented to show the influences of using psilocybin on the cognitive part of the body, same as the brain components. Hebbian learning used to show the adaptivity and learning section of the presented model.

Keywords: Network-oriented modeling · Extreme emotion · Psilocybin · Temporal-causal network

1 Introduction

An intense stimulus, life-threatening phenomena such as an accident, war, leads to a well-known impact on the brain called stress. Based on the psychological peace of researches, 5 to 8% of men and 10 to 14% of women and up to 25% of combat veterans are involved in post-traumatic stress disorder (PTSD) in their lives. Extreme emotions such as stress have a significant impact on both cognitive and psychological functionality on the patient's brains [27].

In this paper, the influence of using psilocybin on patients with anxiety and depression disorders has been investigated. Neuroscience literature review shows a significant role of using psilocybin on two main brain components namely thalamus and cerebral cortex. The aim of this paper is to show the effect of using this drug on the brain components and having prediction of activity of those on patient. In [9] the definition and performance of using aforementioned drug has been mentioned as follows.

'Psilocybin is the prodrug of psilocybin (4-hydroxy-dimethyltryptamine), the primary hallucinogenic component of magic mushrooms, and a classic psychedelic ('mind-manifesting') drug. Psilocybin has been used for centuries in healing

© Springer Nature Switzerland AG 2019
R. Silhavy et al. (Eds.): CoMeSySo 2019, AISC 1047, pp. 176–186, 2019.
https://doi.org/10.1007/978-3-030-31362-3_18

ceremonies (1) and more recently in psychotherapy (2); it is capable of stimulating profound existential experiences (3), which can leave a lasting psychological impression (4).'

The use of psilocybin goes back to many years ago in religious rituals and medical usages [11]. There are several network-oriented modelings via the temporal-causal network for therapies or drugs taking to decrease the extreme emotion [13–26]. In this paper, the effect of using psilocybin has been used to demonstrate this event.

The paper is organized as follows. In Sect. 2 the underlying neurological principles concerning the parts of the brain involved in stress and in the suppression of stress are addressed. In Sect. 3 the integrative temporal-causal network model is introduced. In Sect. 4 the results of the simulation model are discussed, in Sect. 5 the mathematical analysis of the model is presented and eventually in the last section a conclusion is presented.

2 Underlying Neurological Principles

Psilocybin remarkably decreases brain blood flow and venous oxygenation in a way that related to the subjective impacts on the brain. Psilocybin based on the dose of use has different effects in other words it is dose-dependently and has a huge influence on changing the mood, perception, self-experience and thought and based on the result appeared in [10] experienced pleasurable moments, and non-threatening situations.

Neuroscience research shows that psilocybin has a significant agonist 5-HT (serotonergic) pathways and receptors located in the Thalamus and the cerebral cortex [1]. Triggering receptors such as 5-HT$_{2A}$ in the thalamus by using, which is responsible for sensory input, decreases the activity of the thalamic performance. Having decreased the thalamic activity, leading to hallucinations [2, 3]. As far as anxiety disorders and depression cause abnormalities in sensory perception of individuals, psilocybin has an impact on these areas in line with serotonin hormone. In [4] the moderate advantages of the effect of using psilocybin on patients in the Netherlands have been discussed. In [5] also the influence of periodical use of psilocybin and the psychological disorders investigated and also the playing role of psilocybin on decreasing the sense of committing suicide and improvement of moods have been highlighted. In comparison to methylphenidate the effective role of psilocybin on increasing social activity, behavior and positive behavior have been noted [6]. Some medical experiments [7, 8] have been done to show the positive effect of using psilocybin on patients with depression and anxiety due to tough diseases such as cancer. The reduction in an anxiety in patients with cancer was remarkable during the period between 1 to 3 months of posttreatment with psilocybin and had lower negative effect comparing to patients who used niacin [6].

Decreased activity in thalamus, cerebral cortex has been mentioned in [1, 9]. Cerebral cortex contains some brain components such as gustatory cortex, visual cortex, somatosensory cortex, and auditory cortex. These cortices received information, and input from thalamus when a patient uses psilocybin. Other components of cerebral cortex involved in decreasing anxiety consists of primary motor and premotor cortex.

The timing process of the effect of using psilocybin mentioned in [9] precisely:

'The subjective effects began toward the end of the infusion period and reached a sustained peak after ∼ 4 min (5). The first level results were entered into higher level analysis, contrasting cerebral blood flow (CBF) after placebo for all 15 subjects.'

In [10] acute, short and long-term subjective effect of using psilocybin in healthy individuals have been examined. The remaining cognitive parts of the model namely sensory representation state, preparation state, feeling state are based on the cognitive behavior of mind during extreme emotion.

3   The Temporal-Causal Network Model

First the Network-Oriented Modelling approach used to model the integrative overall process is briefly explained. As discussed in detail in [11, Ch 2] this approach is based on temporal-causal network models which can be represented at two levels: by a conceptual representation and by a numerical representation. A conceptual representation of a temporal-causal network model in the first place involves representing in a declarative manner states and connections between them that represent (causal) impacts of states on each other, as assumed to hold for the application domain addressed. The states are assumed to have (activation) levels that vary over time. In reality, not all causal relations are equally strong, so some notion of *strength of a connection* is used. Furthermore, when more than one causal relation affects a state, some way to *aggregate multiple causal impacts* on a state is used. Moreover, a notion of *speed of change* of a state is used for timing of the processes. These three notions form the defining part of a conceptual representation of a temporal-causal network model:

- **Strength of a connection** $\omega_{X,Y}$. Each connection from a state X to a state Y has a *connection weight value* $\omega_{X,Y}$ representing the strength of the connection, often between 0 and 1, but sometimes also below 0 (negative effect) or above 1.
- **Combining multiple impacts on a state** $c_Y(..)$. For each state (a reference to) a *combination function* $c_Y(..)$ is chosen to combine the causal impacts of other states on state Y.
- **Speed of change of a state** η_Y. For each state Y a *speed factor* η_Y is used to represent how fast a state is changing upon causal impact.

Combination functions can have different forms, as there are many different approaches possible to address the issue of combining multiple impacts. Therefore, the Network-Oriented Modelling approach based on temporal-causal networks incorporates for each state, as a kind of label or parameter, a way to specify how multiple causal impacts on this state are aggregated by some combination function. For this aggregation a number of standard combination functions are available as options and a number of desirable properties of such combination functions have been identified.

Figure 1 represents the conceptual representation of the temporal-causal network mode. The components of the conceptual representation shown in Fig. 1 are explained here. The state ws_c shows the world state of the contextual stimulus c. The states ss_c and ss_{ee} are the sensor state for the context c and sensor state of the body state ee for the extreme emotion. The states srs_c and srs_{ee} are the sensory representation of the contextual stimulus c and the extreme emotion, respectively. The state srs_{ee} is a stimulus influencing the activation level of the preparation state. Moreover, ps_{ee} is the preparation state of an extreme emotional response to the sensory representation srs_c of the context c, and fs_{ee} shows the feeling state associated to this extreme emotion. The state es_{ee} indicates the execution of the body state for the extreme emotion. All these relate to the affective processes. The (cognitive) goal state shows the goal for using psilocybin in the body. The (cognitive) state ps_{psil} is the preparation state of taking a psilocybin. The state es_{psil} is the execution state of taking psilocybin. The other states relate to biological brain parts, cerebral cortex which consists of two major components namely sensory areas (gustatory cortex, visual cortex, somatosensory cortex, auditory cortex) and motor areas (primary motor cortex, premotor cortex), thalamus, hippocampus, medial prefrontal cortex, and amygdala which are involved in the stress condition, and in the influence of the psilocybin applied (Table 1).

Table 1. Explanation of the states in the model

X_1	ws_{ee}	World (body) state of extreme emotion ee
X_2	ss_{ee}	Sensor state of extreme emotion ee
X_3	ws_c	World state for context c
X_4	ss_c	Sensor state for context c
X_5	srs_{ee}	Sensory representation state of extreme emotion ee
X_6	srs_c	Sensory representation state of context c
X_7	fs_{ee}	Feeling state for extreme emotion ee
X_8	ps_{ee}	Preparation state for extreme emotion ee
X_9	es_{ee}	Execution state (bodily expression) of extreme emotion ee
X_{10}	goal	Goal of using psilocybin
X_{11}	ps_{psil}	Preparation state of using psilocybin
X_{12}	es_{psil}	Execution of using psilocybin
X_{13}	Gustatory cortex	Brain part
X_{14}	Visual cortex	Brain part
X_{15}	Somatosensory cortex	Brain part
X_{16}	Auditory cortex	Brain part
X_{17}	Thalamus	Brain part
X_{18}	Hippocampus	Brain part
X_{19}	Medial prefrontal cortex	Brain part
X_{20}	Amygdala	Brain part
X_{21}	Primary motor cortex	Brain part
X_{22}	Premotor cortex	Brain part

The connection weights ω_i in Fig. 1 are as follows. The sensor states ss_{ee}, ss_{cc} have two incoming connections from ws_{ee} and ws_c (weights ω_1, ω_2). The world state of extreme emotion ws_{ee} has one arriving connection from es_{ee}, ω_{11} as a body-loop with weight. The sensory representation state of an extreme emotion srs_{ee} has an incoming connection weights ω_8 from state preparation state of an extreme emotion ps_{ee}. The feeling state fs_{ee} has one outgoing connection weight ω_5 from srs_{ee} and also five other connections from sensory areas of cerebral cortex such as gustatory cortex, visual cortex, somatosensory cortex, and auditory cortex, ω_{22}–ω_{25}, respectively, these connection weights are considered as the hallucination connections as it has been mentioned earlier in Sect. 2 as a neurological principles. The preparation state of an extreme emotion ps_{ee} has three incoming connection weights ω_6, ω_7, and ω_{55} from states srs_c and fs_{ee}, Primary cortex, respectively.

The goal has one arriving connection weight from the sensory representation srs_{ee} (ω_{12}) and preparation state ps_{psil} has an entering connection from the goal with weight ω_{13}. The execution of taking the drug (psilocybin) is called es_{pil}, and has an entering connection weight ω_{14} from preparation state of taking ps_{psil}. The state Thalamus has two entering connection weights ω_{15}, ω_{17} from es_{psil}, and Hippocampus, respectively. The sensory area in the cerebral cortex has four components, gustatory cortex, visual cortex, somatosensory cortex, and auditory cortex. All aforementioned components have outgoing connection to srs_{ee}, with connection weights, ω_{26}–ω_{29}. The Hippocampus in the brain has one incoming connection weight, ω_{16} from Thalamus. The Amygdala has two incoming connection weights from Hippocampus, and medial Prefrontal cortex, with the weights ω_{31}, ω_{33} and it has two outgoing connection weights ω_{32}, ω_{36}, ω_{37} to mPFC, Premotor cortex, and Primary cortex, respectively. Note that the connection weight between states medial Prefrontal Cortex and Hippocampus is adaptive and using Hebbian learning means through time will be changed, in this case the patient learned to cope with the stress and after some period of time she does not need to intake psilocybin. The state es_{psil} has also two outgoing connection weights to Premotor Cortex state and Primary cortex with ω_{37} and ω_{38}. Eventually, the state Primary motor cortex has an outgoing connection weight to ps_{ee}.

This conceptual representation was transformed into a numerical representation as follows [11, Ch 2]:

- at each time point t each state Y in the model has a real number value in the interval [0, 1], denoted by $Y(t)$
- at each time point t each state X connected to state Y has an impact on Y defined as $\mathbf{impact}_{X,Y}(t) = \omega_{X,Y} X(t)$ where $\omega_{X,Y}$ is the weight of the connection from X to Y
- The *aggregated impact* of multiple states X_i on Y at t is determined using a *combination function* $\mathbf{c}_Y(..)$:

$$\mathbf{aggimpact}_Y(t) = \mathbf{c}_Y(\mathbf{impact}_{X_1,Y}(t), \ldots, \mathbf{impact}_{X_k,Y}(t))$$
$$= \mathbf{c}_Y(\omega_{X_1,Y}X_1(t), \ldots, \omega_{X_k,Y}X_k(t))$$

where X_i are the states with connections to state Y
- The effect of $\mathbf{aggimpact}_Y(t)$ on Y is exerted over time gradually, depending on speed factor η_Y:

$$Y(t + \Delta t) = Y(t) + \eta_Y[\mathbf{aggimpact}_Y(t) - Y(t)]\Delta t$$
$$\text{or} \quad \mathbf{d}Y(t)/\mathbf{d}t = \eta_Y[\mathbf{aggimpact}\mathbf{Y(t)} - Y(t)]$$

- Thus, the following *difference* and *differential equation* for Y are obtained:

$$Y(t + \Delta t) = Y(t) + \eta_Y[\mathbf{c}_Y(\omega_{X_1,Y}X_1(t), \ldots, \omega_{X_k,Y}X_k(t)) - Y(t)]\Delta t$$

$$\mathbf{d}Y(t)/\mathbf{d}t = \eta_Y[\mathbf{c}_Y(\omega_{X_1,Y}X_1(t), \ldots, \omega_{X_k,Y}X_k(t)) - Y(t)]$$

For states the following combination functions $\mathbf{c}_Y(\ldots)$ were used, the identity function **id(.)** for states with impact from only one other state, and for states with multiple impacts the scaled sum function $\mathbf{ssum}_\lambda(\ldots)$ with scaling factor λ, and the advanced logistic sum function $\mathbf{alogistic}_{\sigma,\tau}(\ldots)$ with steepness σ and threshold τ.

$$\mathbf{id}(V) = V$$
$$\mathbf{ssum}_\lambda(V_1, \ldots, V_k) = (V_1 + \ldots + V_k)/\lambda$$

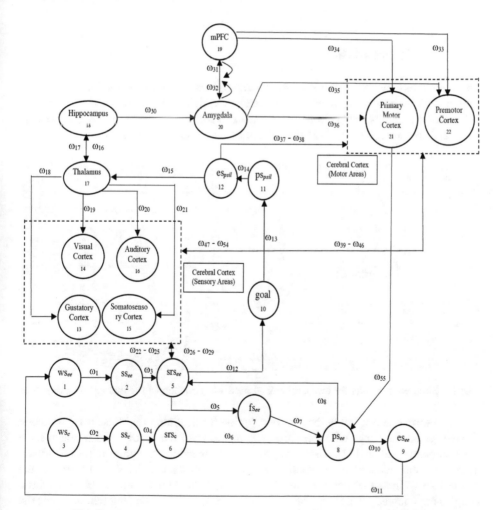

Fig. 1. Conceptual representation of the temporal-causal network model

$$\textbf{alogistic}_{\sigma,\tau}(V_1,\ldots,V_k) = [(1/(1 + e^{-\sigma(V_1 + \ldots + V_{k-\tau})}))-1/(1+e^{\sigma_\tau})]\,(1+e^{-\sigma_\tau})$$

Here first the general Hebbian Learning is explained which is applied to ω_{40} and ω_{41} for (X_5, X_{23}). In a general example model it is assumed that the strength ω of such a connection between states X_1 and X_2 is adapted using the following Hebbian Learning rule, taking into account a maximal connection strength 1, a learning rate $\eta > 0$ and a persistence factor $\mu \geq 0$, and activation levels $X_1(t)$ and $X_2(t)$ (between 0 and 1) of the two states involved [10]. The first expression is in differential equation format, the second one in difference equation format:

$$d\omega(t)/dt = \eta[X_1(t)X_2(t)(1 - \omega(t) - (1 - \mu)\omega(t))]$$
$$\omega(t+\Delta t) = \omega(t) + \eta[X_1(t)X_2(t)(1 - \omega(t))(1 - \mu)\omega(t)]\Delta t$$

4   Example Simulation

The simulation results of the cognitive temporal causal network model, which was constructed based on the neurological science which contains qualitative empirical information (such as fMRI) both for the mechanism by which the brain components work and for emerging result of the processes, has been shown in Fig. 2.

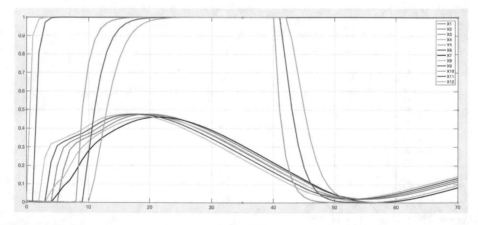

Fig. 2. Simulation results for temporal-causal network modeling of the therapy by psilocybin

Therefore, one can imply that the best option for declining the stress level has been chosen, given the usage of psilocybin. The model used the Matlab codes which have been implemented in [12]. Using appropriate connections weights make the model numerical and adapted to qualitative empirical information. Table 2 illustrates the connection weights that has been used, where the values for are initial values as these weights are adapted over time. The time step was $\Delta t = 1$. The scaling factors λ_i for the nodes with more than one arriving connection weights are mentioned in Table 2. At first, an external world state of an extreme emotion-stimuli context c (represented by ws_c) will influence the affective internal states of the individual by influencing the emotional response es_{ee} (via ss_c, srs_c, and ps_{ee}) conducted to show the extreme emotion by body state ws_{ee}. As a consequence, the stressed individual senses the extreme emotion (and at the same time all the biological brain components increased over time), so as a cognitive process, as a next step the goal (using psilocybin) becomes active to decrease this stress level by using psilocybin at time around 10.

Table 2. Connection weights and scaling factors for the example simulation

Connection weight	ω_1	ω_2	ω_3	ω_4	ω_5	ω_6	ω_7	ω_8	ω_{10}	ω_{11}	ω_{12}	ω_{13}
Value	1	1	1	1	1	1	1	1	1	1	1	1
Connection Weight	ω_{14}	ω_{15}	ω_{16}	ω_{17}	ω_{18}	ω_{19}	ω_{20}	ω_{21}	ω_{22}	ω_{23}	ω_{24}	ω_{25}
Value	1	-0.5	1	1	1	1	1	1	0.01	0.01	0.01	0.01
Connection Weight	ω_{26}	ω_{27}	ω_{28}	ω_{29}	ω_{30}	ω_{31}	ω_{32}	ω_{33}	ω_{34}	ω_{35}	ω_{36}	ω_{37}
Value	1	1	1	1	1	0.4	0.4	1	1	0.1	0.1	-0.5
Connection Weight	ω_{38}	ω_{39}	ω_{40}	ω_{41}	ω_{42}	ω_{43}	ω_{44}	ω_{45}	ω_{46}	ω_{47}	ω_{48}	ω_{49}
Value	−0.5	1	1	1	1	1	1	1	1	1	1	1
Connection Weight	ω_{50}	ω_{51}	ω_{52}	ω_{53}	ω_{54}							
Value	0.5	1	1	1	1							

state	X_5	X_8	X_{14}	X_{15}	X_{17}	X_{18}	X_{19}	X_{20}	X_{21}
λ_i	2.04	3	3.4	3	1.4	2	2	2	2

As a biological process, the goal and in further steps, execution of taking psilocybin activates the alterations and suppression of execution of stress at the first state and this affects other brain components to be less active around time 10. The environment is constant, external input for the model is only the constant world state ws_c. Therefore, based on the simulation results it is illustrated that the model for the drug therapy psilocybin works as expected.

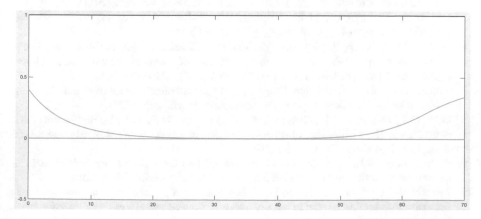

Fig. 3. Simulation results for adaptivity connection weight between amygdala and medial Prefrontal Cortex

The adaptivity connection (Hebbian learning) between two brain parts; Amygdala and medial Prefrontal Cortex (mPFC) is shown in Fig. 3. As it is indicated in the figure the adaptivity, learning to cope with stress and decreasing that over time starts at time around 50 and continues until time 70 to remain constant.

5 Conclusion

In this paper, an adaptive cognitive temporal causal-network model for extreme emotion extinction using psilocybin for patients under extreme emotion is presented. The proposed model can be used to test different hypothesis and neurological principles about the influences of the brain components and hormones and the effects that different brain components have the extinction of extreme emotion, but also on other processes such as fear.

Some simulations have been implemented, one of which was presented in the paper. This model can be used to predict human behavior dynamically in different conditions such as extreme emotion to get insight in such processes and to bring up a certain cure or treatment of individuals to perform the therapies of extreme emotions for post-traumatic disorder individuals. Future of this work can be analysis of other treatments together with this therapy.

References

1. Daniel, J., Haberman, M.: Clinical potential of psilocybin as a treatment for mental health conditions. Ment. Health Clin. **7**(1), 24–28 (2017)
2. Cart-Harris, RL., Erritzoe, D., Williams, T., Stone, J.M., Reed, L.J., Colasanti, A., et.al.: Neural correlates of the psychedelic state as determined by fMRI studies with psilocybin. Proc. Natl. Acad. Sci. U S A. **109** (6), 2138–2143 (2012)
3. Nicholas, D.E.: Hallucinogens. Pharmacol. Tehr. **101**(2), 131–181 (2004)
4. Van Amsterdam, J., Opperhuizen, A., van den Brink, W.: Harm potential of magic mushroom use: a review. Regul. Toxicol. Pharmacol. **59**(3), 423–429 (2011)
5. Hendricks, P.S., Johnson, M.W., Griffiths, R.R.: Psilocybin, psychological distress, and suicidality. J. Psychopharmacol. **29**(9), 1041–1043 (2015)
6. Griffiths, R., Richards, W., Johnson, M., McCann, U., Jesse, R.: Mystical-type experiences occasioned by psilocybin mediate the attribution of personal meaning and spiritual significance 14 months later. J. Psychopharmacol. **22**(6), 621–632 (2008)
7. American Psychiatry Association. Diagnostic and and statistical manual of mental disorders. 4th text revision. American Psychiatric Association, Washington (2000)
8. Grob, C.S., Danforth, A.L., Chopra, G.S., Hagerty, M., McKay, C.R., Halberstadt, A.L., et al.: Pilot study of psilocybin treatment for anxiety in patients with advanced-stage cancer. Arch. Gen. Psychiatry **68**(1), 71–88 (2011)
9. Carhart-Harris, R.L., Erritzoe, D., et al.: Neural correlates of the psychedelic state as determined by fMRI studies with psilocybin. PNAS **109**(6), 2138–2143 (2012)
10. Studerus, E., Kometer, M., Halser, F., Vollenweider, F.X.: Acute, subacute and long-term subjective effects of psilocybin in healthy humans: a pooled analysis of experimental studies. J. Psychopharmacol. **25**(11), 1434–1452 (2011)

11. Treur, J.: Network-Oriented Modeling: Addressing Complexity of Cognitive, Affective and Social Interactions. Springer, Heidelberg (2016)
12. Mohammadi Ziabari, S.S., Treur, J.: A modeling environment for dynamic and adaptive network models implemented in Matlab. In: Proceedings of the 4th International Congress on Information and Communication Technology (ICICT2019), 25–26 February 2019. Springer, London (2019)
13. Treur, J., Mohammadi Ziabari, S.S.: An adaptive temporal-causal network model for decision making under acute stress. In: Nguyen, N.T. (ed.) Proceedings of the 10th International Conference on Computational Collective Intelligence, ICCCI 2018. LNCS, vol. 11056, pp. 13–25. Springer, Berlin (2018)
14. Mohammadi Ziabari, S.S., Treur, J.: Computational analysis of gender differences in coping with extreme stressful emotions. In: Proceedings of the 9th International Conference on Biologically Inspired Cognitive Architecture (BICA2018). Elsevier, Czech Republic (2018)
15. Mohammadi Ziabari, S.S., Treur, J.: Integrative biological, cognitive and affective modeling of a drug-therapy for a post-traumatic stress disorder. In: Proceedings of the 7th International Conference on Theory and Practice of Natural Computing, TPNC 2018. Springer, Berlin (2018)
16. Mohammadi Ziabari, S.S., Treur, J.: An adaptive cognitive temporal-causal network model of a mindfulness therapy based on music. In: Proceedings of the 10th International Conference on Intelligent Human-Computer Interaction, IHCI 2018. Springer, India (2018)
17. Mohammadi Ziabari, S.S., Treur, J.: Cognitive modelling of mindfulness therapy by autogenic training. In: Proceedings of the 5th International Conference on Information System Design and Intelligent Applications, INDIA 2018. Advances in Intelligent Systems and Computing. Springer, Berlin (2018)
18. Lelieveld, I., Storre, G., Mohammadi Ziabari, S.S.: A temporal cognitive model of the influence of methylphenidate (ritalin) on test anxiety. In: Proceedings of the 4th International Congress on Information and Communication Technology (ICICT2019), 25–26 February. Springer, London (2019)
19. Mohammadi Ziabari, S.S., Treur, J.: An adaptive cognitive temporal-causal network model of a mindfulness therapy based on humor. NeuroIS Retreat, 4–6 June, Vienna, Austria (2019)
20. Mohammadi Ziabari, S.S.: Integrative cognitive and affective modeling of deep Brain stimulation. In: Proceedings of the 32nd International Conference on Industrial, Engineering and Other Applications of Applied Intelligent Systems (IEA/AIE 2019), Graz, Austria (2019)
21. Andrianov, A., Guerriero, E., Mohammadi Ziabari, S.S.: Cognitive modeling of mindfulness therapy: effects of yoga on overcoming stress. In: Proceedings of the 16th International Conference on Distributed Computing and Artificial Intelligence (DCAI 2019), 26–28 June, Avila, Spain (2019)
22. de Haan, R.E., Blankert, M., Mohammadi Ziabari, S.S.: Integrative biological, cognitive and affective modeling of caffeine use on stress. In: Proceedings of the 16th International Conference on Distributed Computing and Artificial Intelligence (DCAI 2019), 26–28 June, Avila, Spain (2019)
23. Mohammadi Ziabari, S.S.: An adaptive temporal-causal network model for stress extinction using fluoxetine. In: Proceedings of the 15th International Conference on Artificial Intelligence Applications and Innovations (AIAI 2019), 24–26 May, Crete, Greece (2019)
24. Mohammadi Ziabari, S.S., Gerritsen, C.: An adaptive temporal-causal network model using electroconvulsive therapy (ECT) for PTSD patients. In: 12th International Conference on Brain Informatics (BI 2019) (2019, submitted)

25. Mohammadi Ziabari, S.S., Treur, J.: An adaptive cognitive temporal-causal model for extreme emotion extinction using psilocybin. In: 3rd Computational Methods in Systems and Software, 3–5 October 2019
26. Mohammadi Ziabari, S.S.: A cognitive temporal-causal network model of hormone therapy. In: Proceedings of the 11th International Conference on Computational Collective Intelligence, ICCCI 2019. LNCS, Springer, Heidelberg (2019)
27. Taghva, A., Oluigbo, C., Corrigan, J., Rezai, A.: Posttraumatic stress disorder: neurocircuitry and implications for potential deep brain stimulation. Sterotact. Funct. Neurosurg. **2013**(91), 207–219 (2013)

The Simulation Study of Recursive ABC Method for Warehouse Management

Milan Jemelka$^{(\boxtimes)}$ and Bronislav Chramcov

Faculty of Applied Informatics, Tomas Bata University in Zlín,
Nad Stráněmi 4511, 760 05 Zlín, Czech Republic
{mjemelka, chramcov}@fai.utb.cz

Abstract. The paper deals with a complex warehouse simulation to accomplish a competent solution. It belongs to a group of articles where we are constantly trying to explore the use of warehouses and add further extensions. Greater consideration is concentrated on the use of recursive ABC method for warehouse management in extended concept. The aspiration of the simulation study is to prove whether recursive ABC method returns additional benefits in optimizing the warehouse in this case at a warehouse of different sizes. The complete simulation and the mathematical calculations are accomplished in the Witness Lanner simulation program. The goal of this simulation study is to observe a better solution using recursive ABC method in each part of the model multiple times. Both warehouses are established first on the ABC method, secondary are based on the recursion method. The focus is on two very different layouts of warehouses. Further, the simulation study contributes to propositions that can enhance warehouse management and thus decrease costs. The Witness simulation environment is used for modelling and experimenting. All mathematical computations and simulations are evaluated and measured, as well as all settings of input and output values. Description of the proposed simulation experiments and evaluation of achieved results are presented in tables.

Keywords: Simulation · Random solutions algorithm ·
All combinations algorithm · Optimization · Logistics · Warehouse ·
ABC classification · Inventory

1 Introduction

In our current production organizations, every business concern is currently exposed to competitive pressures. Operations efficiency is the key success of every business concern that processes inventories. Management is under constant pressure to optimize the stock operations every day. Therefore, it is important for any business to pay great attention to the effectiveness of each operation and to continually improve their performance in the light of rapidly changing market conditions. Companies must use effective inventory management process to reduce this time to minimum [1]. When the overall warehouse management efficiency is low, material supply is at risk and the overall primary and secondary costs are high. Companies are constantly finding new ways to optimize processes [2]. Current science uses several ways to increase the

© Springer Nature Switzerland AG 2019
R. Silhavy et al. (Eds.): CoMeSySo 2019, AISC 1047, pp. 187–196, 2019.
https://doi.org/10.1007/978-3-030-31362-3_19

efficiency of storage devices using mathematical methods. ABC analysis is one of the basic mathematical methods used in storage management and in some other areas. It is based on the principles of Vilfredo Pareto, who first noted that 20% of Italians own 80% of Italian land [3]. Then he realized that this division 20/80 also applies to the whole economy. Based on Pareto's observations, Paret's law was also defined: "In many projects, 20% of the effort produces 80% of the outputs."

ABC analysis is also based on Pareto's law, and it is often also called Pareto's analysis [4]. This analysis divides the stored units into three categories: A, B and C per their percentage of the total value of the selected parameter. For example, if it is needed to analyse the total value of a storage location, sorting into categories A, B, C and D [5].

Category A includes products that represent about 15-20% of the assortment and represent about 80% of the value of the warehouse. Category B covers products that represent 30–35% of the assortment and approximately 15% of the total value of the warehouse. Category C includes 50% of products with a 5% share in the value of the warehouse. Sometimes, this classification also includes group D, which has a 0% share [6]. This group includes goods that are no longer used or sold. These goods should be discounted, written off or disposed of, their storage is being a net loss [7].

XYZ analysis is a modification of ABC analysis [8]. The essence of XYZ analysis is the classification of products per the stability of their sales. Sales stability is measured as a variance of sales or consumption over a given time horizon. The XYZ analysis is a way to classify inventory items per variability of their demand. X means „Very little variation". X items are characterized by steady turnover over time. Future demand can be reliably predicted. Y means "Some variation". Although demand for Y items is not steady, variability in demand can be predicted to an extent. This is usually because demand fluctuations are caused by known factors, such as seasonality, product lifecycles, competitor action or economic factors. It's more difficult to forecast demands accurately. Z means "The most variation". Demand for Z items can fluctuate strongly or occur sporadically. There is no trend or predictable causal factors [9].

ABC and XYZ analysis alone may be sufficient, but their combination has a much better predictive value [10]. Therefore, the results of both analyses should be combined. For this purpose, the ABCXYZ matrix is used, where categories A, B and C are set on one axis, and on the other X, Y and Z. The control of the AX, BX, CX, AY, BY, CY, CY, AY, BY, CY products can be fully or partially automated, the demand forecast for these products is sufficiently accurate [11]. The management of AZ, BZ, CZ products needs to be given additional attention, it cannot be fully automated, the predictions are not reliable in relation to them. AZ and BZ products have a significant share of the company's profit, but are characterized by unstable turnover, meaning that caution should be exercised in handling this product group [12]. CY and CZ products should be discounted or their sales should be guaranteed in some other way. The products of the AZ, BZ, CZ, AY, BY, CY groups could be suitable for JIT supply.

Recursive ABC method is a new approach to the ABC method where the same mathematical method is used for smaller parts of the model [13]. It is a mathematical method, where the effort is to recalculate all areas in the mathematical model recursively. The goal is to re-calculate all these areas again. All the results need to be measured [27]. To measure results, it is best to build a model that is as close to reality as possible. This model is then debugged, and multiple random processes are started. All results are measured and entered into tables for further comparison. This work refers to the CoDIT conference, but is focused on another subject of research [28].

2   Problem Formulation

The Witness Lanner simulation program is used for complete simulation and the mathematical calculations. The aim of this simulation is to discover a better solution using recursive ABC method in each part of the model for two different types of warehouse. This recursive method has a finer resolution of part of the model, unlike the basic ABC method. One can assume this recursive method by applying the mathematical distribution again to the already divided parts of the model. The common warehouse operation is usually based on the ABC method. It is important to create the model of the current state and the model (warehouse per the ABC method) and the model using the recursive ABC model. Both approaches (models) are compared on the base of some simulation experiments. The model for this observation has a fixed warehouse stock size to make the results as close as possible to reality. The simulation analysis includes a total of 9 simulation experiments for each type of warehouse. Each simulation experiment has a different storage area of a different size. The overall experiments are measured by the total distance travelled by the operator during the storage process.

The focus of this article is experimentation with the Pareto method ratio on two typologically different warehouses. Regarding the logical distribution of the warehouse, the default setting is variable ratio. For all the various ratios, two different warehouse layouts are tested. For ease of comprehension, the total size of the warehouse is of a fixed size. The basis of a good experiment is a quality simulation that is done for multiple test cases in the simulation tool. In the next step, the calculation moves further, and the next ratio is advantageous and the overall results of the model are measured. As the unit of measurement is meters, it is the total distance that an operator must travel to handle 100,000 storage keeping units (SKU). Storage keeping unit is a universal storage unit that is used globally in warehouse management. The best solution in the model is that the total distance travelled by the operator is as low as possible.

3   Model of Warehouse and the Objective Function

The aim of this paper is to simulate and compare ABC method with the recursive ABC method for warehouse management. The entire experiment is created in the Witness simulation tool. There are two models to explore created in Witness environment. One model is created for basic ABC method (layout is shown in the Fig. 1) and the other one is created for a recursive ABC method.

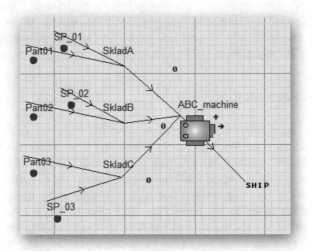

Fig. 1. Model layout in the Witness environment

The basic element of the model is the operator who manages the supply truck. The simulation experiments are evaluated by the total distance travelled by the operator during the storage process. All the measured results are recorded in the excel tables. For the simulation study, the number of SKU movements must be determined in advance. The total number of the supply truck movements is 100,000 SKUs. For all simulations, the rectangular shape of the warehouse is used, which includes two warehouse layouts. SKU is an international term for storage keeping unit. The total path is determined by the sum of the supply trucks movements in each store section (see the formula (1)).

$$D^{total} = \sum_{n_A=1}^{N_A} D_A + \sum_{n_B=1}^{N_B} D_B + \sum_{n_C=1}^{N_C} D_C \tag{1}$$

The distance, which supply truck needs totally for picking up the SKUs in numerous storage sections is different. The letter D is to indicate the total distance. Average distance is used for each warehouse area. The average distance of the supply truck path is used and calculated using the Pythagoras theorem (see the formula (2)).

$$D_A = \sqrt{\left(\frac{W}{2}\right)^2 + (P_A * L)^2} \tag{2}$$

This formula presents the calculation of the average distance of the supply truck path for the A section, where D is the total average distance in a section, W is the width of the warehouse, P is the percentage size (length) of section A (In this simulation study, it is 70%.) and L is the length of the warehouse. This is the average distance, which the operators select in the A section of the warehouse during the storage process of 1 SKU. Average distance is calculated analogously, which the operators select in the section B or C of the warehouse. It is possible to substitute other letters in the formula for calculating other areas.

The universal picture of the warehouse layout is shown in the Fig. 2. Input means where the SKU comes and goes in the warehouse. The letter P indicates the dividing line of the individual storage locations.

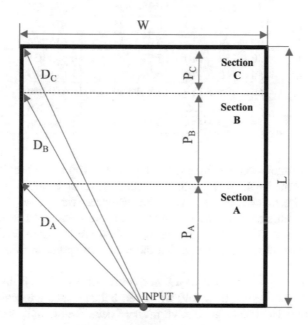

Fig. 2. Storage location layout

4 Results of Simulation Experiments

The results of all experiments are displayed in well-arranged tables. The whole ABC model and recursive ABC models are set to variable ratios and tested on two warehouses with different layouts. The variable ratios are all summarized in tables. The amount of storage keeping units is 100,000 for each experiment. The size of the warehouses is fixed. First warehouse is 100 by 120 m^2 and experiment values are displayed in the Table 1. Second warehouse is 50 by 240 m^2 and experiment values are displayed in the Table 2. Nine different experiments with the same storage size are selected and simulated. To prove the calculation expectations, a further change is chosen and variable ratio is different for the same warehouse each time. A random variable is selected to run the model to provide the most accurate data. It has 1000 different streams available. All simulations share the fact that the model is always tested to relocate 100,000 SKUs. The Table 1 presents handling of 100,000 SKUs as well as Table 2. The purpose of verification is to discover, if recursive calculations in the same warehouse size manage higher storage efficiency, while the different storage ratio is used.

Table 1. Results of simulation experiments I.

No.	Pareto method Ratio			Distance of stock lift [m]	
	A	B	C	ABC method	Recursive ABC method
1	80	15	5	5 508 003	5 078 787
2	70	20	10	6 134 996	5 334 506
3	65	25	10	6 278 885	5 418 892
4	60	25	15	6 874 214	5 815 304
5	50	30	20	7 721 702	6 533 031
6	50	40	10	6 997 456	5 821 489
7	50	35	15	7 341 829	6 169 999
8	40	35	25	8 655 597	7 491 166
9	35	40	25	9 001 009	7 854 056

The Table 1 follows an experiment in stock that is 100 by 120 m^2. There are nine marked experiments in the table with a number. In other columns, we can see the distribution per Pareto method. In a group of distances, we can see the overall comparison of the resulting value. The resulting values are measured in meters. The total results of ABC method and Recursive ABC method are in meters. According to the results of both methods, it is obvious that the recursive approach to inventory brings savings.

The same approach is applied to the next series of tests. The Table 2 follows an experiment in stock that is 50 by 240 m^2. Even better results are achieved with this warehouse layout. The results are shown in more detail below in summary of results.

Table 2. Results of simulation experiments II.

No.	Pareto method Ratio			Distance of stock lift [m]	
	A	B	C	ABC method	Recursive ABC method
1	80	15	5	4 228 752	2 848 593
2	70	20	10	6 370 887	3 839 791
3	65	25	10	6 866 845	4 146 746
4	60	25	15	8 736 053	5 501 863
5	50	30	20	11 219 663	7 770 586
6	50	40	10	9 052 119	5 535 082
7	50	35	15	10 101 550	6 655 241
8	40	35	25	13 756 108	10 566 769
9	35	40	25	14 632 895	11 573 989

5   Summary of Results

Summary of results focuses on the difference between two different methods and even compare both methods when using other types of warehouses. The purpose of this work is to verify the overall performance of the warehouse when changing the storage ratio in the warehouse managed storage area, changing the storage layout and storage strategy using two different methods. Such a verification requires the use of a simulation tool. Witness Lanner simulation tool was used for testing the complex managed warehouse. The Witness simulation program allowed us to quickly and efficiently display the computational results. All test results are presented in Tables 3 and 4. For all simulations, the rectangular shapes of the warehouse are used. The dimensions are 100 by 120 m^2 and 50 by 240 m^2.

Table 3. Comparison of results of simulation experiments I

No.	Distance of stock lift [m]		Difference [m]
	ABC	Recursive ABC	
1	5 508 003	5 078 787	−429 215
2	6 134 996	5 334 506	−800 490
3	6 278 885	5 418 892	−859 993
4	6 874 214	5 815 304	−1 058 910
5	7 721 702	6 533 031	−1 188 671
6	6 997 456	5 821 489	−1 175 967
7	7 341 829	6 169 999	−1 171 830
8	8 655 597	7 491 166	−1 164 431
9	9 001 009	7 854 056	−1 146 953

The first column in the table presents us the experiment number. The second column presents the measured distance in the ABC method. The third column display us the distance measured from the recursive ABC method. In the fourth column, there is difference of basic ABC method and recursive method. The negative number in the fourth column represents the total distance savings in favour of the recursive model. These descriptions are valid for both Tables 3 and 4.

Table 4. Comparison of results of simulation experiments II

No.	Distance of stock lift [m]		Difference [m]
	ABC	Recursive ABC	
1	4 228 752	2 848 593	−1 380 159
2	6 370 887	3 839 791	−2 531 096
3	6 866 845	4 146 746	−2 720 099
4	8 736 053	5 501 863	−3 234 190
5	11 219 663	7 770 586	−3 449 077
6	9 052 119	5 535 082	−3 517 037
7	10 101 550	6 655 241	−3 446 309
8	13 756 108	10 566 769	−3 189 339
9	14 632 895	11 573 989	−3 058 906

In both tables, it is apparent that the recursive method has brought savings in both warehouses. Comparing Tables 3 and 4 results, it is also apparent that the warehouse dimension of 50 by 240 m² is preferable to. At the same time, the mathematical simulation demonstrates that the lowest costs are achieved when the distribution of the warehouse is as close as possible to the basic Pareto theorem. The best result of Tables 1 and 2 is achieved by the ratio of 80:15:5. This ratio could also be evaluated as the ideal ratio where storage companies should aim their intention.

6   Conclusion

The conclusion of the work was to deal with a complex warehouse simulation to accomplish a competent solution. Greater consideration was concentrated on the use of recursive ABC method for warehouse management in extended concept. This benefit was expressed at a total distance that is lower the operator had to drive. The aspiration of the simulation study proved that recursive ABC method returned additional benefits in optimizing the warehouse in this case at a warehouse of different sizes. The complete simulation and the mathematical calculations were accomplished in the Witness Lanner simulation program. Both tested warehouses were established first on the ABC method, secondary were based on the recursion method. The warehouse dimensions were of various ratios and dimensions. The method showed that if the warehouse was divided according to the Pareto rule and mainly the basic percentage distribution, the warehouse achieved greater savings than when the warehouse was divided in a ratio other than that defined in the initial theory. Furthermore, the shape of the warehouse had a major impact on overall warehouse performance and overall savings of manipulation in the warehouses. These two factors were then confirmed by being tested by the recursive ABC method, which showed additional savings in all variants.

Acknowledgment. This work was supported by the Ministry of Education, Youth and Sports of the Czech Republic within the National Sustainability Programme Project No. LO1303 (MSMT-7778/2014) and also by the European Regional Development Fund under the project CEBIA-Tech No. CZ.1.05/2.1.00/03.0089 and also by the Internal Grant Agency of Tomas Bata University under the project No. IGA/FAI/2017/003.

References

1. Davendra, D., Zelinka, I., Senkerik, R., Bialic-Davendra, M.: Chaos driven evolutionary algorithm for the traveling salesman problem. In: Davendra, D. (ed.) Traveling Salesman Problem, Theory and Applications. InTech Europe, Rijeka (2010)
2. Davendra, D.: Evolutionary algorithms and the edge of Chaos. In: Zelinka, I., Celikovsky, S., Richter, H., Chen, G. (eds.) Evolutionary Algorithms and Chaotic Systems. Studies in Computational Intelligence, vol. 267, pp. 145–161. Springer, Berlin Heidelberg (2010)
3. Davendra, D., Bialic-Davendra, M., Senkerik, R.: Scheduling the lot-streaming flowshop scheduling problem with setup time with the chaos-induced enhanced differential evolution. In: 2013 IEEE Symposium on Differential Evolution (SDE), pp. 119–126 (2013)

4. Deugo, D., Ferguson, D.: Evolution to the Xtreme: evolving evolutionary strategies using a meta-level approach. In: CEC2004 Congress on Evolutionary Computation, 19–23 June 2004, pp. 31–38 (2004)
5. Eiben, A.E., Michalewicz, Z., Schoenauer, M., Smith, J.: Parameter control in evolutionary algorithms. Parameter Setting in Evolutionary Algorithms, pp. 19–46. Springer, Heidelberg (2007)
6. Hilborn, R.C.: Chaos and Nonlinear Dynamics: an Introduction for Scientists and Engineers. Oxford University Press, Oxford (2000)
7. Kennedy, J., Eberhart, R.: Particle swarm optimization. In: IEEE International Conference on Neural Networks, November/December 1995, pp. 1942–1948 (1995)
8. May, R.M.C.: Stability and Complexity in Model Ecosystems. Princeton University Press, Princeton (2001)
9. Oplatkova, Z.: Metaevolution: Synthesis of Optimization Algorithms by Means of Symbolic Regression and Evolutionary Algorithms. Lambert Academic Publishing, Saarbrücken (2010)
10. Pluhacek, M., Senkerik, R., Zelinka, I.: Multiple choice strategy based PSO algorithm with chaotic decision making, a preliminary study. In: Herrero, Á., Baruque, B., Klett, F., et al. (eds.) International Joint Conference SOCO 2013-CISIS 2013-ICEUTE 2013. Advances in Intelligent Systems and Computing, vol. 239, pp. 21–30. Springer, Heidelberg (2014)
11. Price, K.V.: An introduction to differential evolution. In: Corne, D., Dorigo, M., Glover, F. (eds.) New Ideas in Optimization. McGraw-Hill Ltd., London (1999)
12. Price, K.V., Storn, R.M., Lampinen, J.A.: Differential Evolution—a Practical Approach to Global Optimization. Natural Computing Series. Springer, Heidelberg (2005)
13. Senkerik, R., Davendra, D., Zelinka, I., Pluhacek, M., Oplatkova, Z.: An investigation on the chaos driven differential evolution: an initial study. In: 5th International Conference on Bioinspired Optimization Methods and Their Applications, BIOMA 2012, pp. 185–194 (2012)
14. Sprott, J.C.: Chaos and Time-Series Analysis. Oxford University Press, Oxford (2003)
15. Zelinka, I.: SOMA—Self-organizing Migrating Algorithm. New Optimization Techniques in Engineering, vol. 141. Studies in Fuzziness and Soft Computing, pp. 167–217. Springer, Heidelberg (2004)
16. Zelinka, I., Raidl, A.: Evolutionary synchronization of chaotic systems. In: Zelinka, I., Celikovsky, S., Richter, H., Chen, G. (eds.) Evolutionary Algorithms and Chaotic Systems. Studies in Computational Intelligence, vol. 267, pp. 385–407. Springer, Heidelberg (2010)
17. Bottani, E., Montanari, R., Rinaldi, M., Vignali, G.: Intelligent algorithms for ware-house management. Intell. Syst. Ref. Libr. **87**, 645–667 (2015)
18. Curcio, D., Longo, F.: Inventory and internal logistics management as critical factors affecting the Supply Chain performances. Int. J. Simul. Process Model. **5**(4), 278–288 (2009)
19. Karasek, J.: An overview of warehouse optimization. Int. J. Adv. Telecommun. Electrotech. Sig. Syst. **2**(3), 111–117 (2013)
20. Muppani, V.R., Adil, G.K., Bandyopadhyay, A.: A review of methodologies for class-based storage location assignment in a warehouse. Int. J. Adv. Oper. Manag. **2**(3–4), 274–291 (2010)
21. Raidl, G., Pferschy, U.: Hybrid optimization methods for warehouse logistics and the reconstruction of destroyed paper documents. Dissertation paper. Vienna University of Technology, Austria (2010)
22. Jemelka, M., Chramcov, B., Kříž, P.: ABC analyses with recursive method for warehouse. In: CoDIT 2017 (2017)
23. Jemelka, M., BChramcov, M.: The use of recursive ABC method for warehouse management. In: 8th Computer Science On-line Conference (2019)

24. Markt, P.L., Mayer, M.H.: WITNESS simulation software: a flexible suite of simulation tools. In: Proceedings of the 1997 Winter Simulation Conference, pp. 711–717 (1997)
25. Waller, A.P.: Optimization of simulation experiments. http://www2.humusoft.cz/download/white-papers/witness/optimization_white_paper.pdf. Accessed 3 Oct 2011
26. Jirsa, J.: Environments for modelling and simulation of production processes. In: Proceedings of the Conference Tvorba softwaru 2004, pp. 65–70. VSB-TU Ostrava, Ostrava, Czech Republic (2004). (in Czech)
27. Chramcov, B., Daníček, L.: Simulation study of the short barrel of the gun manufacture. In: 23rd European Conference on Modelling and Simulation, pp. 275–280. European Council for Modelling and Simulation, Madrid (2009)
28. Chramcov, B., Beran, P., Daníček, L., Jašek, R.: A simulation approach to achieving more efficient production systems. Int. J. Math. Comput. Simul. 5(4), 299–309 (2011). http://www.naun.org/multimedia/NAUN//mcs/20-786.pdf. 20 Apr 2013

Influence of Digital Transformation on the Customer Relationship

Sarah Myriam Lydia Hahn[✉]

Computers and Information Technology, Technical University Cluj Napoca,
Cluj-Napoca, Romania
sarah-hahn@gmx.net

Abstract. The importance of digital technologies is continuously increasing for the customers and thereby also for the companies. Examples for the digital transformation are the usage of smartphones and other gadgets as well as social media channels as Twitter and Facebook. The changing behavior of the customers has an impact on the relationship between a customer and a company. In order to adapt the clients care accordingly the changed attitude itself has to be understood. Therefore, the customer behavior without digital technologies is compared to the one influenced by the digital transformation and differences are accentuated. Following it is pointed out how to deal with these differences in the customer relationship. Also, the question will be answered what the consequences are if notwithstanding the digital transformation the marketing strategy is not changed.

Keywords: Customer relationship · Digital transformation ·
Customer journey · Campaign automation · User identification

1 Introduction

According to Hagens [14], customers want a comprehensive communication to a company across different channels, regardless of the chosen channel. The challenge for a company is to identify a customer via several touchpoints and bring the interactions together. The number of touchpoints is continuously increasing due to the digital transformation [5]. Digital media get more and more importance [15]. Examples for the digital transformation are the usage of smartphones and other gadgets in private as well as in business. Especially social media channels as Twitter and Facebook are continuously gaining more influence.

Digital transformation is described as the financing of technologies or business models to interact with customers in a digital way. This includes not only the point of sale but also the whole customer journey, beginning from the first touchpoint of a customer with the company to the customers win-back [25]. It can be understood as an extension of the traditional marketing themes and channels. New technologies besides the conventional ones as print, radio or television in the context of the digital transformation are electronic media such as computers, smartphones, tablets, digital billboards, game consoles or wearables [28].

© Springer Nature Switzerland AG 2019
R. Silhavy et al. (Eds.): CoMeSySo 2019, AISC 1047, pp. 197–207, 2019.
https://doi.org/10.1007/978-3-030-31362-3_20

The objective of the customer relationship management is to improve customer loyalty and profitability. This can be achieved by the attraction of new customers as well as the optimization of existing customer relationships. The loss of an existing customer is extremely expensive and it is even more difficult to win a lost customer back. Besides, it could have a very negative signal to the market. The most economic way to improve profitability is the optimization of the purchase behavior of customers. This could be done e.g. with cross- or up-selling [26].

2   Influence of Digital Transformation

2.1   ASIDAS-Model

In the traditional way there are four phases defined until a customer buys a product – the so-called AIDA-model. This model was created by E St. Elmo Lewis and its name is an acronym for the steps attention, interest, desire and action [27]. Figure 1 shows the step of the AIDA-model extended with the steps search and share.

Fig. 1. ASIDAS-model

1. **Attention:** The first thing is to gain the attention of the potential customers. If they are not aware of the product, they cannot buy it. There are many different ways in advertising to gain the viewer's attention such as the right placement or use of celebrities for promotion purposes.
2. **Interest:** Once the customer is aware of the product an interest for the product itself has to be created. For creating such an interest, the content of the advertising is important and has to be emotive.
3. **Desire:** Compared to the interest phase, in the phase of desire the customer really wants to get the product. Therefore, there has to be a need for a customer to own the product.
4. **Action:** In the last step, the customer actually buys the product.
   In the course of the digital transformation, the behaviour of the customers changed. Customers are now able to use the internet to get additional information on a specific product. This implicates that not only an advertisement is responsible and important for an order but also the function of the product itself. The traditional AIDA-model has to be extended by the phases search and share – the so-called ASIDAS-model [16].
5. **Search:** After gaining the attention of the customer, the customer itself often searches extensively for more information about the product and experiences from other customers. Thus, it is possible to get product reviews before using it. In this second phase the steps interest and desire have an interdependency with the search. To get more information about a product by searching for it, it increases the interest and the desire for it.

6. **Share:** The last step is sharing the own experience. Platforms can be special review sites such as hotel portals, forums for special subjects as well as social media channels like Facebook. The shared experience can be used from other customers in the search phase.

2.2   Touchpoints Customer Journey

There are different channels, which can be used for the search and the share phase and for other touchpoints at the customer journey. Table 1 shows, which channels are usually used for what kind of information.

Table 1. Overview channels in customer journey

	Attention	Information	Advice	Product comparison	Purchase	Delivery status	After-sales
E-Mail	Mass mailings	–	Personal mailings	–	Personal mailings	Status mailings	Follow-up mailings/personal mailings
Shop/Company's website	Content	Product site	Product details	–	Online purchase	Personal section	Personalized content
Search engine	Search results	Search results	Search results	Search results	–	–	–
Social media	Product placement	Product reviews	Direct communication	Product reviews	–	–	Product placement
Review sites	–	Product reviews	–	Product reviews	–	–	Creating review
Forums	Product placement	Product reviews	Direct communication	Product reviews	–	–	–
Website	Product placement	–	–	–	–	–	Product placement
Suggestion							

As it can be seen in Fig. 1 there are many channels where content is generated which is linked to the customer relationship on the several steps of the customer journey [16]. For example, the e-mail channel can be used for creating attention through mass mailings such as newsletter. Requirement for sending mass e-mails is the opt-in of the recipient. Often the application for e-mail communication can be made on the website or on forms.

This is just one way to place advertisement. As a follow up during the customer journey, it is possible to utilize direct communication with the customer via e-mail for example by detail questions about the product. Questions about a product can usually be sent to the supplier via e-mail or contact forms on the website. Depending on the product and the seller's service, it is possible to order it directly. Normally this is related to products with a high value or a huge amount. With automatically generated status mails, the customer could then be informed about the delivery status. Status mails can be seen as an extra, value-adding service to the customer. To follow up, the respective company could then send additional, personalized content in context to the purchased product in the consequent mailings. This could include follow-up products, special service offers to the purchase or product recommendations based on the purchase [1].

Another channel is the online shop or the website of a company. If the customer is on the website they can be attracted with special content such as teaser or content on a specific theme for example in a blog [2]. The product site itself gives more information to the customer and could contain reviews and advices from other customer who already bought the product. For getting some additional advice regarding the product there can be a FAQ-site or personal advice in terms of a direct chat with the customer service. FAQ-sites answer questions to the company or product, which are often asked. The supplier does not have to answer the same questions repeatedly and saves costs. For the customer it is easier to get the required information without putting an extra question. Finally, the purchase can be made on the website. In a personal section of the website with a tracking number or a login the status of the order can be reviewed. Other services are the delivery status, the possibility to modify or cancel the order or buy some additional services or products. The data of the purchase can be used for showing product recommendations, follow up products or other personalized content on the website.

A search engine helps accessing the required information. In the case of the customer journey this could be information on a special subject without knowing which product could be the solution, searching for detailed information or advice about a specific product [20]. A search engine only supports by finding the information but does not have information itself. The result of a search is a link to a website where the searched information can be found. The most known-one is Google.

Via social media, many touchpoints can be created with the customer. By placing advertisement or product placement, paid and unpaid, through other users, the user can be attracted. Paid advertisement are posts in the name of the company which are on top of the social feed or influencers which show a product or a company. Unpaid advertisement is content of other users regarding a product or company that is created by the user himself of his own accord. More information about a product can be obtained from other users by directly contacting them or by reading product reviews. Other users can be customers as well as the company itself. An example therefore are sites on Facebook. With such a site a company can create their own space in the network for placing messages and getting in contact with its customers. The direct communication can be used for getting advice from other customers. Different product reviews help to compare different suppliers. Based on the purchase special advertisement can be placed on the platforms [8, 29].

Review sites can be directly included in the online shop or they can be independent platforms. Product reviews can support gaining detailed product information or compare different suppliers [4]. Important is that it is possible that a product review is not a real product review because it is written from the supplier or a competitor. Customers do not always recognize that a review is not real. Therefore, suppliers should also check the reviews and delete them if they are not legitimate. After a purchase, the customer can write his own review. By doing so, they reflect on the product and their relationship with the supplier. Regarding to the customer journey this could either be in a positive as well as in a negative way [17].

Another channel in customer journey is forums. A forum often discusses a specific topic such as fashion or technologies. A user can be attracted by reading posts from other users. However, posts can be product reviews and granting additional information to a product or compare different products. Every user can open a new topic, answer to other users or just read their posts. The direct communication with other users can help with getting advice [23].

In general, there can be banners on every website for a specific product. If it is possible to place a banner on a website depends on whether the operator sells space to an ad server or not. Such banners can attract the user to a specific product and be adapted based on the purchase behavior of the user. With a click on the banner, the user is referred to the online shop where he can buy the product directly [18].

In this context, there are several challenges. The first one is to get the data on what the user viewed and the content they created across all channels. To gather this data various methods can be implemented. Different channels partly have different methods for data collection. Nevertheless, it is not only collecting the data but also using it in a meaningful way. To do so, the second step after the data collection is to identify the user over all channels. Otherwise, there is no holistic view of the customer. This includes the connection between the digital media channels and the traditional back-end systems such as a CRM- or ERP-system. In addition, the identification of the user depends on the channel.

The easiest channel for data collection is e-mail. By sending an e-mail, the recipient for the e-mail is known. The challenge here is, if it is possible that customers have the same e-mail address, to identify who really read this e-mail. Because it cannot be determined who really opens the e-mail, it is necessary to have business rules in place. It has to be kept in mind that these business rules are not totally correct. An example for such a business rule is that the response is always related to the user with the latest order. To know whether the e-mail is opened, clicked, bounced or unsubscribed a pixel or image should be integrated in the mailing. When the e-mail is opened, the pixel or image is loaded. The pixel itself is in the same color as the background so the user cannot see it. Implemented pixel and links should be unique for all recipients. By tracking the requests to the webserver, it is tracked who opened, clicked or unsubscribed. The webserver gets a request for every item that should be loaded in the mailing – a pixel or a link. If the mailing is forwarded to someone else, this is not tracked and the response is assigned to the original recipient. The webserver gets the same request as if the original recipient would open the mail and could not make any difference [13].

For collection the data from an online shop or the own website a web tracking system such as Adobe Analytics or Google Analytics has to be implemented. With such a system, a cookie is set by visiting the website the first time. A cookie is a small file saved on the local file system. With the aid of this cookie a user can be identified throughout a session and when he visits the website again. By deleting the cookie, it is hard to identify the user again [7]. One possibility is the so-called digital fingerprint. The concept here is to create an artificial identity by the combination of different properties of the used web browser. Properties are the version of the browser, the language or the downloaded extensions. It is assumed that the combination of various properties is almost unique, but it cannot be guaranteed. Thus, this method is not totally correct if more than one user is using the same browser with the same properties [21]. In addition, apps on tablets or smartphones are a good method to identify the user across several sessions. Each app download has a unique identity. This identity is saved in the app. The app has to be deleted and downloaded again to get another identity. When the app is used the identity can always be tracked [6]. Another challenge is to connect the user to a customer in the back-end system. There are several possibilities to achieve this: login, orders, forms, mailings or personalized codes.

1. **Login:** By logging in on the website the username can be tracked. This information can be used for a connection to the back-end system. The objective is that all users are using the login. To encourage the user interesting, value-adding services could be offered in the personal section of the website or the user gets an extra coupon for the sign-up.
2. **Order:** If a purchase in an online shop is made an order number is generated. By tracking this number, a connection to the transaction data in the back-end system can be made. From the transaction data the buyer is known.
3. **Forms:** By filling in a form on a website the content of the various fields can be tracked. Depending on the fields, these can be used for comparison to the back-end systems. Useful information could be the e-mail address, an order number or the postal address. This method has a lack of definition because the data is not verified and could include misspellings.
4. **Mailing:** When sending a mailing the links should be personalized. If a link is clicked, the link itself or a parameter out of it can be tracked. The simplest way is to use the hashed e-mail address of the recipient as a parameter. In the back-end system the e-mail address has to be hashed with the same technique. For the identification, only the values have to be compared. In addition, here are business rules necessary for linking an e-mail address to a customer.
5. **Personalized code:** Personalized codes such as QR-codes on a print mailing could be generated. In the back-end system, it has to be saved which customer received which code. By entering the code or opening the website via a QR-code the code has to be tracked. Afterwards the inserted code has to be compared to the sent code. If the code is forwarded to someone else and used the identification of the customer is wrong.

In general, almost everything can be tracked on a website. To do so a pixel has to be implemented on the website for the events which should be tracked. Examples for events are adding a product to the shopping cart, clicking on a teaser or filling in a form. The objective is to have an idea, which actions are interesting and which are not before implementing a web tracking system. Afterwards the web tracking has to be configured so that the relevant data for the use cases are collected.

When using a search engine there is not a strict definition when a search is relevant for the customer journey and when it is not. For example, the search for a topic, which is relevant for the business of the company, is an interesting information. It is in the eye of the beholder which search topics are interesting. It is hard to implement an algorithm evaluating the interest of a search topic. The provider of the search engine has the data of all searches from a user. However, this data is not accessible for a company. With an implemented web tracking on the company's site the referrer can be tracked. This enables a company to know from which website the customer was referred to the company's website. Referrers are not only search engines but can be other websites such as a special-interest website as well. If the referrer is a search engine, it is possible to get the search phrase which was used. Search topics which are not referred in the company's website are not available.

A lot data is generated on social media platforms. To get the relevant data the platform itself has to be searched through. Most of the platforms offer application programming interfaces (APIs) to get the data in an automated way. Facebook for example offers the so-called Graph API and Twitter has a Search API [12, 31]. These APIs are uniquely developed by each platform and cannot be called in a standardized way. It is not possible to retrieve and analyse the complete data of the platform with these APIs. Only data that is accessible to the public is given back, no private messages. It is necessary to explicitly search for something and as a result, the content generated fitting to the search is returned. Examples for such a search can be all content from a specific user or all content including a search phrase. This means that it is necessary to have a clear vision what exactly one is searching for before getting the data. If a search is not defined well it will return useless or not all relevant and necessary data. Another point is the identification of the user. Depending on the platform, different data from the user is submitted. This could be a username, the full name of the person or the e-mail address. For example, Facebook will return the full name [11]. Nevertheless, it is not verified that the full name is the correct one. It is possible that there are a few persons with the same name. Twitter will deliver a username [31]. For a connection to the back-end, the username of a customer has to be known. An idea is to have special marketing campaigns where a customer could get something extra or special when telling their username. The simplest way is to identify the customer with the e-mail address. Here are the same challenges as with e-mail channel regarding the assignment of an e-mail address to a customer. Another challenge is if the user has another e-mail address for social media than in the back-end system.

Depending on the review site or forum there are APIs similar to the APIs for social media platforms to get data. For example, the touristic review platform Tripadvisor has the Content API and Amazon the Product Advertising API [3, 30]. The challenge here lies in the identification of the users. The possibilities are the same as for the other channels. If there is no API, it is possible to get the data from the website with a self-developed crawler. A Crawler is a program which automatically scans a website and analyzes it. With such a program it is possible to search a website for defined search terms [22].

Last but not least an ad server can be used for showing advertisements on different websites. Therefore, a cookie is set when visiting the first website with the ad server. With this cookie, it is possible to track all banners displayed across different websites if the used ad server is the same. It depends on the supplier if this collected data is accessible. Besides that, the user has to be identified. This is possible if the user has the same cookie on the company's website. Then they can be identified with the web tracking methods. Overall, it will be difficult to get this data from the supplier of the ad server and connect it to a customer [19].

Which channel in detail is important for a company and should be listened to depends on the offered product. For example, for a fashion retailer Tripadvisor as a touristic review platform will not be useful but maybe Amazon as a retailer instead. It is important to choose the right channels in advance before starting the data collection.

Once the customer is identified and the data is collected, it is important to get the relevant data and draw the right conclusions from the huge amount of collected data. With this data, the knowledge about the customers can be expanded. This additional information can be used in two ways. The first one is the more detailed segmentation in marketing campaigns. For example, when sending a newsletter there could be a space for a personal recommendation. Such a recommendation can be based on the customer's behavior or on an analytical model. With knowledge of the whole customer journey there is more data for such a recommendation and so it could be more granular and precise. All in all a higher personalization and segmentation in marketing campaigns is possible. Another point is that because of the digital media the customer and their need is known earlier. This can be used to get in contact earlier with them to get a conversion before they choose another supplier.

In both ways, the customer gets advertisement which is more relevant to them than in the traditional way. This results in a better customer experience and the advertisement is more recognized because it is a benefit [15]. The result is not only a rise in sales but also an emotional tie to the customer is built up. The effect could be word-of-mouth propaganda or recommendations, which have indirect influence on the sales, too.

2.3   Automatization of Data Processing

Because of the huge amount of data, a manual processing of it is not possible and a partly processing of the data will be too laborious and lasts too long. The objective is to get and process the data and use it for marketing campaigns in an automated way in near-time. This could be partitioned in the following three steps:

1. **ETL:** ETL stands for extract, transform and load in the context of data preparation. First, the data has to be extracted from the source. Depending on the source this can be an API, flat files or a database. Important is that there is an interface to the source. After the extraction the data has to be transformed in a defined way. Only prepared data can be used in an easy and wise way for constitutive use cases. One transformation is the identification of the customer. Another one is the aggregation and filtering of the data, for example to identify all shopping cart abandonments from the web tracking data. In the last step the prepared data is loaded to the target such as a database. For the implementation of such ETL-processes there are several tools from different suppliers like Microsoft, Oracle and also open-source alternatives like Pentaho. The challenge is the performance of these processes in the face of the increasing data volume. The more data has to be loaded the longer the process takes. This is against the objective that the data should be available in near-time. The fast provision of data is important to react in an appropriate time concerning the digital media [26].

2. **Analytics:** With the aid of analytics, customers and their behaviour can be analysed based on their historical data. Without a connection to the customer, this data can

give information about the customer journey itself. An example is the analysis of the website and the click paths. As a result, it is possible to identify pages where the user often leaves the website. With this knowledge, the page itself has to be proofed for its content [16]. Besides, data from different sources can be used in combination to get an overview across the different touchpoints of the customer journey. It is only important, that the data has a connection via the customer to connect the different sources. Historical Data can not only be analysed but it can also be used for forecasting such as a next best offer or churn prevention. Therefore, the data is split into a group of customers as a reference and a group for which the forecast should be made. From the reference group the expected behaviour is deviated and projected on to the forecast group. In the case of next best offer the customer journey and the transaction data is the basis to forecast which product a customer of the forecasting group will buy next with which probability based on their customer journey. The required data for a churn prevention is the customer journey as well as the churn data. An attempt is made to predict which customer will resign next. Such analytics can be done with so-called data mining models. With different algorithms patterns in the data are searched and used for the forecasting. Once a pattern has been found the reference group is no longer necessary but the model should be retrained constantly [32]. Another part of analytics is the presentation of the data in the form of reports. With such reports the data can be illustrated for persons which are not involved in analytics, for example the management.

3. **Campaign Automation:** The data from the customer journey as well as the analytic key performance indicators should be used for personalized marketing campaigns. Personalized communications have a higher relevance to the customer and get more attention [10]. This is visible in the response quote of personalized e-mails [24]. It is also important in the case of the digital transformation to react faster to customers' actions. Examples for such marketing campaigns are an e-mail when the user have a shopping cart abandonment in the online shop. The mailing could include a reminder for the shopping cart with its products or next best offers based on the shopping cart and the viewed products on the website. Another example is a complaint post on a social media platform. Based on this post the user could be called directly from a call centre agent with a solution for their complaint and an additional service offer. With marketing automation software, marketing campaigns can be created and controlled. Therefore, the different data sources as the CRM and Enterprise Resource Planning (ERP) system has to be connected to the tool as well as the outgoing channels like e-mail, call centre, letter shop or social media platforms. In workflows campaigns can be defined and automatically executed [9].

3 Conclusion

In the course of the digital transformation more and more data is generated. With this data the information about a customer can be expanded and used for marketing campaigns with the objective of an increase of sales. However, there are several challenges, which are not solved yet.

First all relevant touchpoints have to be defined. Because of the volume of possible touchpoints, it is not possible to think of all of them. The focus should be to define the most important and common touchpoints in the customer journey. After defining the touchpoints an interface is needed to gather the data automatically. There are interfaces for mailings, web tracking on the own website and the common social media and review platforms. However, there is a lack of interfaces on channels like search engines, more unknown platforms and other websites such as special interest websites. This goes along with a lack of information. A holistic view on the customer is not possible, only an approximation.

Depending on the volume of data, the data preparation with standard ETL-processes can be very slow. Because of the digital transformation it is important to get the data in near-time to react very close to the customer's action. A solution can be the optimization of the existing ETL-process or using new technologies and hardware such as the processing in the cloud.

Another point is the identification of the customer. Without an identification the collected data cannot be used directly for a personalized communication. The data itself is not useless because it can be used for a general analysis of the customer behaviour and data mining models for forecasting. Nevertheless, the objective is to connect the data with the customer data to have a higher benefit. One solution could be to collect more customer data such as the e-mail address with special marketing campaigns. It is important to offer an incentive with those campaigns.

After extracting the data, it is another challenge to prepare the data in a meaningful way. From the huge amount of data, the important information has to be defined. This definition should be frequently checked with the aid of a reporting of the marketing actions. The prepared data could then be used for analytics. For developing data mining models, the requirements and objectives have to be defined as well. For forecasting a lot of historical data is necessary. In the beginning of the data collection the conclusion of forecasting models has to be watched critically. It is important to retrain these models regularly and to check the forecasts for correctness.

References

1. Baggott, C.: Email Marketing by the Numbers: How to Use the World's Greatest Marketing Tool to Take Any Organization to the Next Level. Wiley, New York (2011)
2. Bleier, A., Eisenbeiss, M.: Personalized online advertising effectiveness: the interplay of what, when, and where. Mark. Sci. **34**(5), 669–688 (2015)
3. Amazon: Welcome – Product Advertising API. http://docs.aws.amazon.com/AWSECommerceService/latest/DG/Welcome.html
4. Van Bommel, E., Edelman, D., Ungerman, K.: Digitizing the consumer decision journey. McKinsey Q. (2014)
5. Bonner Management Forum: Management Summary of Conference "Die Customer Journey – vom Interessenten zum Käufer" (2014)
6. Boyle, S.S., et al.: Method and system for personalized venue marketing. U.S. Patent No. 8,433,342 (2013)
7. Butler, E., Teddy, J., Waugh, M.: First-party cookie for tracking web traffic. U.S. Patent Application No. 11/437,989 (2012)

8. Carter, B., Levy, J.R.: Facebook Marketing: Leveraging Facebook's Features for Your Marketing Campaigns. Que Publishing (2012)
9. Chakraborty, A.: Decision points on the path to unified customer engagement. https://www.oracle.com/webfolder/mediaeloqua/documents/White+Paper+-+Decision+Points+on+the+path+to+Unified+Customer+Engagement.pdf
10. Goldsmith, R.E., Freiden, J.B.: Have it your way: consumer attitudes toward personalized marketing. Mark. Intell. Plann. **22**, 228–239 (2004)
11. Facebook: Graph API User. https://developers.facebook.com/docs/graph-api/reference/user
12. Facebook: Graph API – Übersicht. https://developers.facebook.com/docs/graph-api/overview/
13. Foulger, M.G., et al.: System and method related to generating and tracking an email campaign. U.S. Patent No. 7,065,555 (2006)
14. Hagen, P.: Beyond CRM: manage customer experiences. Forrester Research (2011)
15. Kalyanaraman, S., Sundar, S.S.: The psychological appeal of personalized content in web portals: does customization affect attitudes and behavior? J. Commun. **56**(1), 110–132 (2006)
16. Kodali, S.: US Cross-Channel Retail Sales Forecast: 2014 to 2018. Forrester Research (2014)
17. Kreutzer, R.T.: Praxisorientiertes Online-Marketing, Konzepte - Instrumente – Checklisten. Springer Gabler (2014)
18. Luca, M., Zervas, G.: Fake it till you make it: reputation, competition, and Yelp review fraud. Manag. Sci. **62**, 3412–3427 (2016)
19. Manchanda, P., et al.: The effect of banner advertising on internet purchasing. J. Mark. Res. **43**, 98–108 (2006)
20. Mayer, J.R., Mitchell, J.C.: Third-party web tracking: policy and technology. In: IEEE Symposium on Security and Privacy (SP). IEEE (2012)
21. Molenaar, C.: Shopping 3.0: Shopping, the Internet or Both. Routledge, Abingdon (2016)
22. Nikiforakis, N., et al.: Cookieless monster: exploring the ecosystem of web-based device fingerprinting. In: IEEE Symposium on Security and Privacy (SP). IEEE (2013)
23. Olston, C., Najork, M.: Web crawling. Foundations and Trends® in Information Retrieval 4.3 (2010)
24. Rosenbaum, M.S., Otalora, M.L., Ramírez, G.C.: How to create a realistic customer journey map. Bus. Horiz. **60**, 143–150 (2017)
25. Sahni, N.S., Wheeler, S.C., Chintagunta, P.K.: Personalization in email marketing: the role of non-informative advertising content. Mark. Sci. **37**, 236–258 (2016)
26. Simitsis, A., Vassiliadis, P., Sellis, T.: Optimizing ETL processes in data warehouses. In: Proceedings of 21st International Conference on Data Engineering, ICDE 2005. IEEE (2005)
27. Solis, B.: The 2014 State of Digital Transformation. Altimeter Group (2014)
28. Strauss, R.E.: Digital Business Excellence: Strategien und Erfolgsfaktoren im E-Business; mit über 250 Unternehmensbeispielen, Stuttgart (2013)
29. Strong, E.: The Psychology of Selling and Advertising, New York (1925)
30. Tehrani, K., Andrew, M.: Wearable technology and wearable devices: everything you need to know. Wearable Devices Magazine, March 2014
31. Tuten, T.L., Solomon, M.R.: Social Media Marketing. Sage, Thousand Oaks (2017)
32. Witten, I.H., et al.: Data Mining: Practical Machine Learning Tools and Techniques. Morgan Kaufmann Publishers, Burlington (2016)

The Comparison of Machine-Learning Methods XGBoost and LightGBM to Predict Energy Development

Martin Nemeth[✉], Dmitrii Borkin, and German Michalconok

Faculty of Materials Science and Technology in Trnava,
Institute of Applied Informatics, Automation and Mechatronics,
Slovak University of Technology in Bratislava, Bratislava, Slovakia
{martin.nemeth, dmitrii.borkin,
german.michalconok}@stuba.sk

Abstract. This paper follows the paper named Comparison of methods for time series data analysis for further use of machine learning algorithms. In previous paper we were dealing with the initial analysis of the time series data from a thermal plant. In this paper we aim at the design of a prediction model with the use of machine learning methods XGBoost and LightGBM. In the first part of this paper we focus at feature engineering where we add supplemental parameters to the existing dataset. In the second part of this paper we are dealing with the performance comparison of the mentioned machine-learning methods XGBoost and LightGBM.

Keywords: XGBoost · LightGBM · Energy prediction · Machine learning

1 Introduction

Prediction is special task of data mining. It is used to predict future values of a target variable based on the past values of this variable. Various methods can be used for prediction type of tasks. We can use for example classical methods of statistical analysis like exponential smoothing methods. These methods can provide precise predictions, however their performance decreases when trying to predict values more in the future. Another approach is to use the machine-learning methods and algorithms. These algorithms can learn from past data to be able to predict future values of the target variable. In machine-learning we can solve two main tasks. First task is called classification. The goal of classification is to train the classification model on the past data in a way, it will be able to classify new object into the given set of classes. The second task is regression. The main objective of this task is to predict future continuous values of the target variable based on the training set of past values of this target. In this and also in previous paper we are dealing with the data from a thermal plant. Our original dataset contains the timestamp information and also the thermal power output in MW (megawatts). For this particular dataset we are going to solve the regression type of task to predict the future values of the thermal power output.

© Springer Nature Switzerland AG 2019
R. Silhavy et al. (Eds.): CoMeSySo 2019, AISC 1047, pp. 208–215, 2019.
https://doi.org/10.1007/978-3-030-31362-3_21

2 Feature Analysis

For the purpose of building a prediction model we have decided to add supplemental parameters/features into our data set. We have derived these new parameters from the timestamp information. New parameters are *hour, year, weekday and is_weekend*. When adding new supplemental parameters to the dataset, it is needed to assess their correlation with the target variable and also if there is some kind of correlation between these parameters. This step is important for further use of the machine learning methods. For assessing the correlation between new parameters and the target variable we have chosen the Pearson correlation method. The computation of the Pearson correlation coefficient was performed using Python v3.6 and corresponding libraries.

2.1 Computation of Pearson Correlation Coefficient

The Pearson correlation coefficient is used to measure the linear relationship between two data sets. The calculation of the p-value relies on the assumption that each dataset is normally distributed. The correlation coefficient value varies between -1 and $+1$. If the value of the Pearson correlation coefficient is equal to 0, it means that there is no correlation. However, if the value of the coefficient is -1 or $+1$, it means that there is an exact linear relationship. Positive values of correlation coefficient imply that as x increases, so does y. Negative correlation values imply that as x increases, y has decreasing character as well.

The p-value is indicating the probability of an uncorrelated system producing datasets that have a Pearson correlation at least as extreme as the one computed from these datasets.

The correlation coefficient is calculated as follows [1–3]:

$$r = \frac{\sum (x - m_x)(y - m_y)}{\sqrt{\sum (x - m_x)^2 \sum (y - m_y)^2}} \tag{1}$$

where mx is the mean of the vector x and m_y is the mean of the vector y.

We assume, that x and y are picked from independent normal distributions and we can also say, that the population correlation coefficient is equal to 0, and the probability density function of the sample correlation coefficient r is defined as follows [1, 2]:

$$f(r) = \frac{(1 - r^2)^{\left(\frac{n}{2} - 2\right)}}{B\left(\frac{1}{2}, \frac{n}{2} - 1\right)} \tag{2}$$

where n is the sample count, B is the beta function. This function is can be also represented to as the exact distribution of the sample correlation coefficient r. This distribution is used in Pearson to calculate the p-value. This can be represented as a beta distribution on the interval $[-1, 1]$, and equal shape parameters a = b = n/2 − 1. According to the SciPy's implementation of the beta distribution, the distribution of r is defined as follows [2, 3]:

$$dist = scipy.stats.beta\left(\frac{n}{2} - 1, \frac{n}{2} - 1, loc = -1, scale = 2\right) \tag{3}$$

Pearson returns the p-value as a two-sided p-value. For a given sample with correlation coefficient r, the p-value is described as the probability, that $abs(r')$ of a random sample x' and y' drawn from the population with zero correlation would be greater than or equal to $abs(r)$. Considering the object $dist.$, which is shown in previous equation, the p-value for a given r and length n are computed as follows [2, 3]:

$$p = 2 * dist.cdf(-abs(r)) \tag{4}$$

If n is equal to 2, the continuous distribution described above is not well defined. The limit of the beta distribution can be interpreted as the shape parameters a and b, and they are approaching $a = b = 0$ as a discrete distribution with equal probability masses at $r = 1$ and $r = -1$. More directly, one can observe that for given data $x = [x1, x2]$ and $y = [y1, y2]$, and assuming $x1! = X2$ and $y1! = Y2$, the only possible values for r are from the interval 1 and -1. The two-sided p-value for a sample of length 2 is always 1, because $abs(r')$ for any sample x' and y' with length 2 will be equal to 1.

The Table 1 shows the computed values of correlations of each new supplemental data parameter. The coefficient describes the positive or negative correlation and the p-value the dependency level. It is assumed that when this value is less than 0.001, it means that the correlation between the observed parameters is strong. According to the results of the correlation analysis we can tell that the correlation between each parameter and the target variable is strong and it is reasonable to include these parameters into the prediction model.

Table 1. The computed values of correlations of each new supplemental data parameter

	Coefficient	p-value
hour	−0.01474928394	4.647606075e−09
year	0.04156915041	2.631213412e−61
weekday	−0.0246343476	1.277480079e−22
is_weekend	−0.03030450926	2.165167817e−33

The Fig. 1 shows the example of the distribution of supplemental parameter weekday. The x-axis is representing the actual day of the week from Monday (represented by 0) to Sunday (represented by 6). The y-axis shows the values of the thermal power output. There are slight differences in ranges of the values for each day.

In the next step we have evaluated the possible correlations between new supplemental parameters. The character of the new parameters suggests, that there cannot be any correlation between these parameters. However, we have tested the correlation to ensure we have derived these parameters correctly. Figure 2 shows the heat-map of the mutual correlations between derived parameters. We can see that there is correlation equal to 1 only on the diagonal of the heat map. The only significant correlation is obviously between parameter is_weekend and weekday and is equal to 0,79008.

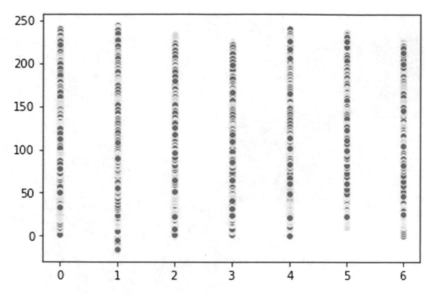

Fig. 1. The example of the distribution of supplemental parameter weekday, where on the y-axis is represented thermal power output in MW

From the heat-map it is clear that the weekday parameter has strong correlation with the is_weekend parameter. Following table shows the basic statistical indicators of the weekday parameters (Table 2).

3   Building the Prediction Model

3.1   XGBoosST Method

In this section of the paper we discuss and compare two chosen methods for building the prediction model. Both algorithms were implemented in Python v3.6 with the use corresponding libraries.

First machine learning method is a gradient boosting machine. We have implemented this method with the use of the XGBoost library in Python envirnomnet. Gradient boosting is used for supervised learning. The XGBoost library works with two following methods [5, 6].

In gradient descent we follow the cost function. This function measures the closeness of the predicted values to the corresponding actual values. The main goal is to minimize the difference between these 2 values. We can understand the gradient descent as an iterative optimization algorithm. In its first iteration it assigns the model initial weights. In next iterations, the cost function is minimized by updating the weights.

Boosting can be understood as a set of weak learners. The records that are not classified correctly are boosted by assigning a greater weight to them. These weak

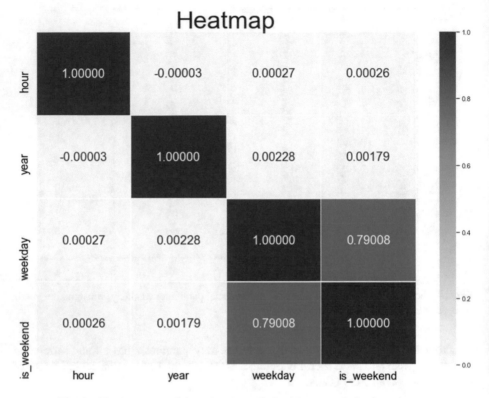

Fig. 2. The heat-map of the mutual correlations between derived parameters

Table 2. Basic statistical indicators of the derived parameter weekday

	Mean	Std	Amax	Amin
Monday	72.83	53.98	240.6á	1.51
Tuesday	73.28	54.67	243.95	−16.42
Wednesday	74.34	54.22	234.07	0.01
Thursday	73.57	53.65	224.79	2.04
Friday	72.54	53.01	239.71	0.00
Saturday	69.97	51.31	236.39	10.76
Sunday	69.52	51.17	224.86	0.00

learners are later combined to form one single strong learner. The XGBoost algorithm produces a tree-based model [7, 8].

Following figure shows the result of the model based on the XGBoost algorithm. The blue line represents the actual values and the green line represents the prediction produced by the trained model (Fig. 3).

We have chose to calculate the Mean absolute percentage error (MAPE) as a performance indicator. This value has the same nature as the mean absolute error. It is

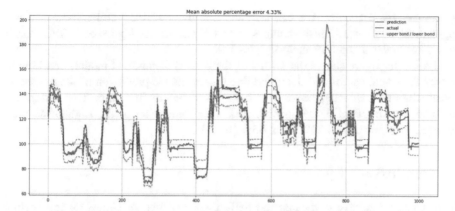

Fig. 3. The result of the model based on the XGBoost algorithm

computed as a percentage. This approach is convenient to comprehensibly explain the quality of presented model. In our case the MAPE was equal to 4.33%, which can be considered as a sufficient result. The mean absolute percentage error is computed as follows [7, 8]:

$$MAPE = \frac{100}{n} \sum_{i=1}^{n} \frac{\left| y_1 - \hat{y_2} \right|}{y_t} \tag{5}$$

3.2 LigthGBM Method

The LightGBM is a framework for gradient boosting. It uses a learning algorithm that has a tree structure. In contrast to other tree based learning algorithms, the LightGBM

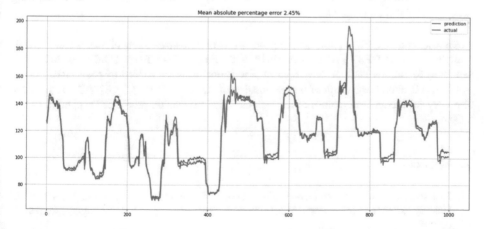

Fig. 4. The results of the trained LightGBM model

grows the tree vertically. We can also say that it grows the tree leaf-wise and other algorithms level-wise. The advantages of this algorithm are high speed, GPU learning and takes lower amount of memory [10–12].

Following figure shows the results of the trained LightGBM model. The blue line shows the actual values and the green line represents the prediction of the model. The mean absolute percentage error was also calculated in this case. The performance showed the MAPE of 2.45%, which makes the prediction of the LightGBM model more accurate than the XGBoost model (Fig. 4).

4   Conclusion

In this paper we have compared the performance of two machine-learning methods XGBoost and LightGBM for building the prediction model. In our paper we were working with real life data from a thermal plant as well as in the previous paper Comparison of methods for time series data analysis for further use of machine learning algorithms.

The aim of this paper was the initial comparison of the performance of two machine-learning methods for prediction of the thermal power output. Both methods were used to build a prediction model to predict the thermal power output. These models were trained on real life data from a thermal plant. The dataset and the structure of these data is described in detail in mentioned previous paper. In this paper we have decided to derive new additional parameters from the existing dataset to enhance the prediction performance of the prediction models. The analysis of these additional parameters is discussed in the first section of this paper.

Both machine-learning methods provided accurate prediction performance on real life data. However, the LightGBM method predicted values of the target variable with smaller value of the mean absolute percentage error, which was equal to 2.45%.

Our future work will be focused on optimizing the LightGBM prediction model by optimizing the parameterization of this method. The main objective of our research leads to building a machine-learning based decision support system for optimizing processes of the thermal plant.

Acknowledgments. This publication is the result of implementation of the project: "UNI-VERSITY SCIENTIFIC PARK: CAMPUS MTF STU - CAMBO" (ITMS: 26220220179) supported by the Research & Development Operational Program funded by the EFRR.

This publication is the result of implementation of the project VEGA 1/0673/15: "Knowledge discovery for hierarchical control of technological and production processes" supported by the VEGA.

References

1. Benesty, J., et al.: Pearson correlation coefficient. In: Cohen, I., Huang, Y., Chen, J., Benesty, J. (eds.) Noise Reduction in Speech Processing, pp. 1–4. Springer, Heidelberg (2009)

2. Lawrence, I., Lin, K.: A concordance correlation coefficient to evaluate reproducibility. Biometrics **45**, 255–268 (1989)
3. Hauke, J., Kossowski, T.: Comparison of values of Pearson's and Spearman's correlation coefficients on the same sets of data. Quaestiones Geographicae **30**(2), 87–93 (2011)
4. Mukaka, M.M.: A guide to appropriate use of correlation coefficient in medical research. Malawi Med. J. **24**(3), 69–71 (2012)
5. Chen, T., Guestrin, C.: XGBoost: a scalable tree boosting system. In: Proceedings of the 22nd ACM SIGKDD International Conference on Knowledge Discovery and Data Mining, pp. 785–794. ACM (2016)
6. Gumus, M., Kiran, M.S.: Crude oil price forecasting using XGBoost. In: 2017 International Conference on Computer Science and Engineering (UBMK), pp. 1100–1103. IEEE (2017)
7. Ren, X., et al.: A novel image classification method with CNN-XGBoost model. In: Kraetzer, C., Shi, Y.Q., Dittmann, J., Kim, H. (eds.) International Workshop on Digital Watermarking, pp. 378–390. Springer, Cham (2017)
8. Sheridan, R.P., et al.: Extreme gradient boosting as a method for quantitative structure–activity relationships. J. Chem. Inf. Model. **56**(12), 2353–2360 (2016)
9. Ke, G., et al.: LightGBM: a highly efficient gradient boosting decision tree. In: Advances in Neural Information Processing Systems, pp. 3146–3154 (2017)
10. Xie, Y., et al.: Evaluation of machine learning methods for formation lithology identification: a comparison of tuning processes and model performances. J. Pet. Sci. Eng. **160**, 182–193 (2018)
11. Cao, Y., Gui, L.: Multi-step wind power forecasting model using LSTM networks. Similar Time Series and LightGBM. In: 2018 5th International Conference on Systems and Informatics (ICSAI), pp. 192–197. IEEE (2018)
12. Ju, Y., et al.: A model combining convolutional neural network and LightGBM algorithm for ultra-short-term wind power forecasting. IEEE Access **7**, 28309–28318 (2019)

C4.5 Decision Tree Enhanced with AdaBoost Versus Multilayer Perceptron for Credit Scoring Modeling

Thitimanan Damrongsakmethee[1,2]($\boxtimes$) and Victor-Emil Neagoe[1]

[1] Department of Applied Electronics and Information Engineering,
Faculty of Electronics, Telecommunications and Information Technology,
Polytechnic University of Bucharest, Bucharest, Romania
manancc@gmail.com, victoremil@gmail.com
[2] Department of Business Information System, Suratthani Rajabhat University,
Suratthani, Thailand

Abstract. Within this paper, we evaluate two main machine learning techniques for credit scoring. The first algorithm consists of a cascade with two steps: (a) C4.5 decision tree; (b) AdaBoost for binary classification (credit accepted or rejected). The second technique corresponds to choosing a neural network classifier implemented by Multilayer Perceptron (MLP). For evaluation of the proposed models, we have used the German credit dataset and the Australian credit dataset. For the German dataset, MLP leads to the best result corresponding to an accuracy of 81.0%, versus C4.5 enhanced with AdaBoost that leads to an accuracy of 78.67%. For the Australian credit dataset, we found that MLP is also the best classifier with an accuracy of 90.85%, versus C4.5 followed by AdaBoost obtaining an accuracy of 89.00%. At the same time, one can remark that C4.5 enhanced by AdaBoost has led to a better performance than a simple C4.5.

Keywords: Decision tree · C4.5 · AdaBoost · Multilayer perceptron · Credit scoring

1 Introduction

Credit scoring is concerned with the financial risks and informing managerial decision making in the money lending business. Crook et al. [5] explained credit scoring as a model-based estimate of the probability that a borrower will show some undesirable behavior in the future. The classic example is application scoring to decide on an application for some credit product, the bank employs the predictive models, called a scorecard, to estimate how likely the applicant is to default. The term of credit scoring refers to credit decisions in the businesses. The management of collaborative or national credit is treated in the literature on business failure prediction or credit rating analysis, respectively [11]. Credit scoring [6, 13] is a client's credit worthiness assessment (credit risks), based on numerical statistical methods - a well-known method of applying mathematical methods in a banking field. For increasing the profitability of credit operations, a bank has to qualitatively assess credit risks. Based

© Springer Nature Switzerland AG 2019
R. Silhavy et al. (Eds.): CoMeSySo 2019, AISC 1047, pp. 216–226, 2019.
https://doi.org/10.1007/978-3-030-31362-3_22

on clients' classification into risk groups, a bank decides whether to issue a loan to a client or not, and what lending limit and interest should be set. The task of classifying clients into risk groups is solved by a scoring system. The main task of scoring is not only in finding out whether the client is able to pay the loan or not, but also in finding out a degree of reliability and compulsion of the client. In other words, scoring estimates how much the customer is "worthy" of the loan. The scoring model uses a set of specific features. The result is a real valued score; the higher it is the higher the client's reliability, and the bank can streamline its customers by the degree of credit enhancement. The indicator of each customer is compared with a certain numerical threshold, which can be called the break-even line. The complexity of constructing a model resides in the determination of which characteristics should be included in a model. Characteristics of loan application (factors) can be either discrete (for example, the gender of a borrower, and level of education) or continuous (for example, the borrower's age, work experience, income, expenses, loan amount). Credit scoring systems play an important role in banks, since such systems allow for the reducing of costs and minimizing of operational risk by automated decision-making, reducing the processing time for loan applications, enabling banks to pursue their credit policy centrally and providing additional protection for financial organizations from fraud [14]. In addition, credit scoring is a tool that reduces the risk of credit management for financial institutions as well [11]. Therefore, the reason for using a credit scoring model is to find a classifier that separates the good credit sample from the bad credit sample. This endeavor is considered as one of the best application fields for both machine learning and operational research techniques [22]. Many models of machine learning classification have been examined to pursue credit scoring accuracy. Classification models for the credit scoring include K-Nearest Neighbors (K-NN), Support Vector Machine (SVM), Multilayer Perceptron (MLP), Naive Bayes, Random forests, Decision trees, Logistic [2, 6, 19, 28]. For the future, using Deep Learning as a MLP extension for credit scoring seems to be a promising challenge [15, 16, 18].

In this paper, we have proposed two machine learning techniques to improve the accuracy for credit scoring: C4.5 decision tree cascaded with AdaBoost as a first variant versus Multilayer Perceptron (MLP) as a second variant. For performance evaluation of the proposed algorithms, we have used two datasets: German credit dataset and Australian credit dataset in a public benchmark from the UCI machine learning repository [26, 27].

The paper is structured as follows: Sect. 2 presents the details of related works. Section 3 explains the details of our proposed methods. The experimental results which show accuracy improvements of the proposed models are given in Sect. 4, while the conclusions are presented in Sect. 5.

2 Related Works

We will specify related works only for the algorithms applied on German and Australian credit datasets from UCI repository. Zhang et al. [29] proposed a model to classify the dataset based on a decision trees model by learning historical data and combined technique of the Genetic algorithm (GA) and K-means clustering technique to improve the accuracy of credit scoring approval. This research pointed out the advantage of applying a hybrid technique to improve the classification accuracy for credit scoring. Jiang proposed a new credit scoring model based on applying C4.5 and Simulated Annealing algorithm to improve the accuracy of credit scoring [10]. The research of Bastos [2] used boosted decision trees (BDT) which is a learning technique that combines several decision trees to build a classifier which is obtained from a weighted majority vote of the classifications given by individual trees; the boosted decision provides an elegant way to rank the attributes that most significantly indicate the likelihood of default. In the research of Paleologo et al., they have evaluated the results of applying bagging to decision trees, linear SVM, RBF SVM, MLP, K-Nearest Neighbor (KNN), AdaBoost and also of using subagging; they found that subagging decision trees achieve an improved performance for credit scoring and keep the model simple and reasonably interpretable [19]. In addition, Shrivas et al. [24] developed a robust binary classifier as an Intrusion Detection System (IDS); they used decision trees based techniques on C4.5, Random forests, ID3, CART and REF tree (meaning to build a decision or regression tree using information gain/variance reduction and prunes to select critical features). The research paper of Davoodabadi and Moeini [7] found that using feature selection methods with decision tree algorithms (hybrid models) make more accurate models than models without feature selection. They proposed hybrid models by combination of three methods of feature selection (GA +ReliefF+Gainratio) and three decision tree algorithms which are ID3, C4.5 and CART. In [15] Lyn, applied the feature selection using Principal Component Analysis (PCA) to reduce the attributes and then to classify the selected attributes in the transformed space C4.5 using decision trees. This method showed the best experimental results on C4.5 when compared with other techniques. Lakshmi et al. [12] has applied a C4.5 decision tree classifier with parameter selection to predict the risk levels during pregnancy. In [20] Pandya and Pandya proposed the technique of feature selection for decision tree method to reduce error pruning. The results of this research achieved the objective of classifying on credit scoring datasets. Wang et al. [28] presented the combination model to use random forests with bagging technique. In this research, they applied the parallel random forests with feature selection method to evaluate the credit risk, while, Al-hroot [1] developed a MLP model for prediction of bankruptcy in Jordanian companies. He found that MLP can be investigated as a potential model for bankruptcy prediction and it achieves the best overall prediction results. Shukla et al. [25] applied an artificial neural network (ANN) model on the Australian credit dataset based on choosing some features of the applicant. It is clear that this model has shown the graph of the accuracy is being achieved. Byanjanka et al. [3] developed a MLP model to find promising results in classifying credit applications. They have proved that the MLP model could be effective in the screening of the worst applications. The research of Ilgun et al. [9] used a MLP algorithm model to classify the Australian credit dataset. The experimental results of this research were done by choosing a MLP with two and three hidden layers, and using the backpropagation training method. In the last year, several researchers have pointed out the power of deep learning neural networks to manage credit scoring. Neagoe et al. [16] presented a deep learning convolutional neural network (DCNN) and MLP model for credit scoring on German and Australian datasets; this research found that the DCNN has led to the best overall accuracy. Munkhdalai et al. [15] have applied a deep learning approach combined with grid search to get the best accuracy on credit scoring; they found that a bigger sample dataset can lead to a higher performance. In [18] Niimi has applied deep neural networks to predict the status of credit cards (good or bad) and the results have confirmed the effectiveness of the proposed model.

3   The Proposed Methods

3.1   C4.5

C4.5 [21] is an evolution of ID3, and uses gain ratio as splitting criteria. The splitting ceases when the number of instances to be split is below a certain threshold. Error–based pruning is performed after the growing phase. C4.5 can handle the numeric of attributes. It can induce from a training set that incorporates missing values by using corrected gain ratio criteria as presented above.

3.2   AdaBoost

As a method to reduce errors in the predictions made by the decision trees, a boosting algorithm can be implemented. The fundamental of the boosting algorithms [22] is to perform the following 5 steps:

Step 1 choose and implement a classifier.
Step 2 split the dataset into a training set (of size n) and a validation set (of size m). The samples in these sets are chosen randomly in the first iteration. Both sets need to contain samples from all classes.
Step 3 train the classifier using the training set. This classifier is said to be a weak classifier. Test the classifier by classifying the testing set and create weights to indicate the flaws in the weak classifier.
Step 4 repeat steps 2 and 3, for given number of iterations, storing each weak classifier constructed this way. Each iteration, the selection of samples is reflected by the assigned weights.
Step 5 combine all weak classifiers into one strong classifier.

In C4.5 the particular boosting algorithm implemented is based on the idea of adaptive Boosting or AdaBoost, which was introduced in [8]. This work made great impact in the field of machine learning due to the ability of combining several weak classifiers into a stronger one, while retaining the robustness to overfitting of the weak classifiers. The main idea of adaptive boosting is to weigh the data points in each successive boosting iteration during the construction of a classifier. These weights for each sample in the training data are distributed such that the algorithm will be focused to correctly classify the data points which were missclassified by the previous classifiers. The AdaBoost algorithm [12] for binary classification contains the following steps:

Step 1 set n observations with $(x_1, y_1), \ldots, (x_n, y_n)$ where x_j is a vector of attributes and y_j is the class.

Step 2 set $w_t(i)$ is a weighting function as $\frac{1}{n}$ for $i = \{1..n\}i = \{1..n\}$.

Step 3 set T is a number of trials.

Step 4 loop t 1 to T to find weak classifier. The equation to find a weak classifier as:

$$f_t(x_i) \tag{1}$$

$$\in_t = \sum_{y_{i \neq ft(x_i)}} w_t(i) \tag{2}$$

$$\alpha_t = \frac{1}{2} \log\left(\frac{1 - \in_t}{\in_t}\right) \tag{3}$$

Step 5 loop i 1 to n to correct classified and incorrect classified as:

$$w_{t+1}(i) = w_t e^{-\alpha_t}, \text{ correctly classified} \tag{4}$$

$$w_{t+1}(i) = w_t e^{\alpha_t}, \text{ incorrectly classified} \tag{5}$$

Step 6 normalize w_t as:

$$w_t(i) = \frac{w_t(i)}{\sum_{i=1}^{n} w_t(i)} \tag{6}$$

Step 7 final classifier as:

$$F(x) = \text{sign}\left(\sum_{t=1}^{T} \alpha_t f_t(x)\right) \tag{7}$$

AdaBoost algorithm is the binary classification which considers all observations of a variable xi including the true class association of that observation yi. The data set is given by (xi, yi), where $i = \{1..n\}$. In the process to set the variable, all weights are set to be equal for all samples. These weights are normalized such that their sum equals one. The constant T affects the number of weak classifiers to use in the final boosted model.

The main loop builds a weighted sum of weak classifiers to produce a stronger final classifier. This is achieved by sum value of the weights of all misclassified observations and constructing a modifier for the updated weights. The boosting procedure will end if the value of $\sum_{\{i:y_i \neq F_t(x_i)\}} w_i(i) < 0.1$ or $\frac{\sum_{\{i:y_i \neq F_t(x_i)\}} w_i(i)}{|W_m|} > 0.5$, where $|w_m|$ is the measure of the set of weights associated with misclassified observations. This are 3 steps of the algorithm of depicting the weighting procedure in C4.5 [4]:

Step 1 setting the variables; the number of samples in the training set, the number of misclassified samples, the number of boosting iterations, the weight of i-th sample during t-th round of boosting, sum correct classified and sum incorrect classified samples.

Step 2 loop t 1 to T, step to build a decision tree.

Step 3 loop i 1 to N, step to compute the midpoint as:

$$\text{midpoint} = \frac{1}{2}\left[\frac{1}{2}(S_+ + S_-) - S_-\right] \tag{8}$$

weight if correctly classified:

$$w_{i,t} = w_{i,t-1}\frac{S_+ - \text{midpoint}}{S_+} \tag{9}$$

weight if misclassified:

$$w_{i,t} = w_{i,t-1} + \frac{\text{midpoint}}{N_-}, \tag{10}$$

where N is Number of the sample in training set, N_- is Number of the misclassified sample, T is Number of boosting iterations, $w_{i,t}$ is weight of i-th sample during t-th round of boosting, S_+ is Sum over all weights associated with the correct classified sample, S_- is Sum over all weights associated with the misclassified sample.

3.3 Multilayer Perceptron

Multilayer Perceptron (MLP) is a feed-forward neural network with multiple layers [19]. There are three types of layers in MLP: the input layer, the hidden layers and the output layer. Each neuron computes an activation function. The input layer receives the signal or the independent variables in the dataset [2]. The input layer doesn't have an activation function. The hidden layers are the true computational engine of the MLP. The output layer takes the decision about the input. Then the number of output neurons depends on the way the target values of the training patterns are described [19]. The training process of MLP occurs by continuing adjustment of the weights of the connection after each processing. This adjustment is based on the error in the output layer which is the difference between the expected result and the output. This continuous adjustment of the weights by supervised learning technique is called "backpropagation" training [17, 23]. The backpropagation consists of two parts: forward pass and backward pass. In the forward pass, the expect output correlated with the given input is evaluated. In the backward pass, partial derivatives of the cost function with respects to the different parameters are propagated back over the network. A general MLP active neuron can be simply described by the following formula [23].

$$y_j = \varphi\left(\sum_{i=1}^{n}(\omega_{ij}.x_i + b_j)\right), \tag{11}$$

for $j = 1...n$, where $\omega_{1j}, ..., \omega_{nj}$ are weights connecting the input neurons with hidden neuron j, $\{x_1, ..., x_n\}$ are input features, b_j is the bias term, φ is a nonlinear function and y_j is the output variable. The scheme of an active neuron of MLP model shown in Fig. 1.

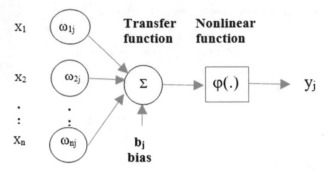

Fig. 1. Neuron of the MLP model

3.4   Proposed Techniques

In this paper, we propose two algorithms in order to compare the performances of a credit scoring model by using improved C4.5 decision tree versus MLP.

Regarding enhanced C4.5 algorithm, we integrated AdaBoost with the decision tree in order to improve the accuracy of credit scoring. This model consists of two steps: (a) C4.5 decision tree; (b) AdaBoost for binary classification (credit accepted or rejected). It iteratively trains the AdaBoost machine learning model by selecting the training set based on the accurate prediction of the last training. It assigns the higher weight to wrong classified observations so that in the next iteration these observations will get the high probability for classification. Also, it assigns the weight to the trained classifier in each iteration according to the accuracy of the classifier. The more accurate classifier will get a higher weight. In this process, we iterate until all the training data fits without any error or until the specified maximum number of estimators is reached. In this paper, we used base learner bagging with 100 iterations as shown in Fig. 2.

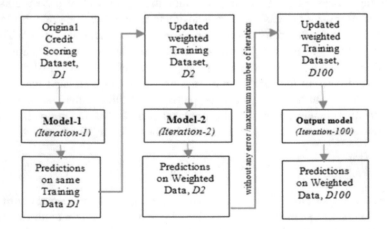

Fig. 2. Proposed credit scoring of C4.5+AdaBoost model.

MLP algorithm, In order to get the best accuracy, we have designed a MLP with the following characteristics. The input layer has a number "n" of input neurons, where n is the number of credit applicant's features corresponding to the considered datasets. We have chosen a MLP with H hidden layers. The number of hidden neurons for each hidden layer stands for h_j, where $j = 1 \ldots H$. The number N of epochs (corresponding to the training phase, where an epoch corresponds to passing by the whole training set) has a significant influence on the MLP performance. The output layer consists of two neurons corresponding to two classes of credit cards: "good" or "bad".

3.5 Performance Evaluation

We used the prediction accuracy rate defined by the equation:

$$\text{Accuracy} = \frac{TP + TN}{TP + TN + FP + FN} \times 100, \tag{12}$$

where TP is the number of "bad" applicants predicted as 'bad", TN is the number of "good" applicants predicted as "good", FP is the number of good" applicants predicted as "bad", and FN is the number of "bad" applicants predicted as "good". The variables TP, TN, FP, and FN are considered on the test lot.

3.6 Datasets

To evaluate the techniques proposed in this paper, we have used two credit datasets: the German credit dataset and Australian credit dataset, both taken from the UCI repository of machine learning databases.

The German credit dataset classifies people described by a set of attributes as good or bad credit risks [27]. The data characterizing a potential client contain 25 attributes including the label ("good' or "bad" credit risk). There are 1000 loan applicants divided into 700 with "good credit" label and 300 with "bad credit" label. There are no missing values in this dataset.

The Australian credit approval dataset [26] is dedicated to credit card applications. The data contains 15 attributes (including the label). This dataset is more interesting because there is a good mix of attributes of various structures: continuous, nominal with a small number of values, as well as nominal with a large number of values. The total number of instances is 690; 307 instances have "good credit" label and 383 correspond to "bad credit" label. The details of the datasets are shown in Table 1.

Table 1. Characteristics of the considered datasets.

Dataset name	Number of attributes	Good credit	Bad credit	Total	Classes
German	25	700	300	1000	1 = good 2 = bad
Australian	15	307	383	690	1 = good 0 = bad

4 Experimental Results

For the German dataset, we have considered 500 training data and 250 test data. For the Australian data set, we have chosen 344 training data and 173 test data. The results from Tables 2, 3, and 4 show the performances of credit scoring for German credit dataset and respectively, for the Australian credit dataset using the considered classification techniques of C4.5 (simple and enhanced with AdaBoost) and MLP.

Table 2. Experimental results of the C4.5 decision tree-based credit scoring models.

Dataset	Model	Accuracy (%)
German	C4.5	72.68
	C4.5+AdaBoost	**78.67**
Australian	C4.5	83.42
	C4.5+AdaBoost	**89.00**

Table 2 shows one can point out that the technique using C4.5 decision tree followed by AdaBoost (bagging with 100 iterations) leads to best predictive performances for both datasets. On German credit, its classification accuracy is of 78.67% versus the simple C4.5 method that leads to 72.68% accuracy. On Australian dataset, the C4.5 enhanced with AdaBoost leads to 89.00% accuracy by comparison to simple C4.5 that obtains 83.42% accuracy.

Table 3 shows best performances of the MLP for credit scoring. On the German credit dataset, the best accuracy is 81.20%. For the Australian credit dataset, MLP leads to the maximum classification accuracy of 90.85%.

Table 3. Experimental performances of the MLP credit scoring model (H = 6 hidden layers).

Dataset	Number of input neurons	Optimum number of neurons for any hidden layer $h_1 = \ldots = h_6$	Optimum number of training epochs	Accuracy (%)
German	24	18	8	81.20
Australian	14	41	12	90.85

Table 4 shows the global accuracy evaluation by our two main proposed models. For both datasets, the MLP outperforms the C4.5 decision tree model (in its two variants: without and with AdaBoost). MLP obtains the best accuracy of 81.20% for the German dataset and that of 90.85% for the Australian dataset.

Table 4. Accuracy comparisons of C4.5 (simple C4.5 and C4.5 enhanced with AdaBoost) versus MLP.

Model	German Credit (% correct)	Australian Credit (% correct)
C4.5	72.68	83.42
C4.5 +AdaBoost	78.67	89.00
MLP	**81.20**	**90.85**

5   Conclusions

This paper is an approach for design, implementation and accuracy evaluation of two machine learning credit scoring models: C4.5 decision tree enhanced with AdaBoost versus MLP classifier. The experiments have used a German credit dataset and an Australian dataset as well. For the German dataset, MLP leads with the accuracy of 81.20%, versus C4.5 followed by AdaBoost that has obtained an accuracy of 78.67%. For the Australian credit dataset, we have found that MLP has led to the best performance of 90.85%, versus C4.5 enhanced with AdaBoost that has led to an accuracy of 89.00%. For the future, we intend to evaluate a deep learning neural network performance for credit scoring modeling.

References

1. Al-hroot, Y.-A.-K.: Bankruptcy prediction using multilayer perceptron neural networks in Jordan. Eur. Sci. J. **12**(4), 425–435 (2016)
2. Bastos, J.: Credit scoring with boosted dcision trees. MPRA Paper (2008)
3. Byanjanka, A., Heikkilä, M., Mezei, J.: Predicting credit risk in peer-to-peer lending: a neural network approach. In: IEEE Symposium Series on Computational Intelligence, pp. 719–725 (2015)
4. Christmas, B.: AdaBoost and the super bowl of classifiers a tutorial introduction to adaptive boosting, 1–6 (2009)
5. Crook, J.N., Edelman, D.B., Lyn, C.T.: Recent developments in consumer credit risk assessment. Eur. J. Oper. Res. **183**, 1447–1465 (2007)
6. Damrongsakmethee, T., Neagoe, V.-E.: Data mining and machine learning for financial analysis. Indian J. Sci. Technol. **10**(39), 1–7 (2017)
7. Davoodabadi, Z., Moeini, A.: Building customers' credit scoring models with combination of feature selection and decision tree algorithms. Int. J. Adv. Comput. Sci. **4**(2), 97–103 (2015)
8. Freund, Y., Schapire, R.E.: A decision-theoretic generalization of on-line learning and an application to boosting. J. Comput. Syst. Sci. **139**(55), 119–139 (1997)
9. Ilgun, E., Mekic, E., Mekic, E.: Application of Ann in Australian credit card approval. Eur. Res. **69**(2), 334–342 (2014)
10. Jiang, Y.: Credit scoring model based on the decision tree and the simulated annealing algorithm (2009)
11. Jingqiao, Z., Viswanath, A., Raj, S.: Evolutionary optimization of transition probability matrices for credit decision-making. Eur. J. Oper. Res. **200**(2), 557–567 (2010)
12. Lakshmi, B.N., Indumathi, T.S., Ravi, N.: A study on C.5 decision tree classification algorithm for risk predictions during pregnancy. Procedia Technol. **24**, 1542–1549 (2016)
13. Li, X.-L., Zhong., Y.: An overview of personal credit scoring: techniques and future work. Int. J. Intell. Sci. (2012). ISSN 2163-0283
14. Lyn, C.T.: A survey of credit and behavioural scoring: forecasting financial risk of lending to consumers. Int. J. Forecast. **16**(2), 149–172 (2000)
15. Munkhdalai, L., Namsrai, O., Ryu, KH.: Credit scoring with deep learning. In: 4th International Conference on Information, System and Convergence Applications, pp. 1–5 (2018)

16. Neagoe, V.-E., Ciotec, A.-D., Cucu, G.-S.: Deep convolutional neural networks versus multilayer perceptron for financial prediction. In: 2018 12th International Conference on Communications (COMM), Bucharest, p. 201. IEEE (2018)
17. Neagoe, V.-E.: Retele neurale pentru explorarea datelor (Neural Networks for Data Mining). Matrix rom, Bucharest (2018)
18. Niimi, A.: Deep learning for credit card data analysis. In: 2015 World Congress on Internet Security (WorldCIS), Dublin, Ireland, pp. 73–77. IEEE (2015)
19. Paleologo, G., Elisseeff, A., Antonini, G.: Subagging for credit scoring models. Eur. J. Oper. Res. **201**(2), 490–499 (2010)
20. Pandya, R., Pandya, J.: C5.0 algorithm to improved decision tree with feature selection and reduced error pruning. Int. J. Comput. Appl. **117**(16), 18–21 (2015)
21. Quinlan, J.R.: C4.5: Programs for Machine Learning. USA (1993)
22. Schapire, R.E.: The boosting approach to machine learning: an overview. In: Denison, D.D., Hansen, M.H., Holmes, C.C., Mallick, B., Yu, B. (eds.) Nonlinear Estimation and Classification, vol. 171, pp. 149–171. Springer, New York (2003)
23. Shai, S.-S., Shai, B.-D.: Understanding Machine Learning: From Theory to Algorithms. Cambridge University Press, Cambridge (2014)
24. Shrivas, A.-K., Singhai, S.-K., Hota, H.-S.: An efficient decision tree model for classification of attacks with feature selection. Int. J. Comput. Appl. **84**(14), 42–48 (2013)
25. Shukla, A., Gwalior, M., Mishra, A.: Design of credit approval system using artificial neural network: a case study. Int. J. Eng. Res. Comput. Sci. Eng. **4**(6), 337–341 (2017)
26. UCI: Australian Credit Approval. http://archive.ics.uci.edu/ml/datasets/statlog+(australian +credit+approval). Accessed 10 Jan 2019
27. UCI: German Credit Data. https://archive.ics.uci.edu/ml/datasets/statlog+(german+credit +data). Accessed 10 Jan 2019
28. Wang, G., Ma, J., Huang, L., Xu, K.: Two credit scoring models based on dual strategy ensemble trees. Knowl. Based Syst. **26**, 61–68 (2012)
29. Zhang, D., Leung, S.C.H., Ye, Z.: A decision tree scoring model based on genetic algorithm and K means algorithm. In: Third International Conference on Convergence and Hybrid Information Technology, pp. 1043–1047 (2008)

Stability and Dissipativity of Multistep Runge–Kutta Methods for Nonlinear Delay Differential Equations with Constrained Grid

Yuhuan Fan, Haiyan Yuan$^{(\boxtimes)}$, and Peichen Wang

Department of Mathematics, Heilongjiang Institute of Technology,
Harbin 150050, China
yhy82_47@163.com

Abstract. This paper is devoted to the stability and dissipativity of multistep Runge–Kutta methods with constrained grid for a class of nonlinear neutral delay differential equations. Nonlinear stability and dissipativity are introduced and proved. We discuss both the $GR(l)$-stability, $GAR(l)$-stability, and the weak $GAR(l)$-stability on the basis of (k, l)-algebraically stable of the multistep Runge–Kutta methods, we also prove that an algebraically stable, irreducible multistep Runge–Kutta method is finite-dimensional dissipative.

Keywords: Stability · Dissipativity · (k, l) algebraically stability ·
Nonlinear neutral delay differential equation · Multistep Runge–Kutta methods

1 Introduction

Neutral delay differential equations (NDDEs) arise in a variety of fields as biology, economy, control theory, electrodynamics and so on (see, for example, [1–5]). The stability and dissipativity properties of numerical methods for linear NDDEs have been widely researched by many authors (see, for example, [6–11]). For the case of nonlinear delay differential equations, this kind of methodology had been first introduced by Torelli [12, 13] and then developed by Bellen and Zennaro [14], Bellen [15] and Zennaro [16, 17]. 1997, Koto proved the asymptotic stability of natural Runge–Kutta method for a class of nonlinear delay differential equations in [18], Bellen [19] gave a discussion of the stability of continuous numerical methods for a special class of nonlinear neutral delay differential equations. In 2009, Yang, Gao and Shi, gave a novel robust stability criteria for stochastic Hopfield neural networks with time delays in [23]. Yang, Zhang, and Shi [24] studied the exponential stability on stochastic neural networks with discrete interval and distributed delays in 2010. 2011, Liu [25] gave the robust stability for Neutral Time-Varying Delay Systems with Non-Linear Perturbations. On the stability, Akio studied the Values of Random Zero-Sum Games in [26], and in [27] Basin and Calderon-Alvarez gave the delay-dependent stability studies for Vector Nonlinear Stochastic Systems with Multiple Delays.

However, these important convergence results are based on the classical Lipschitz conditions. The studies focusing on the stability and convergence of the numerical method for nonlinear NDDEs based on a one-sided Lipschitz condition have not yet

© Springer Nature Switzerland AG 2019
R. Silhavy et al. (Eds.): CoMeSySo 2019, AISC 1047, pp. 227–242, 2019.
https://doi.org/10.1007/978-3-030-31362-3_23

been seen in literature until now. By means of a one-sided Lipschitz condition, in the present paper we discuss the stability and convergence of two-step Runge–Kutta (TSRK) methods for nonlinear NDDEs. Thanks to the one-sided nature of the Lipschitz condition, the error bounds obtained in the present paper are sharper than those given in the references mentioned.

Many problems arising in the fields of physics, control theory, biosciences, medicine, physics and so on [1, 2] are modeled by dissipative dynamical systems. They are characterized by possessing a bounded positively invariant absorbing set which all trajectories starting from any bounded set enter in a finite time and thereafter remain inside (see Humphries and Stuart [3, 4]). In the study of dissipative systems it is often the asymptotic behavior of the system that is of interest, and so it is important to analyze whether or not numerical methods inherit the dissipativity of the dynamical systems when considering the applicability of numerical methods for these systems.

Humphries and Stuart [3, 4] first studied the dissipativity of Runge–Kutta methods for initial value problems (IVPs) of ordinary differential equations (ODEs), and proved that an algebraically stable, irreducible method can inherit the dissipativity of finite-dimensional systems, in 1994. Later, many results on the dissipativity of numerical methods for ODEs have already been found [5–7]. For the delay differential equations (DDEs) with constant delay, Huang [8] gave a sufficient condition for the dissipativity of theoretical solution, and investigated the dissipativity of (k, l)-algebraically stable Runge–Kutta methods. Huang and Chen [9] and Huang [10], subsequently, obtained some results about the dissipativity of linear θ-methods and $G(c, p, 0)$-algebraically stable one-leg methods, respectively. In addition, Huang [11] further discussed the dissipativity of multistep Runge–Kutta methods, and proved that an algebraically stable, irreducible multistep Runge–Kutta methods with linear interpolation procedure is finite-dimensional dissipative. In 2004, Tian [12] studied the dissipativity of DDEs with a bounded variable lag and the dissipativity of θ-method. Moreover, Wen [13] discussed the dissipativity of Volterra functional differential equations, and further investigated the dissipativity of DDEs with piecewise delays and a class of linear multistep methods. In recent years, a number of works on the dissipativity of numerical methods have been carried out. Gan [14–16] studied the dissipativity of numerical methods for nonlinear integro differential equations (IDEs), nonlinear delay-integro-differential equations (DIDEs) and nonlinear pantograph equations, respectively. As to nonlinear Volterra delay-integro-differential Equations, it was shown that for $\theta \in [1/2, 1]$, any linear θ-method and one-leg method can inherit the dissipativity property, which was obtained by Gan [15]. In addition, Cheng and Huang [19], Wen et al. [20] and Wang et al. [21] considered the dissipativity for nonlinear neutral delay differential equations (NDDEs).

This paper pursues this, and further investigates the dissipativity of multistep Runge–Kutta methods for nonlinear neutral delay differential equations. The motivations are as follows. Multistep Runge–Kutta methods are a wider class of methods which has as special cases not only one-leg methods, linear multistep methods, and Runge–Kutta methods, but also a wide range of hybrid methods. In particular, there exist algebraically stable multistep Runge–Kutta methods with only real eigenvalues such that they not only possess very good stability, but also can be performed in parallel.

2 The Description of the Problem and Numerical Methods

Let H be a real or complex, finite dimensional or infinite-dimensional Hilbert space with the inner product $\langle \cdot, \cdot \rangle$ and the corresponding induced norm $\|\cdot\|$, and the matrix norm is subordinated to the vector norm. X be a dense continuously imbedded subspace of H. Consider the following initial value problems (IVPs) of nonlinear NDDEs:

$$\begin{cases} \frac{d}{dt}[y(t) - Ny(t - \tau)] = f(y(t), y(t - \tau)), t \geq 0, \\ y(t) = \varphi(t), -\tau \leq t \leq 0. \end{cases} \tag{2.1}$$

where τ is a given constant delay, $N \in X \times X$ stands for a constant matrix with $\|N\| < 1$, $\varphi : [-\tau, 0] \to X$ is a continuous function, and $f : X \times X \to H$ is a locally Lipschitz continuous function and f satisfies the following conditions:

$$\mathrm{Re}\langle u - Nv, f(u, v) \rangle \leq \beta_0 + \beta_1 \|u\|^2 + \beta_2 \|v\|^2, u, v \in X, \tag{2.2}$$

where β_0, β_1, β_2 are real constants.

Throughout this paper, we assume that the problem (2.1) has unique exact solution $y(t)$. For the study of solvability, we refer the reader to [2].

Remark 2.1. When $N = 0$, the problem (2.1) degenerates into an IVP of DDEs, the number of papers dealing with this case amounts to several hundreds.

Proposition 2.2 [15]. Condition (2.2) implies that $\beta_0 \geq 0$, $\beta_2 \geq 0$.

Next, let us consider the adaptation of s-stage multistep Runge–Kutta methods for solving problem (2.1) based on the formula

$$\begin{cases} Y_i^{(n)} - N\bar{Y}_i^{(n)} = \sum_{j=1}^{r} a_{ij}(y_{n+j-1} - N\bar{y}_{n+j-1}) + h \sum_{j=1}^{s} b_{ij} f(Y_j^{(n)}, \bar{Y}_j^{(n)}), \ i = 1, 2, \cdots, s. \\ y_{n+r} - N\bar{y}_{n+r} = \sum_{j=1}^{r} \theta_j(y_{n+j-1} - N\bar{y}_{n+j-1}) + h \sum_{j=1}^{s} \gamma_j f(Y_j^{(n)}, \bar{Y}_j^{(n)}). \end{cases}$$

$$\tag{2.3}$$

where $h > 0$ is the fixed stepsize, the parameters a_{ij}, b_{ij}, θ_j and γ_j are real constants, $Y_i^{(n)}$ and y_n are approximation to $y(t_n + c_i h)$ and $y(t_n)$, respectively, and $t_n = nh$. The argument $\bar{Y}_i^{(n)}$ and $\bar{y}_n$ denotes an approximation to $y(t_n + c_i h - \tau)$ and $y(t_n - \tau)$, those are obtained by a specific interpolation procedure using values $Y_i^{(k)}$ and y_{k+r-1} $(k \leq n)$. The initial values $y_n = \varphi(t_n)$, $\bar{y}_n = \varphi(t_n - \tau)$ for $t_n \leq 0$, and $Y_i^{(n)} = \varphi(t_n + c_i h)$, $\bar{Y}_i^{(n)} = \varphi(t_n + c_i h - \tau)$ for $t_n + c_i h \leq 0$. Following the referee's suggestion, we assume that $0 \leq c_i \leq 1$, $i = 1, 2, \cdots, s$.

As to the computation of the delay terms $\bar{Y}_i^{(n)}$, $i = 1, 2, \cdots, s$, we use the constrained stepsize h satisfying $hm = \tau$ with a positive integer m.

Let

$$\bar{Y}_i^{(n)} = Y_i^{(n-m)}, \quad i = 1, 2, \cdots, s, \tag{2.4a}$$

$$\bar{y}_n = y_{n-m}. \tag{2.4b}$$

Remark 2.3. We consider the procedure (2.4a and 2.4b) here because, in the case that the order of the method is more than 2, there will be no order reduction. But it must be noticed that the stepsize h is limited by $mh = \tau$.

The used values $Y_i^{(n)}$ and y_n with $n < -m < 0$ are assumed to be 0. Here, we do not discuss other details.

It is well known that multistep Runge–Kutta methods are a subclass of a general linear methods. Let

$$C_{11} = [b_{ij}] \in R^{s \times s}, \quad C_{12} = [a_{ij}] \in R^{s \times r}, \tag{2.5a}$$

$$C_{21} = \begin{bmatrix} 0 & \cdots & 0 \\ \vdots & & \vdots \\ 0 & \cdots & 0 \\ \gamma_1 & \cdots & \gamma_s \end{bmatrix} \in R^{r \times s}, C_{22} = \begin{bmatrix} 0 & 1 & 0 & \cdots & 0 \\ 0 & 0 & 1 & \cdots & 0 \\ \vdots & \vdots & \ddots & \ddots & \vdots \\ 0 & 0 & \cdots & 0 & 1 \\ \theta_1 & \theta_2 & \cdots & \theta_{r-1} & \theta_r \end{bmatrix} \in R^{r \times r} \tag{2.5b}$$

For any given $k \times l$ real matrix $Q = [q_{ij}]$, we define the corresponding linear operator

$$Q : X^l \to X^k,$$
$$QU = V = (v_1, v_2, \cdots, v_k) \in X^k, U = (u_1, u_2, \cdots, u_l) \in X^l, u_j \in X,$$

with

$$v_i = \sum_{j=1}^{l} q_{ij} u_j, \quad i = 1, 2, \cdots, k.$$

Then, method (2.3) can be rewritten in the form of general linear method

$$\begin{cases} G^{(n)} = hC_{11}F(Y^{(n)}, \bar{Y}^{(n)}) + C_{12}g^{(n-1)}, \\ g^{(n)} = hC_{21}F(Y^{(n)}, \bar{Y}^{(n)}) + C_{22}g^{(n-1)}. \end{cases} \tag{2.6}$$

with the following notational conventions:

$$Y^{(n)} = (Y_1^{(n)}, Y_2^{(n)}, \cdots, Y_s^{(n)})^T, \bar{Y}^{(n)} = (\bar{Y}_1^{(n)}, \bar{Y}_2^{(n)}, \cdots, \bar{Y}_s^{(n)})^T,$$
$$G^{(n)} = (Y_1^{(n)} - N\bar{Y}_1^{(n)}, Y_2^{(n)} - N\bar{Y}_2^{(n)}, \cdots, Y_s^{(n)} - N\bar{Y}_s^{(n)})^T,$$
$$g^{(n)} = (y_{n+1} - N\bar{y}_{n+1}, y_{n+2} - N\bar{y}_{n+2}, \cdots, y_{n+r} - N\bar{y}_{n+r})^T,$$
$$F(Y^{(n)}, \bar{Y}^{(n)}) = (f(Y_1^{(n)}, \bar{Y}_1^{(n)}), f(Y_2^{(n)}, \bar{Y}_2^{(n)}), \cdots, f(Y_s^{(n)}, \bar{Y}_s^{(n)}))^T.$$

We introduce the following notations for brevity, for any real symmetric $p \times p$ matrix $Q = [q_{ij}]$, and $Q \geq 0 (> 0)$ means that Q is nonnegative definite (positive definite). For any $Q \geq 0$, define a pseudo inner product on H^p by

$$\langle Y, Z \rangle_Q = \sum_{i,j=1}^{p} q_{ij} \langle Y_i, Z_j \rangle, \, Y = (Y_1, Y_2, \cdots, Y_p) \in H^p, Z = (Z_1, Z_2, \cdots, Z_p) \in H^p,$$

and the corresponding pseudo norm on H^p by

$$\|Y\|_Q = \langle Y, Y \rangle_Q^{1/2}.$$

Especially $\|\cdot\|$ is the simplicity for $\|\cdot\|_Q$ when Q is identity matrix.

Definition 2.4. Let k, l be real constants. A multistep Runge–Kutta method (2.3) is said to be (k, l)-algebraically stable if there exists a real symmetric $r \times r$ matrix $G > 0$ and a diagonal matrix $D = diag(d_1, d_2, \cdots, d_s) \geq 0$ such that $M = [M_{ij}] \geq 0$, where

$$M = \begin{bmatrix} kG - C_{22}^T G C_{22} - 2l C_{12}^T D C_{12} & C_{12}^T D - C_{22}^T G C_{21} - 2l C_{12}^T D C_{11} \\ DC_{12} - C_{21}^T G C_{22} - 2l C_{11}^T D C_{12} & C_{11}^T D + DC_{11} - C_{21}^T G C_{21} - 2l C_{11}^T D C_{11} \end{bmatrix}. \tag{2.7}$$

As an important special case, a $(1, 0)$-algebraically stable method is called algebraically stable for short.

Definition 2.5. Let l be a real constant, and H be a finite-dimensional (or infinite-dimensional) space. A multistep Runge–Kutta method (2.3) with an interpolation procedure is said to be finite-dimensionally (or infinite-dimensionally) $D(l)$ dissipative if, when the method is applied to problem (2.1) in H with stepsize h satisfying $d(\beta_1 h + \beta_2 h) < l d_{min}(1 - \|N\|)^2$, $(d(\beta_1 h + \beta_2 h) < l d(1 - \|N\|)^2)$,   and   constraint $mh = \tau$, where $d_{min} = \min_{1 \leq j \leq s} d_j$, $(d = \sum_{j=1}^{s} d_j)$, there exists a constant C such that, for any initial values, there exists an n_0, dependent only on initial values, such that

$$\|y_n\| \leq C, n \geq n_0,$$

holds. As an important special case, a $D(0)$-dissipative method is called D-dissipative for short.

$GD(l)$-and GD-dissipativity are defined by dropping restriction $\tau = mh$.

Definition 2.6 [11]. A multistep Runge–Kutta method (2.3) is said to be stage-reducible if, for some nonempty index set $T \subset \{1, 2, \cdots, s\}$,

$$\gamma_j = 0, \quad for \, \, j \in T,$$
$$b_{ij} = 0, \quad for \, \, j \in T, i \notin T.$$

Otherwise, it is said to be stage-irreducible.

Definition 2.7. A multistep Runge–Kutta method (2.3) is said to be step-reducible if polynomials $\{\sigma_i(x)\}_{i=0}^s$ have common divisor where

$$\sigma_0(x) = x^r - N\bar{x}^r - \sum_{j=1}^r \theta_j(x^{j-1} - N\bar{x}^{j-1}),$$

$$\sigma_i(x) = \sum_{j=1}^r a_{ij}(x^{j-1} - N\bar{x}^{j-1}), i = 1, 2, \cdots, s.$$

Otherwise, it is said to be step-irreducible.

Definition 2.8 [11]. A multistep Runge–Kutta method (2.3) is said to be reducible if it is stage-reducible or step-reducible.

3   Finite-Dimensional Numerical Dissipativity

In this section, we focus on the dissipativity analysis of (k, l)-algebraically stable multistep Runge–Kutta methods with respect to nonlinear NDDEs in finite-dimensional spaces. We always assume that $H = X = C^N$.

Lemma 3.1 [11]. Suppose $\{\xi_i(x)\}_{i=1}^r$ are a basis of polynomials for P^{r-1}, the space of polynomials of degree strictly less than r and E is the translation operator: $Ey_n = y_{n+1}$. Then there is always a unique solution $y_n, y_{n+1}, \ldots, y_{n+r-1}$ to the system of equations

$$\xi_i(E)y_n = \Delta_i, \Delta_i \in C^N, i = 1, 2, \cdots, r.$$

and there exists a constant χ, independent of Δ_i, such that

$$\max_{0 \le i \le r-1} \|y_{n+i}\| \le \chi \max_{0 \le i \le r-1} \|\Delta_i\|.$$

Lemma 3.2 [11]. Suppose that a multistep Runge–Kutta method (2.3) is step-irreducible. Then, there exist real constants $v_i, i = 1, 2, \cdots, s$, such that $\sigma_0(x)$ and $\sum_{i=1}^s v_i\sigma_i(x)$ have no common divisor.

Now we state and prove the main results.

Theorem 3.3. Assume that a step-irreducible multistep Runge–Kutta method (2.3) is (k, l)-algebraically stable, $D > 0, l > 0$ and $k \le 1$, the problem (2.1) satisfies (2.2) with $d(\beta_1 h + \beta_2 h) < l d_{\min}(1 - \|N\|)^2$. Then the method (2.3) with (2.2) and (2.4a and 2.4b) is finite-dimensionally $D(l)$-dissipative.

Proof. From (2.4a and 2.4b) we have

$$\left\| \bar{Y}^{(j)} \right\|^2 = \left\| Y^{(j-m)} \right\|^2 \tag{3.1a}$$

$$\|\bar{y}_n\| = \|y_{n-m}\| \tag{3.1b}$$

As in [17] and [15], by means of (k, l)-algebraically stability of the method, we can easily obtain that

$$
\begin{aligned}
&\left\|g^{(n)}\right\|_G^2 - k\left\|g^{(n-1)}\right\|_G^2 - 2\mathrm{Re}\left\langle G^{(n)}, hF(Y^{(n)}, \bar{Y}^{(n)})\right\rangle_D + 2l\left\|G^{(n)}\right\|_D^2 \\
&= \left\langle C_{21}hF(Y^{(n)}, \bar{Y}^{(n)}) + C_{22}g^{(n-1)}, G(C_{21}hF(Y^{(n)}, \bar{Y}^{(n)}) + C_{22}g^{(n-1)})\right\rangle + \\
&\quad \left\langle g^{(n-1)}, -kGg^{(n-1)}\right\rangle + 2\mathrm{Re}\left\langle C_{11}hF(Y^{(n)}, \bar{Y}^{(n)}) + C_{12}g^{(n-1)}, -DhF(Y^{(n)}, \bar{Y}^{(n)})\right\rangle \quad (3.2) \\
&\quad + \left\langle C_{11}hF(Y^{(n)}, \bar{Y}^{(n)}) + C_{12}g^{(n-1)}, 2lD(C_{11}hF(Y^{(n)}, \bar{Y}^{(n)}) + C_{12}g^{(n-1)})\right\rangle \\
&= -\left\langle\left\langle g^{(n-1)}, hF(Y^{(n)}, \bar{Y}^{(n)})\right\rangle, M\left\langle g^{(n-1)}, hF(Y^{(n)}, \bar{Y}^{(n)})\right\rangle\right\rangle \leq 0.
\end{aligned}
$$

Considering (2.2) and $k \leq 1$, we have

$$
\begin{aligned}
\left\|g^{(n)}\right\|_G^2 &\leq \left\|g^{(n-1)}\right\|_G^2 + 2hd\beta_0 + 2\beta_1 h\left\|Y^{(n)}\right\|_D^2 + 2h\beta_2\left\|\bar{Y}^{(n)}\right\|_D^2 - 2l\left\|G^{(n)}\right\|_D^2, \\
&\leq \left\|g^{(n-1)}\right\|_G^2 + 2hd\beta_0 + 2d\beta_1 h\left\|Y^{(n)}\right\|^2 + 2hd\beta_2\left\|\bar{Y}^{(n)}\right\|^2 \\
&\quad - 2ld_{\min}\left(\left\|Y^{(n)}\right\|^2 + \|N\|^2\left\|\bar{Y}^{(n)}\right\|^2 - 2\|N\|\left\langle Y^{(n)}, \bar{Y}^{(n)}\right\rangle\right)
\end{aligned}
$$

Using Cauchy-Schwarz inequality, we have

$$
\begin{aligned}
&\left\|g^{(n)}\right\|_G^2 \leq \left\|g^{(n-1)}\right\|_G^2 + 2hd\beta_0 + 2\beta_1 dh\left\|Y^{(n)}\right\|^2 + 2hd\beta_2\left\|\bar{Y}^{(n)}\right\|^2 \\
&- 2ld_{\min}(\left\|Y^{(n)}\right\|^2 + \|N\|^2\left\|\bar{Y}^{(n)}\right\|^2) + 2ld_{\min}\|N\|(\left\|Y^{(n)}\right\|^2 + \left\|\bar{Y}^{(n)}\right\|^2) \\
&\leq \left\|g^{(n-1)}\right\|_G^2 + 2hd\beta_0 + [2\beta_1 dh - 2ld_{\min}(1 - \|N\|)]\left\|Y^{(n)}\right\|^2 + [2hd\beta_2 + 2ld_{\min}\|N\|(1 - \|N\|)]\left\|\bar{Y}^{(n)}\right\|^2
\end{aligned}
$$

$$(3.3)$$

Where

$$
d_{\min} = \min_{1 \leq j \leq s} d_j, \quad d = \sum_{j=1}^{s} d_j \tag{3.4}
$$

By induction, we can easily obtain

$$
\begin{aligned}
\left\|g^{(n)}\right\|_G^2 &\leq \left\|g^{(-1)}\right\|_G^2 + 2(n+1)hd\beta_0 + [2d\beta_1 h - 2ld_{\min}(1 - \|N\|)]\sum_{j=0}^{n}\left\|Y^{(j)}\right\|^2 \\
&\quad + [2hd\beta_2 + 2ld_{\min}\|N\|(1 - \|N\|)]\sum_{j=0}^{n}\left\|\bar{Y}^{(j)}\right\|^2.
\end{aligned} \tag{3.5}
$$

When using (2.4a and 2.4b) and (3.1a and 3.1b) on substitution into (3.5) gives

$$\left\| g^{(n)} \right\|_G^2 \le \left\| g^{(-1)} \right\|_G^2 + 2(n+1)hd\beta_0 + [2d\beta_1 h - 2ld_{\min}(1 - \|N\|)] \sum_{j=0}^{n} \left\| Y^{(j)} \right\|^2$$

$$+ [2hd\beta_2 + 2ld_{\min}\|N\|(1 - \|N\|)] \sum_{j=0}^{n} \left\| Y^{(j-m)} \right\|^2.$$

$$\le \left\| g^{(-1)} \right\|_G^2 + 2(n+1)hd\beta_0 + 2\left[d\beta_1 h + d\beta_2 h - ld_{\min}(1 - \|N\|)^2 \right] \sum_{j=0}^{n} \left\| Y^{(j)} \right\|^2$$

$$+ [2hd\beta_2 + 2ld_{\min}\|N\|(1 - \|N\|)]m \max_{-m \le i \le -1} \left\| Y^{(i)} \right\|^2$$

$$(3.6)$$

where we have used that $\beta_2 \ge 0$ and $\tau = mh$. Let λ_1 denote the maximum eigenvalue of the matrix G,

$$a_1 = \max_{0 \le i \le r-1} \left\| y_i \right\|^2, \quad a_2 = \max_{-m \le i \le -1} \left\| Y^{(i)} \right\|^2,$$
$$R_1 = \max(a_1, a_2), \quad \mu = ld_{\min}(1 - \|N\|)^2 - d(\beta_1 h + \beta_2 h).$$

Then, we have $\mu > 0$ and

$$\left\| g^{(n)} \right\|_G^2 + 2\mu \sum_{j=0}^{n} \left\| Y^{(j)} \right\|_D^2 \le \left\| g^{(-1)} \right\|_G^2 + 2(n+1)hd\beta_0 + [2hd\beta_2 + 2ld_{\min}\|N\|(1 - \|N\|)]m \max_{-m \le i \le -1} \left\| Y^{(i)} \right\|^2$$
$$\le r\lambda_1(1 + \|N\|)a_1 + [2hd\beta_2 + 2ld_{\min}\|N\|(1 - \|N\|)]ma_2 + 2(n+1)hd\beta_0$$
$$\le ([2hd\beta_2 + 2ld_{\min}\|N\|(1 - \|N\|))m + r\lambda_1(1 + \|N\|))R_1 + 2(n+1)hd\beta_0$$

$$(3.7)$$

When $\beta_0 = 0$, it follows from (3.7) and $\mu > 0$ that

$$\lim_{n \to \infty} \left\| Y^{(n)} \right\| = 0,$$

which shows that for any $\varepsilon > 0$, there exists $n_0(R_1, \varepsilon) > 0$, such that

$$\left\| Y_j^{(n)} \right\| \le \varepsilon, \left\| \bar{Y}_j^{(n)} \right\| \le \varepsilon \, j = 1, 2, \cdots, s, \, n \ge n_0. \tag{3.8}$$

Hence, (2.3) implies that

$$\left\| \sigma_i(E)(y_n - N\bar{y}_n) \right\| = \left\| \sum_{j=1}^{r} a_{ij}(y_{n+j-1} - N\bar{y}_{n+j-1}) \right\| \le hL \sum_{j=1}^{s} |b_{ij}| + \varepsilon, i = 1, 2, \cdots, s, \, n \ge n_0$$

$$(3.9a)$$

$$\left\| \sigma_0(E)(y_n - N\bar{y}_n) \right\| = \left\| y_{n+r} - N\bar{y}_{n+r} - \sum_{j=1}^{r} \theta_j(y_{n+j-1} - N\bar{y}_{n+j-1}) \right\| \le hL \sum_{j=1}^{s} |\gamma_j|, n \ge n_0 \quad (3.9b)$$

where

$$L = \sup_{\substack{\|u\| \le \varepsilon, \\ \|v\| \le \varepsilon}} \|f(u,v)\|, \, u, v \in X.$$

From Lemma 3.2 it follows that there exist real constants v_i, $i = 1, 2, \cdots, s$, such that $\sigma_0(x)$ and $\sum_{i=1}^{s} v_i \sigma_i(x)$ have no common divisor. Therefore,

$$\left\| \sum_{i=1}^{s} v_i \sigma_i(E)(y_n - N\bar{y}_n) \right\| \le \sum_{i=1}^{s} |v_i| \left[hL \sum_{j=1}^{s} |b_{ij}| + \varepsilon \right], \, n \ge n_0,$$

which further gives

$$\left\| \left[\sigma_0(E) - \sum_{i=1}^{s} v_i \sigma_i(E) \right] (y_n - N\bar{y}_n) \right\| \le hL \sum_{j=1}^{s} |\gamma_j| + \sum_{i=1}^{s} |v_i| \left[hL \sum_{j=1}^{s} |b_{ij}| + \varepsilon \right], \, n \ge n_0.$$

$$(3.10)$$

Since $\sigma_0(x)$ and $\sigma_0(x) - \sum_{i=1}^{s} v_i \sigma_i(x)$ are coprime, and both are of degree r. Hence,

$$\left\{ x^i \sigma_0(x), x^i \left[\sigma_0(x) - \sum_{i=1}^{s} v_i \sigma_i(x) \right] : i = 0, 1, \cdots, r - 1. \right\},$$

form a basis for P^{2r-1}. Considering (3.9a and 3.9b), (3.10) and Lemma 3.1, we have

$$\|y_n - N\bar{y}_n\| \le \chi \left[hL \sum_{j=1}^{s} |\gamma_j| + \sum_{i=1}^{s} |v_i| \left(hL \sum_{j=1}^{s} |b_{ij}| + \varepsilon \right) \right], \text{ for } n \ge n_0, \text{ for } \beta_0 = 0$$

Therefore,

$$\|y_n\| \le \|N\| \|\bar{y}_n\| + \chi \left[hL \sum_{j=1}^{s} |\gamma_j| + \sum_{i=1}^{s} |v_i| \left(hL \sum_{j=1}^{s} |b_{ij}| + \varepsilon \right) \right]$$

$$\le \|N\| \|y_{n-m}\| + \chi \left[hL \sum_{j=1}^{s} |\gamma_j| + \sum_{i=1}^{s} |v_i| \left(hL \sum_{j=1}^{s} |b_{ij}| + \varepsilon \right) \right] \qquad (3.11)$$

$$\le \frac{\chi \left[hL \sum_{j=1}^{s} |\gamma_j| + \sum_{i=1}^{s} |v_i| \left(hL \sum_{j=1}^{s} |b_{ij}| + \varepsilon \right) \right]}{1 - \|N\|} + \max_{-\tau \le \xi \le 0} \|\varphi(\xi)\|$$

where $n \ge n_0'$, $n_0' = m + n_0$.

When $\beta_0 > 0$, let us take

$$n = 2(m+r)q - 1,$$
$$q = \left\lfloor \frac{([2hd\beta_2 + 2ld_{\min}\|N\|(1 - \|N\|)]m + r\lambda_1(1 + \|N\|))R_1}{4(m+r)hd\beta_0} \right\rfloor + 1,$$

where the notation $\lfloor x \rfloor$ means the maximum integer no greater than x, then

$$([2hd\beta_2 + 2ld_{\min}\|N\|(1 - \|N\|)]m + r\lambda_1(1 + \|N\|))R_1 \leq 4(m+r)qhd\beta_0.$$

It follows from (3.7) that

$$\mu \sum_{j=0}^{2(m+r)q-1} \left\| Y^{(i)} \right\|^2 \leq 4(m+r)qhd\beta_0,$$

which gives

$$\sum_{i=0}^{2q-1} \sum_{j=(m+r)i}^{(m+r)(i+1)-1} \left\| Y^{(i)} \right\|^2 \leq 4(m+r)\frac{qhd\beta_0}{\mu}.$$

Hence, there exists an integer $c \in [q, 2q - 1]$ such that

$$\sum_{j=(m+r)c}^{(m+r)(c+1)-1} \left\| Y^{(i)} \right\|^2 \leq 4(m+r)\frac{hd\beta_0}{\mu}. \tag{3.12}$$

Let $p = (m+r)c + m$, then for all $j \in [p - m, p + r - 1]$, we have

$$\left\| Y^{(i)} \right\|^2 \leq a_2', \tag{3.13}$$

where

$$a_2' = 4(m+r)\frac{hd\beta_0}{\mu}.$$

Therefore, by (2.1) and (2.3), for all $n \in [p, p + r - 1]$,

$$\|\sigma_i(E)(y_n - N\bar{y}_n)\| = \left\| \sum_{j=1}^{r} a_{ij}(y_{n+j-1} - N\bar{y}_{n+j-1}) \right\| \leq hL_1 \sum_{j=1}^{s} |b_{ij}| + \sqrt{\frac{a_2'}{d_i}}, i = 1, 2, \cdots, s. \tag{3.14a}$$

$$\|\sigma_0(E)(y_n - N\bar{y}_n)\| = \left\| y_{n+r} - N\bar{y}_{n+r} - \sum_{j=1}^{r} \theta_j(y_{n+j-1} - N\bar{y}_{n+j-1}) \right\| \leq hL_1 \sum_{j=1}^{s} |\gamma_j|. \tag{3.14b}$$

where

$$L_1 = \sup_{\substack{\|y\| \le w \\ \|z\| \le w}} \|f(y,z)\|, y, z \in X,$$

where $w = \sqrt{a_2'/d_{min}}$.

Therefore,

$$\left\| \left[\sigma_0(E) - \sum_{i=1}^{s} v_i \sigma_i(E) \right] (y_n - N\bar{y}_n) \right\| \le hL_1 \sum_{j=1}^{s} |\gamma_j| + \sum_{i=1}^{s} |v_i| \left[hL_1 \sum_{j=1}^{s} |b_{ij}| + \sqrt{\frac{a_2'}{d_i}} \right],$$

(3.15)

with $n \in [p, p+r-1]$.

Considering (3.14a and 3.14b), (3.15), Lemmas 3.1 and 3.2, similar to (3.11), we have

$$\|y_n - N\bar{y}_n\|^2 \le a_1', n \in [p, p+r-1],$$

(3.16)

where

$$a_1' = \chi^2 \left[hL_1 \sum_{j=1}^{s} |\gamma_j| + \sum_{i=1}^{s} |v_i| \left(hL_1 \sum_{j=1}^{s} |b_{ij}| + \sqrt{\frac{a_2'}{d_i}} \right) \right]^2.$$

Let

$$R_2 = \max(a_1', a_2').$$

A repetition of the above analysis implies that there exists a p',

$$p' \in [p + (m+r)q' + m, p + (2q'-1)(m+r) + m]$$

$$q' = \left\lfloor \frac{([2hd\beta_2 + 2ld_{min}\|N\|(1-\|N\|)]m + r\lambda_1(1+\|N\|))R_2}{4(m+r)hd\beta_0} \right\rfloor + 1,$$

such that

$$\left\| Y^{(n)} \right\|^2 \le a_2', n \in [p'-m, p'+r-1],$$

(3.17)

$$\|y_n - N\bar{y}_n\|^2 \le a_1', n \in [p', p'+r-1].$$

(3.18)

Similarly, to (3.3), (3.5) and (3.7), for $n \in [p, p']$, we can obtain

$$\left\|g^{(n-1)}\right\|_G^2 \leq \left\|g^{(p-1)}\right\|_G^2 + 2(n-p)hd\beta_0 + [2hd\beta_2 + 2ld_{\min}\|N\|(1-\|N\|)]m \max_{p-m \leq i \leq p-1} \left\|Y^{(i)}\right\|^2$$
$$\leq 2([2hd\beta_2 + 2ld_{\min}\|N\|(1-\|N\|)]m + r\lambda_1(1+\|N\|))R_2 + 2(2m+r)hd\beta_0.$$
(3.19)

Similar to (3.11), we can obtain

$$\|y_n\| \leq \|N\|\|\bar{y}_n\| + \sqrt{2([2hd\beta_2 + 2ld_{\min}\|N\|(1-\|N\|)]m + r\lambda_1(1+\|N\|))R_2 + 2(2m+r)hd\beta_0}$$
$$\leq \|N\|\|y_{n-m}\| + \sqrt{2([2hd\beta_2 + 2ld_{\min}\|N\|(1-\|N\|)]m + r\lambda_1(1+\|N\|))R_2 + 2(2m+r)hd\beta_0}$$

Hence, by induction, we have

$$\|y_n\| \leq \frac{\sqrt{2([2hd\beta_2 + 2ld_{\min}\|N\|(1-\|N\|)]m + r\lambda_1(1+\|N\|))R_2 + 2(2m+r)hd\beta_0}}{1 - \|N\|} + \max_{-\tau \leq \xi \leq 0}\|\varphi(\xi)\|$$
(3.20)

for $n \geq \frac{([2hd\beta_2 + 2ld_{\min}\|N\|(1-\|N\|)]m + r\lambda_1(1+\|N\|))R_2}{2hd\beta_0} + 2m + r$ and $\beta_0 > 0$.

A combination of (3.11) and (3.20) shows that the method is finite-dimensionally $D(l)$-dissipative.

Theorem 3.4. Assume that a step-irreducible multistep Runge–Kutta method (2.3) is (k,l)-algebraically stable, $D > 0$, $l < 0$ and $k \leq 1$, the problem (2.1) satisfies (2.2) with $d(\beta_1 h + \beta_2 h) < ld(1 - \|N\|)^2$. Then the method (2.3) with (2.2) and (2.4a and 2.4b) is finite-dimensionally $D(l)$-dissipative.

Proof. In the proof of Theorem 3.3, change all $d_{\min}$ into d, we can get the proof of Theorem 3.4.

Theorem 3.5. Assume that a method (2.3) is irreducible and algebraically stable, the problem (2.1) satisfies (2.2) with $d(\beta_1 h + \beta_2 h) < 0$. Then, the method (2.3) with (2.2) and (2.4a and 2.4b) is finite-dimensionally D-dissipative.

Proof. As in [18], we can prove that, if a stage-irreducible method (2.3) is algebraically stable for the matrices G and D, then $D > 0$, therefore, use the proof of Theorem 3.3 for $k = 1, l = 0$, we prove this theorem.

Remark 3.6. Specializing above results for the case that $N = 0$ and $\beta_2 = 0$, we can obtain the finite-dimensional dissipativity of algebraically stable multistep Runge–Kutta methods for dynamical systems without delays. If it is further assumed that $r = 1$, then the corresponding result is just that presented by Humphries and Stuart [4].

4   Infinite-Dimensional Numerical Dissipativity

In this section, we further discuss the dissipativity of (k,l)-algebraically stable multistep Runge–Kutta methods in infinite-dimensional spaces. Here we assume that H is an infinite-dimensional complex Hilbert space instead of $H = X = C^N$.

Theorem 4.1. Assume that a multistep Runge–Kutta method (2.3) is (k, l)-algebraically stable, $D > 0$, $l > 0$ and $k < 1$, the problem (2.1) satisfies (2.2) with $d(\beta_1 h + \beta_2 h) < l d_{\min}(1 - \|N\|)^2$ where $d_{\min} = \min_{1 \le j \le s} d_j$. Then the method (2.3) with (2.2) and (2.4a and 2.4b) is infinite-dimensionally $D(l)$-dissipative.

Proof. Let

$$\mu = d(\beta_1 h + \beta_2 h) - l d_{\min}(1 - \|N\|)^2, \tag{4.1}$$

when $d(\beta_1 h + \beta_2 h) < l d_{\min}(1 - \|N\|)^2$, we have $\mu < 0$. Using (3.1a and 3.1b) and (3.2), we deduce that

$$\|g^{(n)}\|_G^2 \le k \|g^{(n-1)}\|_G^2 + 2hd\beta_0 + [2d\beta_1 h - 2l d_{\min}(1 - \|N\|)] \|Y^{(n)}\|^2 + [2hd\beta_2 + 2l d_{\min} \|N\|(1 - \|N\|)] \|\bar{Y}^{(n)}\|^2. \tag{4.2}$$

Iterating (4.2) and considering (2.4a and 2.4b) and (3.1a and 3.1b), we have

$$\|g^{(n)}\|_G^2 \le k^{n+1} \|g^{(-1)}\|_G^2 + 2\sum_{i=0}^{n} k^{n-i} \left[hd\beta_0 + [d\beta_1 h - l d_{\min}(1 - \|N\|)] \|Y^{(i)}\|^2 \right]$$

$$| 2\sum_{i=0}^{n} k^{n-i} [dh\beta_2 + l d_{\min} \|N\|(1 - \|N\|)] \|\bar{Y}^{(i)}\|^2$$

$$\le k^{n+1} \|g^{(-1)}\|_G^2 + 2\sum_{i=0}^{n} k^{n-i} \left[hd\beta_0 + [d\beta_1 h - l d_{\min}(1 - \|N\|)] \|Y^{(i)}\|^2 \right]$$

$$+ 2\sum_{i=0}^{n} k^{n-i} [dh\beta_2 + l d_{\min} \|N\|(1 - \|N\|)] \|Y^{(i-m)}\|^2$$

$$\le k^{n+1} \|g^{(-1)}\|_G^2 + 2\sum_{i=0}^{n} k^{n-i} \left(hd\beta_0 + \mu \|Y^{(i)}\|^2 \right)$$

$$+ [2dh\beta_2 + 2l d_{\min} \|N\|(1 - \|N\|)] k^{n-m} m \max_{-m \le i \le -1} \|Y^{(i)}\|^2$$

$$\le k^{n+1} \|g^{(-1)}\|_G^2 + \frac{2hd\beta_0}{1 - k} + 2\sum_{i=0}^{n} k^{n-i} \mu \|Y^{(i)}\|^2$$

$$+ [2hd\beta_2 + 2l d_{\min} \|N\|(1 - \|N\|)] k^{n-m} m \max_{-m \le i \le -1} \|Y^{(i)}\|^2$$

$$\le k^{n+1} \|g^{(-1)}\|_G^2 + \frac{2hd\beta_0}{1 - k} + [2hd\beta_2 + 2l d_{\min} \|N\|(1 - \|N\|)] k^{n-m} m \max_{-m \le i \le -1} \|Y^{(i)}\|^2 \tag{4.3}$$

where we have used that $\mu < 0$.

Thus, for any $\varepsilon > 0$ there exists $n_0 > 0$, such that

$$\|g^{(n)}\|_G \le \sqrt{\frac{2hd\beta_0}{1 - k}} + \varepsilon, n \ge n_0. \tag{4.4}$$

Similar to (3.11), we can obtain

$$\|y_n\| \le \|N\|\|\bar{y}_n\| + \sqrt{\frac{2hd\beta_0}{1-k}} + \varepsilon, \quad n \ge n_0,$$

$$\le \|N\|\|y_{n-m}\| + \sqrt{\frac{2hd\beta_0}{1-k}} + \varepsilon \le \frac{\sqrt{\frac{2hd\beta_0}{1-k}} + \varepsilon}{1 - \|N\|} + \max_{-\tau \le \xi \le 0} |\varphi(\xi)|, \quad n \ge n_0'.$$

with $n_0' = m + n_0$, which shows the method (2.3) with (2.2) and (2.4a and 2.4b) is infinite-dimensionally $D(l)$-dissipative.

Theorem 4.2. Assume that a multistep Runge–Kutta method (2.3) is (k, l)-algebraically stable, $D > 0$, $l < 0$ and $k < 1$, the problem (2.1) satisfies (2.2) with $d(\beta_1 h + \beta_2 h) < ld(1 - \|N\|)^2$ where $d = \sum_{j=1}^{s} d_j$. Then the method (2.3) with (2.2) and (2.4a and 2.4b) is infinite-dimensionally $D(l)$-dissipative.

Proof. In the proof of Theorem 4.1, change all $d_{\min}$ into d, we can get the proof of Theorem 4.2.

Theorem 4.3. Assume that there exist nonnegative constants k_1, k_2, $k_1 < k_2$, such that the stage-irreducible method (2.3) is $((1 - k_1 l)/(1 - k_2 l), l)$-algebraically stable for every $l < 0$. Then, the method with (2.6) is infinite-dimensionally D-dissipative.

5 Comparison with Existing Results

When $N = 0$ the problem (2.1) degenerates into an IVP of DDEs. Therefore, the results of Theorems 3.3, 3.4, 4.1 and 4.2 in this paper partially cover the numerical dissipativity of multistep Runge–Kutta for DDEs which is given by Huang in [11].

6 Numerical Examples

Example 6.1. Consider the following problem

$$\begin{cases} \frac{d}{dt}(y_1(t) - 0.1y_2(t-1)) = -6y_1(t) + \sin(y_2(t))\sin(y_1(t-1)), t \ge 0, \\ \frac{d}{dt}(y_2(t) - 0.2y_1(t-1)) = -5.2y_2(t) - \cos(y_1(t))\cos(y_2(t-1)), t \ge 0, \\ y_1(t) = \sin(t), y_2(t) = \cos(t), -1 \le t \le 0. \end{cases} \quad (6.1)$$

For this system $N = \begin{bmatrix} 0 & 0.1 \\ 0.2 & 0 \end{bmatrix}$, with a standard inner product and the corresponding norm we have $\|N\| = 0.04$, $\alpha = 2 \times 1.15$, $\beta_1 = -4.13$, $\beta_2 = 0.63$, and $\tau = 1$.

Consider the two-step two-stage method

$$\begin{cases} Y_1^{(n)} - 0.1\bar{Y}_2^{(n)} = \frac{15}{14}hf(Y_1^{(n)}, \bar{Y}_1^{(n)}) - \frac{5}{14}hf(Y_2^{(n)}, \bar{Y}_2^{(n)}) + \frac{5}{14}y_n + \frac{9}{14}y_{n+1} \\ Y_2^{(n)} - 0.2\bar{Y}_1^{(n)} = -\frac{1}{2}hf(Y_1^{(n)}, \bar{Y}_1^{(n)}) + \frac{3}{2}hf(Y_2^{(n)}, \bar{Y}_2^{(n)}) - \frac{1}{2}y_n + \frac{3}{2}y_{n+1} \\ y_{n+2} = \frac{21}{20}hf(Y_1^{(n)}, \bar{Y}_1^{(n)}) + \frac{3}{20}hf(Y_2^{(n)}, \bar{Y}_2^{(n)}) + \frac{1}{10}y_n + \frac{9}{10}y_{n+1} \end{cases} \quad (6.2)$$

method (6.2) is algebraically stable for

$$G = \begin{bmatrix} \frac{1}{10} & 0 \\ 0 & 1 \end{bmatrix}, D = \begin{bmatrix} \frac{21}{20} & 0 \\ 0 & \frac{3}{20} \end{bmatrix}.$$

Therefore, method (6.2) for (6.1) is finite-dimensionally D-dissipative.

Acknowledgments. This work was supported by Natural Science Foundation of China under Grant 11901173 and Heilongjiang Natural Science Foundation (LH2019A030) and the assisted project by Heilong Jiang Postdoctoral Funds for scientific research initiation and the Research Foundation of Heilongjiang Educational Committee (12531546).

Declare. The authors declare that there is no conflict of interests regarding the publication of this article.

References

1. Bocharov, G.A., Rihan, F.A.: Numerical modelling in biosciences with delay differential equations. J. Comput. Appl. Math. **125**, 183–199 (2000)
2. Xiao, A.G.: On the solvability of general linear methods for dissipative dynamical systems. J. Comput. Math. **18**, 633–638 (2000)
3. Humphries, A.R., Stuart, A.M.: Model problems in numerical stability theory for initial value problems. SIAM Rev. **36**, 226–257 (1994)
4. Humphries, A.R., Stuart, A.M.: Runge–Kutta methods for dissipative and gradient dynamical systems. SIAM J. Numer. Anal. **31**, 1452–1485 (1994)
5. Hill, A.T.: Global dissipativity for A-stable methods. SIAM J. Numer. Anal. **34**, 119–142 (1997)
6. Hill, A.T.: Dissipativity of Runge–Kutta methods in Hilbert spaces. BIT **37**, 37–42 (1997)
7. Xiao, A.G.: Dissipativity of general linear methods for dissipative dynamical systems in Hilbert spaces. Math. Numer. Sin. **22**, 429–436 (2000). (in Chinese)
8. Huang, C.M.: Dissipativity of Runge–Kutta methods for dissipative systems with delays. IMA J. Numer. Anal. **20**, 153–166 (2000)
9. Huang, C.M., Chen, G.: Dissipativity of linear θ-methods for delay dynamical systems. Math. Numer. Sin. **22**, 501–506 (2000). (in Chinese)
10. Huang, C.M.: Dissipativity of one-leg methods for dissipative systems with delays. Appl. Numer. Math. **35**, 11–22 (2000)
11. Huang, C.M.: Dissipativity of multistep Runge–Kutta methods for dynamical systems with delays. Math. Comput. Model. **40**, 1285–1296 (2004)
12. Tian, H.J.: Numerical and analytic dissipativity of the θ-method for delay differential equations with a bounded lag. Int. J. Bifurcat. Chaos **14**, 1839–1845 (2004)
13. Wen, L.P.: Numerical stability analysis for nonlinear Volterra functional differential equations in abstract spaces. Ph.D. thesis, Xiangtan University (2005). (in Chinese)

14. Gan, S.Q.: Dissipativity of linear θ-methods for integro-differential equations. Comput. Math Appl. **52**, 449–458 (2006)
15. Gan, S.Q.: Dissipativity of θ-methods for nonlinear Volterra delay-integro-differential equations. J. Comput. Appl. Math. **206**, 898–907 (2007)
16. Gan, S.Q.: Exact and discretized dissipativity of the pantograph equation. J. Comput. Math. **25**(1), 81–88 (2007)
17. Burrage, K., Butcher, J.C.: Nonlinear stability of a general class of differential equation methods. BIT **20**, 185–203 (1980)
18. Li, S.F.: Theory of Computational Methods for Stiff Differential Equation. Huan Science and Technology Publisher, Changsha (1997)
19. Cheng, Z., Huang, C.M.: Dissipativity for nonlinear neutral delay differential equations. J. Syst. Simul. **19**(14), 3184–3187 (2007)
20. Wen, L.P., Wang, W.S., Yu, Y.X.: Dissipativity of θ-methods for a class of nonlinear neutral delay differential equations. Appl. Math. Comput. **202**(2), 780–786 (2008)
21. Wang, W.S., Li, S.F.: Dissipativity of Runge–Kutta methods for neutral delay differential equations with piecewise constant delay. Appl. Math. Lett. **21**(9), 983–991 (2008)

Monophonic Distance Based Indices of Graphs

V. Kaladevi and G. Kavitha$^{(\boxtimes)}$ ⓘ

Hindustan Institute of Technology and Science, Chennai 603 103, Tamil Nadu,
India
kavithateam@gmail.com

Abstract. The monophonic distance matrix of a simple connected graph G(p, q) is a square matrix of order p whose entries are the monophonic distances. In this paper the monophonic indices of cycle graph, wheel graph, ladder graph and circular ladder graph are determined by forming the monophonic polynomial from the corresponding monophonic distance matrices of graphs.

Keywords: Monophonic distance · Monophonic polynomial · Monophonic index

1 Introduction

Let $G = (V(G), E(G))$ be a simple finite undirected connected graph. For any two vertices $u, v \in V(G)$, the distance $d(u, v)$ from u to v is defined as the length of a shortest u-v path in G. The detour distance $D(u, v)$ from u to v is defined as the length of a longest path between two vertices in G. A chord of a path $v_1, v_2, \ldots, v_n$ in a connected graph G is an edge $v_i v_j$ with $j \geq i + 2$. A u-v path P is called a monophonic path if it is a chordless path. The length of a longest u-v monophonic path is called the monophonic distance from u to v and is denoted by $d_m(u, v)$. The monophonic distance is different from both the shortest distance and the longest distance. The wiener index and the detour index of a graph are respectively defined by $W(G) = \frac{1}{2} \sum\limits_{u,v \in V(G)} d(u, v)$ and $D(G) = \frac{1}{2} \sum\limits_{u,v \in V(G)} D(u, v)$.

The wiener index and detour index of a graph are studied by several authors in the literature [1–11]. Similar to $W(G)$ and $D(G)$, the monophonic index denoted by $MP(G)$ is defined. For the complete graph K_n, $MP(K_n) = W(K_n) = \frac{n(n-1)}{2}$ and for any tree T, $W(T) = D(T) = MP(T)$. In this paper $MP(G)$ is derived for cycle graph C_n, wheel graph W_n and for Cartesian product graphs $P_2 \times P_n$, $P_2 \times C_n$ (when n is odd) and $C_m \circ K_n$.

2 Monophonic Index of C_n and W_n

Definition 2.1. The Monophonic Distance matrix of a graph is a square matrix of order of G whose entries are the monophonic distance $d_m(u, v)$ between every pair of vertices u and v and denoted by MPDM(G)

R. Silhavy et al. (Eds.): CoMeSySo 2019, AISC 1047, pp. 243–255, 2019.
https://doi.org/10.1007/978-3-030-31362-3_24

Definition 2.2. The Monophonic Polynomial of G denoted by $MPP(G : x)$ and defined by $MPP(G : x) = \frac{1}{2} \sum\limits_{u,v \in V(G)} x^{d_m(u,v)}$.

Definition 2.3. The Monophonic Index of G is the first order differentiation of $MPP(G : x)$ at x = 1. That is, $MP(G) = \frac{d}{dx}[MPP(G : x)]_{x=1}$.

Theorem 2.4. Let $G = C_n$ be the cycle graph of order n. Then if n is odd and $n \geq 5$, $MPP(C_n : x) = nx + n(x^{\left(\frac{n+1}{2}\right)} + x^{\left(\frac{n+3}{2}\right)} + x^{\left(\frac{n+5}{2}\right)} + \ldots + x^{(n-2)})$ $MP(C_n = n + \left[\frac{3n(n-3)(n-1)}{8}\right]$ and if n is even and $n \geq 6$,

$$MPP(C_n : x) = nx + \frac{n}{2}x^{\left(\frac{n}{2}\right)} + nx^{\left(\frac{n+2}{2}\right)} + nx^{\left(\frac{n+4}{2}\right)} + \ldots + nx^{(n-2)}$$

$$MP(C_n) = \frac{2(4n + n^2) + (n^2 - 4n)(3n - 2)}{8}$$

Proof: Let $V(C_n) = \{v_1, v_2, \ldots, v_n\}$ and $E(C_n) = \{v_i v_{i+1} : 1 \leq i \leq n - 1\} \cup \{v_n v_1\}$ be the vertex set and edge set of G respectively. The monophonic index of C_n can be found by forming the Monophonic Distance matrix MPDM(G). The proof of the theorem is divided into two cases.

Case i: When n is odd.
Let $n = 5$. The graph C_5 and $MPDM(C_5)$ are given in Figs. 1 and 2 respectively

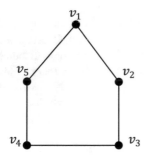

	v_1	v_2	v_3	v_4	v_5
v_1	0	1	3	3	1
v_2	1	0	1	3	3
v_3	3	1	0	1	3
v_4	3	3	1	0	1
v_5	1	3	3	1	0

Fig. 1. Cycle C_5 **Fig. 2.** MPDM (C_5)

$$MPP(C_5 : x) = 5x + 5x^3 \text{ and } MP(C_5) = 5 + 15 = 20$$

Let $n = 7$. Then the graph C_7 and $MPDM(C_7)$ are respectively given in Figs. 3 and 4.

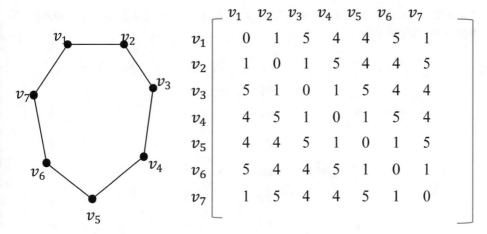

	v_1	v_2	v_3	v_4	v_5	v_6	v_7
v_1	0	1	5	4	4	5	1
v_2	1	0	1	5	4	4	5
v_3	5	1	0	1	5	4	4
v_4	4	5	1	0	1	5	4
v_5	4	4	5	1	0	1	5
v_6	5	4	4	5	1	0	1
v_7	1	5	4	4	5	1	0

Fig. 3. Cycle C_7 **Fig. 4.** MPDM (C_7)

$$MPP(C_7 : x)\ 7x + 7x^4 + 7x^5 \text{ and } MP(C_7) = 7 + 28 + 35 = 70$$

Similarly if n is increased successively by 2, the corresponding monophonic polynomial are given by

$$MPP(C_9 : x) = 9x + 9x^5 + 9x^6 + 9x^7$$
$$MPP(C_{11} : x) = 11x + 11x^6 + 11x^7 + 11x^8 + 11x^9$$
$$MPP(C_{13} : x) = 13x + 13x^7 + 13x^8 + 13x^9 + 13x^{10} + 13x^{11}$$

Hence in general if n is odd and $n \geq 5$, then the monophonic polynomial of C_n is

$$MPP(C_n : x) = nx + n(x^{(\frac{n+1}{2})} + x^{(\frac{n+3}{2})} + x^{(\frac{n+5}{2})} + \ldots + x^{(n-2)})$$

$$MP(C_n) = n + n[\frac{n+1}{2} + \frac{n+3}{2} + \frac{n+5}{2} + \ldots + (n-2)]$$

$$= n + n\left[\left(\frac{n-3}{2}\right)\left(\frac{1}{2}\right)\left(\frac{2(n+1)}{2} + \left(\frac{n-3}{2} - 1\right)1\right)\right]$$

$$= n + n\left[\frac{n-3}{4}\left((n+1) + \frac{n-5}{2}\right)\right]$$

$$= n + n\left[\frac{n-3}{4}\left(\frac{2n+2+n-5}{2}\right)\right]$$

$$= n + n\left[\frac{n-3}{4} \times \frac{3n-3}{2}\right]$$

$$= n + n\left[\frac{3(n-3)(n-1)}{8}\right]$$

$$= n + [\frac{3n(n-3)(n-1)}{9}]$$

Case ii: When n is even. Let $n = 6$. The graph C_6 and MPDM (C_6) is given in Figs. 5 and 6 are respectively

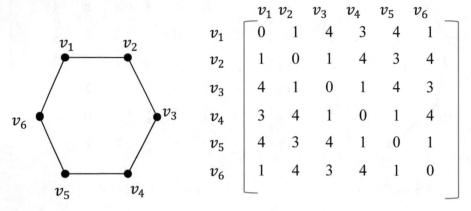

$$\begin{array}{c@{\ }cccccc}
 & v_1 & v_2 & v_3 & v_4 & v_5 & v_6 \\
v_1 & 0 & 1 & 4 & 3 & 4 & 1 \\
v_2 & 1 & 0 & 1 & 4 & 3 & 4 \\
v_3 & 4 & 1 & 0 & 1 & 4 & 3 \\
v_4 & 3 & 4 & 1 & 0 & 1 & 4 \\
v_5 & 4 & 3 & 4 & 1 & 0 & 1 \\
v_6 & 1 & 4 & 3 & 4 & 1 & 0
\end{array}$$

Fig. 5. Cycle C_6 **Fig. 6.** MPDM (C_6)

$$MPP(C_6 : x) = 6x + 3x^3 + 6x^4$$

When $n = 8$, the graph C_8 and $MPDM(C_8)$ is given in Figs. 7 and 8 are respectively

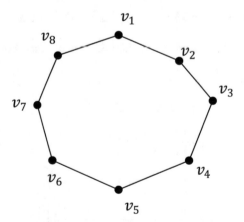

Fig. 7. Cycle C_8

$$
\begin{array}{c}
\begin{array}{cccccccc}
v_1 & v_2 & v_3 & v_4 & v_5 & v_6 & v_7 & v_8
\end{array} \\
\begin{array}{c}
v_1 \\ v_2 \\ v_3 \\ v_4 \\ v_5 \\ v_6 \\ v_7 \\ v_8
\end{array}
\left[
\begin{array}{cccccccc}
0 & 1 & 6 & 5 & 4 & 5 & 6 & 1 \\
1 & 0 & 1 & 6 & 5 & 4 & 5 & 6 \\
6 & 1 & 0 & 1 & 6 & 5 & 4 & 5 \\
5 & 6 & 1 & 0 & 1 & 6 & 5 & 4 \\
4 & 5 & 6 & 1 & 0 & 1 & 6 & 5 \\
5 & 4 & 5 & 6 & 1 & 0 & 1 & 6 \\
6 & 5 & 4 & 5 & 6 & 1 & 0 & 1 \\
1 & 6 & 5 & 4 & 5 & 6 & 1 & 0
\end{array}
\right]
\end{array}
$$

Fig. 8. MPDM (C_8)

$$MPP(C_8 : x) = 8x + 4x^4 + 8x^5 + 8x^6.$$

When n is increased successively by 2,

$$MPP(C_{10} : x) = 10x + 5x^5 + 10x^6 + 10x^7 + 10x^8$$

$$MPP(C_{12} : x) = 12x + 6x^6 + 12x^7 + 12x^8 + 12x^9 + 12x^{10}$$

Hence in general if n is even and $n \geq 6$

$$MPP(C_n : x) = nx + \frac{n}{2}x^{\left(\frac{n}{2}\right)} + nx^{\left(\frac{n+2}{2}\right)} + nx^{\left(\frac{n+4}{2}\right)} + \ldots + nx^{(n-2)}$$

$$MP(C_n) = n + \frac{n}{2}\left(\frac{n}{2}\right) + \frac{n(n+2)}{2} + \frac{n(n+4)}{2} + \ldots + (n-2)$$

$$= \left(n + \frac{n^2}{4}\right) + n\left(\frac{n+2}{2} + \frac{n+4}{2} + \ldots + n - 2\right)$$

$$= \left(n + \frac{n^2}{4}\right) + n\left[\left(\frac{n-4}{2}\right)\left(\frac{1}{2}\right)\left(\frac{2(n+2)}{2} + \left(\frac{n-4}{2} - 1\right)1\right)\right]$$

$$= \left(n + \frac{n^2}{4}\right) + n\left[\left(\frac{n-4}{4}\right)\left(n + 2 + \frac{n-4-2}{2}\right)\right]$$

$$= \left(\frac{4n + n^2}{4}\right) + \frac{n(n-4)}{4}\left(\frac{2n+4+n-6}{2}\right)$$

$$= \frac{2(4n + n^2) + (n^2 - 4n)(3n - 2)}{8}$$

$$= \frac{8n + 2n^2 + (3n^3 - 14n^2 + 8n)}{8}$$

$$= \frac{n(3n^2 - 12n + 16)}{8}$$

Theorem 2.2. Let $G = W_{n+1} = K_1 + C_n$ be the wheel graph. Then

$$MP(G) = \begin{cases} \frac{n}{8}(3n^2 - 12n + 17) & \text{if } n \text{ is even} \\ \frac{n}{8}(3n^2 - 12n + 24) & \text{f } n \text{ is odd} \end{cases}$$

Proof: Let, $V(G) = \{v_o, \ v_1, \ v_2, \ \ldots, \ v_n\}$ and $E(G) = \{v_o v_i : 1 \le i \le n\} \cup \{v_i v_{i+1} : 1 \le i \le n-1\} \cup \{v_n v_1\}$ with v_0 as the center vertex be the vertex set and edge set of G as given in Fig. 9 respectively. The monophonic index of G is defined by

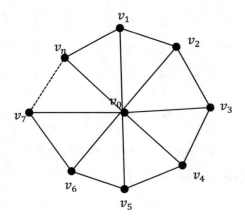

Fig. 9. Wheel Graph W_{n+1}

$$MP(G) = \sum_{0 \le i < j \le n} d(v_i, v_j)$$

If n is even, then

$$MP(G) = \sum_{j=1}^{n} d(v_0, v_j) + \sum_{1 \le i < j \le n} d(v_i, v_j)$$

$$= \sum_{j=1}^{n} d(v_0, v_j) + \sum_{i=1}^{n} \sum_{j=i}^{n} d(v_i, v_j)$$

$$= n + \left[n + \frac{n(n+1)}{2} + \frac{n(n+3)}{2} + \frac{n(n+5)}{2} + \ldots + n(n-3) + n(n-2) \right]$$

$$= \frac{n}{8}(3n^2 - 12n + 17)$$

If n is odd, then

$$MP(G) = \sum_{j=1}^{n} d(v_0, v_j) + \sum_{1 \le i < j \le n} d(v_i, v_j)$$

$$= n + \left[n + \binom{n}{2}\binom{n}{2} + n\left(\frac{n}{2} + 1\right) + n\left(\frac{n}{2} + 2\right) + \ldots + n(n-3) + n(n-2) \right]$$

$$= \frac{n(3n^2 - 12n + 24)}{8}$$

3 Monophonic Index of Cartesian Product of P_2 with P_n and C_n

Definition 3.1. The Cartesian product of the two graphs G_1 and G_2 denoted by $G = G_1 \times G_2$ is a graph [1] whose vertex set is $V(G_1) \times V(G_2)$ and any two vertices $(u_1, v_1), (u_2, v_2)$ in $V(G)$ are adjacent if and only if $u_1 = u_2$ and $v_1 v_2 \in E(G_2)$ or $v_1 = v_2$ and $u_1 u_2 \in E(G_1)$.

Theorem 3.2 Let $G = P_2 \times P_n$ be the Cartesian product of P_2 and P_n. Then

$$MPP(G) = (3n-2)x + (2n-2)x^2 + (2n-4)x^3 + (4n-10)x^4 + (4n-14)x^5$$
$$+ (4n-18)x^6 + (4n-22)x^7 + \ldots + 6x^n + 2x^{n+1})$$

$$MP(G) = \frac{1}{3}\left[2n^3 + 9n^2 - 20n + 12\right]$$

Proof: Let $V(P_2) = \{u_1, u_2\}$ and $V(P_n) = \{v_1, v_2, \ldots, v_n\}$ be the vertex set of P_2 and P_n respectively.

Then $V(P_2 \times P_n) = \{(u_1, v_i), (u_2, v_i) : 1 \le i \le n\}$

For simplicity let us denote $(u_1, v_i) = w_i$; $1 \le i \le n$ and $(u_2, v_i) = w_i$; $n+1 \le i \le 2n$. The edge set

$$E(P_2 \times P_n) = \{w_i w_{i+1} : 1 \le i \le n-1\}$$
$$\cup \{w_i w_{i+1} : n+1 \le i \le 2n-1\} \cup \{w_i w_{n+i} : 1 \le i \le n\}$$

Let $n = 3$. The graph $P_2 \times P_3$ and MPDM $(P_2 \times P_3)$ are given in Figs. 10 and 11 respectively.

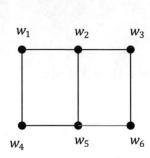

$$
\begin{array}{c|cccccc}
 & w_1 & w_2 & w_3 & w_4 & w_5 & w_6 \\
\hline
w_1 & 0 & 1 & 4 & 1 & 2 & 3 \\
w_2 & 1 & 0 & 1 & 2 & 1 & 2 \\
w_3 & 4 & 1 & 0 & 3 & 2 & 1 \\
w_4 & 1 & 2 & 3 & 0 & 1 & 4 \\
w_5 & 2 & 1 & 2 & 1 & 0 & 1 \\
w_6 & 3 & 2 & 1 & 4 & 1 & 0 \\
\end{array}
$$

Fig. 10. $P_2 \times P_3$ **Fig. 11.** MPDM $(P_2 \times P_3)$

$$MPP(P_2 \times P_3 : x) = 7x + 4x^2 + 2x^3 + 2x^4 \text{ and } MP(P_2 \times P_3) = 29.$$

When $n = 4$, the graph $P_2 \times P_4$ and MPDM $(P_2 \times P_4)$ are given in Figs. 12 and 13 are respectively.

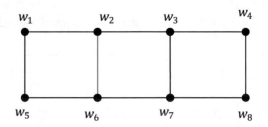

Fig. 12. $P_2 \times P_4$

$$
\begin{array}{c|cccccccc}
 & w_1 & w_2 & w_3 & w_4 & w_5 & w_6 & w_7 & w_8 \\
\hline
w_1 & 0 & 1 & 4 & 5 & 1 & 2 & 3 & 4 \\
w_2 & 1 & 0 & 1 & 4 & 2 & 1 & 2 & 3 \\
w_3 & 4 & 1 & 0 & 1 & 3 & 2 & 1 & 2 \\
w_4 & 5 & 4 & 1 & 0 & 4 & 3 & 2 & 1 \\
w_5 & 1 & 2 & 3 & 4 & 0 & 1 & 4 & 3 \\
w_6 & 2 & 1 & 2 & 3 & 1 & 0 & 1 & 4 \\
w_7 & 3 & 2 & 1 & 2 & 4 & 1 & 0 & 1 \\
w_8 & 4 & 3 & 2 & 1 & 3 & 4 & 1 & 0 \\
\end{array}
$$

Fig. 13. MPDM $(P_2 \times P_4)$

$$MPP(P_2 \times P_4 : x) = 10x + 6x^2 + 4x^3 + 6x^4 + 2x^5;$$

$$MP(P_2 \times P_4) = 68.$$

For $n = 5, 6, 7, \ldots$ The corresponding monophonic polynomial are calculated from MPDM as

$$MPP(P_2 \times P_5 : x) = 13x + 8x^2 + 6x^3 + 10x^4 + 6x^5 + 2x^6;$$

$$MP(P_2 \times P_5) = 129.$$

$$MPP(P_2 \times P_6 : x) = 16x + 10x^2 + 8x^3 + 14x^4 + 10x^5 + 6x^6 + 2x^7;$$

$$MP(P_2 \times P_6) = 216.$$

$$MPP(P_2 \times P_7 : x) = 19x + 12x^2 + 10x^3 + 18x^4 + 14x^5 + 10x^6 + 6x^7 + 2x^8;$$

$$MP(P_2 \times P_7) = 333. \text{ And so on.}$$

Hence in general for any $n \geq 3$,

$$MPP(P_2 \times P_n) = (3n - 2)x + (2n - 2)x^2 + (2n - 4)x^3 + (4n - 10)x^4$$
$$+ (4n - 14)x^5 + \ldots + 6x^n + 2x^{n+1}$$

$$MP(G) = \frac{1}{3}\left[2n^3 + 9n^2 - 20n + 12\right].$$

Theorem 3.3. Let $G = P_2 \times C_n$ be the cartesian product of P_2 and C_n. Then if n is odd, $n \geq 9$, $MP(G) = \frac{n(3n^2 + 6n - 27)}{2}$ and if n is even, $n \geq 10$, $MP(G) = \frac{n(9n^2 + 22n - 40)}{8}$.

Proof: Case (i) when is odd.

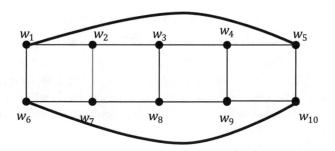

Fig. 14. $P_2 \times C_5$

For $n = 5$, as shown in Fig. 14

$$MPP(G : x) = 2(5x + 5x^5) + (5x + 10x^4 + 10x^5); \; MP(G) = 155$$

For $n = 7$,

$$MPP(G : x) = 2(7x + 7x^6 + 7x^7) + (7x + 14x^7 + 14x^8 + 14x^9);$$

$$MP(G) = 539.$$

For $n = 9$,

$$MPP(G : x) = 2(9x + 9x^7 + 9x^{10} + 9x^{11}) + (9x + 18x^8 + 18x^9 + 18x^{10} + 18x^{11})$$

$$MP(G) = 1215.$$

For $n = 11, 13, 15, 17, 19$, the respective monophonic indices of G are 2211, 3627, 5535, 8007, 11115. Hence for $n \geq 9$ and n is odd, $MP(G) = \frac{n(3n^2 + 6n - 27)}{2}$.

4   Monophonic Index of Corona Product of Graphs

The corona product of any two graphs G and H is defined by $G \circ H$ and is obtained by taking one copy of G and $|V(G)|$ copies of H by joining each vertex of the ith copy of H to the ith vertex of G, where $1 \leq i \leq |V(G)|$. In this section the monophonic polynomial of $C_m \circ K_n$ is studied.

Theorem 4.1. Let C_m and K_n be the cycle graph on m vertices and complete graph on n vertices respectively. Them if m is even, $m \geq 4$ and $n = 2, 3, 4, \ldots$

$$MPP(C_m \circ K_n) = MPP(C_m) + MPP(K_n) + (mnx + 2mnx^2)$$
$$+ mn(x^3 + x^4 + \ldots + x^{\frac{m+2}{2}}) + 2mn(x^5 + x^6 + x^7 + \ldots + x^{\frac{m+4}{2}})$$
$$+ 2mn(x^7 + x^8 + \ldots + x^{\frac{m+6}{2}}) + \ldots + 2mn(x^{m-1}) + \frac{mn^2}{2}x^{\frac{m+4}{2}}$$
$$+ mn^2(x^6 + x^7 + \ldots + x^{\frac{m+6}{2}}) + mn^2(x^8 + x^9 + \ldots + x^{\frac{m+8}{2}})$$
$$+ mn^2(x^9 + x^{10} + \ldots + x^{\frac{m+10}{2}}) + \ldots + mn^2 x^m$$

Proof: Let $V(C_m) = \{v_1, \; v_2, \; \ldots, \; v_m\}$ and $E(C_m) = \{v_i v_{i+1}; \; 1 \leq i \leq m+1\} \cup \{v_m v_1\}$ be the vertex set and edge set of C_m respectively with v_1 as start vertex and v_m as the end vertex. Let $G = C_m \circ K_n$ be the corona product of C_m and K_n with

$$V(G) = \{v_1, v_2, \ldots v_m\} \cup$$
$$\{u_1, u_2, u_n, u_{n+1}, u_{n+2}, \ldots, u_{2n}, u_{2n+1},$$
$$u_{2n+2}, \ldots, u_{3n}, \ldots, u_{mn+1}, u_{mn+2}, \ldots, u_{mn+n}\}$$

be the vertex set of G. Then the edge set of G is

$$E(G) = \{ v_1 u_i; \quad 1 \le i \le n \} \quad \cup \quad \{ v_2 u_i; \quad n+1 \le i \le 2n \}$$
$$\cup \{ v_m u_i; \quad (m-2)n+1 \le i \le (m-1)n \}$$
$$\cup \{ v_i v_{i+1}; \quad 1 \le i \le m-1 \} \quad \cup \quad \{ v_m v_1 \}$$

The corona product graph $C_6 \circ K_4$ is shown in Fig. 15.

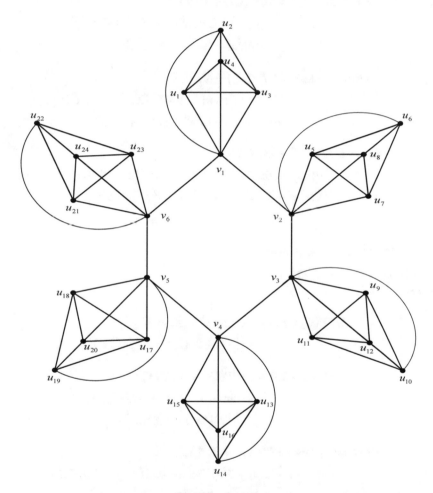

Fig. 15. $C_6 \circ K_4$

The monophonic distance between the vertices of $G = C_m \circ K_n$ can be obtained by forming the monophonic distance matrix of G when m is even and odd.

Case (i): When m is even.
For $n = 2, 3, 4, 5, 6, \ldots$

$$MPP(C_4 \circ K_n) = MPP(C_4) + MPP(K_n)$$
$$+ \left[4(nx) + 4n(2x^2) + 4n(x^3)\right] + \left[(4n^2x^3 + 2n^2x^4)\right]$$

$$MPP(C_6 \circ K_n) = MPP(C_6) + MPP(K_n)$$
$$+ \left[6(nx) + 6n(2x^2) + 6n(x^4) + 6n(2x^5)\right]$$
$$+ \left[(6n^2x^3 + 3n^2x^5 + 6n^2x^6)\right]$$

$$MPP(C_8 \circ K_n) = MPP(C_8) + MPP(K_n)$$
$$+ \left[8(nx) + 8n(2x^2) + 8n(2x^6) + 8n(2x^7)\right]$$
$$+ \left[(8n^2x^3 + 4n^2x^6 + 8n^2x^7 + 8n^2x^8)\right]$$

$$MPP(C_{10} \circ K_n) = MPP(C_{10}) + MPP(K_n)$$
$$+ \left[10(nx) + 10n(2x^2) + 10n(x^6) + 10n(2x^7) + 10n(2x^8) + 10(2x^9)\right]$$
$$+ \left[(10n^2x^3 + 5n^2x^7 + 10n^2x^8 + 10n^2x^9 + 10x^{10})\right]$$

and so on.

Hence for any $m \geq 4$, m even and $n = 2, 3, 4, \ldots$

Case (ii): m is odd, $m \geq 3$.
For $n = 2, 3, 4, 5, \ldots$

$$MPP(C_3 \circ K_n) = MPP(C_3) + MPP(K_n)$$
$$+ \left[3nx + 3n(2x^2)\right] + \left[3n(3x^3)\right]$$

$$MPP(C_5 \circ K_n) = MPP(C_5) + MPP(K_n)$$
$$+ \left[5nx + (5n)(2x^2) + (2n)(x^3)\right]$$
$$+ \left[(5n)(5x^3) + (5n)(5x^5)\right]$$

$$MPP(C_7 \circ K_n) = MPP(C_7) + MPP(K_n)$$
$$+ \left[7(nx) + 7(n)(2x^2) + 7(n)(2x^5) + 7(n)(2x^6)\right]$$
$$+ \left[7(n)(nx^3) + 7(n)(nx^6) + 7(n)(nx^7)\right]$$

$$MPP(C_9 \circ K_n) = MPP(C_9) + MPP(K_n)$$
$$+ \left[9(nx) + 9n(2x^2) + 9(n)(2x^6) + 9n(2x^7) + (9n)(2x^8)\right]$$
$$+ \left[9(n)(nx^3) + 9(n)(nx^7) + 9(n)(nx^8) + 9(n)(nx^9)\right]$$

Hence for any $m \geq 7$, m odd and $n = 2, 3, 4, 5, \ldots$

$$MPP(C_m \circ K_n) = MPP(C_m) + MPP(K_n) + (mnx + 2mnx^2)$$
$$+ 2mn(x^5 + x^6 + \ldots + x^{\frac{n+3}{2}})$$
$$+ 2mn(x^6 + x^7 + \ldots + x^{\frac{n+3}{2}})$$
$$+ 2mn(x^7 + x^8 + \ldots + x^{\frac{n+3}{2}}) + \ldots + 2mnx^{n-1}$$
$$+ mn^2x^3 + mn^2(x^6 + x^7 + \ldots + x^{\frac{m+5}{2}})$$
$$+ mn^2(x^7 + x^8 + \ldots + x^{\frac{m+7}{2}} + \ldots + mn^2x^7)$$

5   Conclusion

In this paper the new topological index known as monophonic index of a graph is introduced and studied for some graphs. These results would help the researchers to develop new results and applications to chemical graph theory. The monophonic distance energy can be found from the monophonic distance matrix will be much useful for the chemists.

References

1. Arul Paul Sudhahar, P., Little Flower, M.: Total edge detour monophonic number of a graph (2019). https://doi.org/10.29055/jcms/1004
2. Arumugam, S.: The upper connected vertex detour monophonic number of a graph. Indian J. Pure Appl. Math. **49**(2), 365–379 (2018)
3. Titus, P., Ganesamoorthy, K.: Upper detour monophonic number of a graph. Electron. Notes Discret. Math. **53**, 331–342 (2016)
4. Titus, P., Ganesamoorthy, K.: On the detour monophonic number of a graph. Ars Comb. **129**, 33–42 (2016)
5. Santhakumaran, A.P., Titus, P.: A note on "monophonic distance in graphs". Discret. Math. Algorithms Appl. **4**(2), 1250018 (2012)
6. Santhakumaran, A.P., Titus, P.: Monophonic distance in graphs. Discret. Math. Algorithms Appl. **3**(2), 159–169 (2011)
7. Kathiresan, K.M., et al.: Detour wiener indices of graphs. Bull. ICA **62**, 33–47 (2011)
8. Kaladevi, V., Backialakshmi, P.: Detour distance polynomial of double star graph and cartesian product of P2 × Cn. Antarct. J. Math. **8**(5), 399–406 (2011)
9. Chantrand, G., Zhang, P.: Distance in graphs, taking the longview AKCE. J. Graphs Combin. **1**, 1–13 (2004)
10. Furtula, B., et al.: Wiener-type topological indices of phenyleness. Indian J. Chem. **41 A**(9), 1767–1772 (2002)
11. Lukovits, L.: The detour index. Croat. Chem. Acta **69**, 873–882 (1996)

A Novel Schema for Secure Data Communication over Content-Centric Network in Future Internet Architecture

M. S. Vidya[1(✉)] and Mala C. Patil[2]

[1] Department of Computer Science and Engineering, BMSIT&M,
Bengaluru, India
rvidyapai@bmsit.in
[2] Department of Computer Science, COHB University of Horticultural Sciences,
Bagalkot, India
malapatil2002@yahoo.co.in

Abstract. The transformation to the existing internet-based communication system to Future Internet Architecture (FIA) has already initiated in the present time. A robust communication system in FIA always calls for a secure communication system to facilitate better scalability with effective resistance towards security threat. Review of existing security approaches in FIA shows that they are not cost effective and uses a complex encryption standard which restricts the applicability towards the efficient, secure transmission. Therefore, the proposed system introduces a novel schema towards secure data transmission, considering the data content-centric approach. Analytical modeling is constructed where block-based access policy is framed for faster accessibility of encrypted data along. The study outcome shows better security performance, along with enhanced data transmission performance.

Keywords: Future Internet of Thing · Secure communication ·
Distributed communication · Access scheme · Authentication

1 Introduction

The conventional form of the internet architecture exhibits a bottleneck of resources that imposes a real-time challenge of scalability when a large amount of internet of things (IoT) nodes to have collaborated for the specific application realization. Apart from the scalability, it also suffers a non-synchronous stage with mobility and high dimension content distribution in advanced communication network due to non-flexibility and lack of control mechanisms [1]. The evolution of 32-bit addressing IPV4 to 128-bit IPV6 to an extend handles the scalability aspect, but the mechanism of host-based IP-network is not suitable for the wide-spread distributed communication systems of modern and advanced collaborative architecture-based applications, where the highly distributed and open architecture of internet is needed [2]. The future generation communication systems like 4G, 4G-LTE and 5G aims to support a delivery of the high-dimension content delivery that makes the future internet as content-centric internet. Irrespective of various solution mechanism like IPsec [3], DNSSec [4] etc.,

© Springer Nature Switzerland AG 2019
R. Silhavy et al. (Eds.): CoMeSySo 2019, AISC 1047, pp. 256–265, 2019.
https://doi.org/10.1007/978-3-030-31362-3_25

adopted towards the security aspect in IPV4 and IPV6 based internet architecture, the legacy of security loopholes and weakness still exits. Another side of the coin carries more risk and threats, where a continuous growth of the media-content is witnessed to be propagating and sharing by the growing user bases where privacy preservation becomes a very vital issue in the security solution domain [5]. Therefore, content integrity in future internet architecture is an open and challenging task that need to be addressed. Since, the content delivery to the user require a huge shift in the architecture of the network, therefore various forms of delivery models are emerging out but yet there is no standard form has established because various underlying technologies like SDN, cloud, virtualization etc. itself are in the emerging stage and many issues of security, capacity of content distribution, interoperability and device heterogeneity etc. to be incorporated [6]. The philosophy of content owner and the content user are reverse to each other, the content owner never wants that an unauthorized or non-subscriber are able to use or view their content as there are huge economy loss to the content service provider [7]. The phase of solution has witnessed handling security paradigms by the third-party vendors as a data security as a service model which is an additional cost as well as there have been many events is witnessed where the third party is held responsible for the breach of security [8–10]. Though there are continuous and significant contributions for this challenge by the wide-spread of academician and researchers with an evolving and dynamic system of the internet architecture new approaches require which shall be efficient, robust and flexible to handle the content security. Though the conventional crucial public-private cryptography (PPKC) offers various complex and hard encryption mechanism with a secret key to offer a higher level of context security but poses a huge computational and time complexity overhead which is not suitable for the energy efficient and resource constraints distributed future internet architecture.

Hence, the proposed system introduces a novel and straightforward mechanism towards safeguarding data transmission in FIA. Section 2 discusses the existing research work followed by problem identification in Sect. 3. Section 4 discusses the proposed methodology, followed by an elaborated discussion of algorithm implementation in Sect. 5. Comparative analysis of the accomplished result is discussed under Sect. 6, followed by conclusion in Sect. 7.

2 Related Work

This section is an extension of our prior review [11]. Most recently, Granjal et al. [12] have used elliptical curve cryptography in order to offer access control in FIA environment. The works of Han et al. [13] have presented a security modeling using software defined network. Approach using shuffling of addresses has been carried out by Nizzi et al. [14] for minimizing the network overhead. The work of Liu et al. [15] has presented a framework that can be used for assessment of security strength connected with IoT. Singh et al. [16] have discussed security approaches using the physical layer in order to incorporate the precise level of secrecy with better adaptability to network alteration. Significance of security and trust is discussed by Ambrosin [17]. Device authentication problem has been discussed by Kim et al. [18] where the physical layer has been used while Meng et al. [19] has used trust factor as a part of the evaluation process. Privacy is another concern in the FIA system, especially when it handles the massive scale of data and the work of Cha et al. [20] has presented a conceptual solution for this. Yu et al. [21] have discussed a key management scheme for promoting the autonomous assessment of security strength. Secure transmission of data using software defined network was also proven to offer better security as seen in the work of Zhang et al. [22]. The agent-based mechanism was also used in IoT and cloud ecosystem as distributed network finds its ultimate deployment over this. The recent works carried out by Prajapati et al. [23] have used agent-based approach for facilitating massive mining data. Another recent work of Ross et al. [24] has discussed how agent-based mechanism can be embedded with smart objects in the IoT environment. The work of Kotenko [25] has presented a technique for facilitating data structurization over IoT media in order to extract the pattern of communication. However, such studies are more data oriented and less on security. Another work carried out by Jing et al. [26] has discussed that hybridizing the characteristics of agents has multiple benefits of communication system assessment in a massive network. The work carried out by Oravec et al. [27] has constructed the design of an agent using the event-driven approach for constructing the communication mechanism in a massive network. The agent-based mechanism was also used in a massive network and cloud ecosystem as specific network finds its ultimate deployment over this. Another recent work of Rufus et al. [28] has discussed how agent-based mechanism can be embedded with smart objects in the IoT environment. The work of Gargees and Peng [29] has presented a technique for facilitating data structurization over IoT media in order to extract the pattern of communication. However, such studies are more data oriented and less on security. The study carried out by Nekrasov et al. [30] has discussed the usage of software agents, along with the incorporation of intelligence. The work carried out in [31] has constructed the design of an agent using the event-driven approach for constructing the communication mechanism in a massive network. Majority of the existing system towards FIA is focused on internet-of-things related securities that are found to affect security features in a massive network. The next section discusses the research problems of existing approaches.

3   Problem Description

Although there is various work being carried out towards securing communication in FIA, the majority of the existing approaches are used sophisticated and agent-based strategy towards security. Developing a successful protocol to be working on large scale FIA is still an open-end problem which calls for more investigation. The conventional usage of encryption-based strategy offers dual problem viz. (i) they are not cost effective for resource constraint nodes, (ii) they are not distributed in nature of work, and (iii) they are specific to resist the only kind of attacks. As FIA is more about scalability, therefore, it is still not known how to make the security algorithm to work in distributed nature and distributed operation. Therefore, the problem statement can be "*Developing a unique, cost-effective solution for securing communication in FIA to offer maximum resistance from threats without affecting data transmission performance.*" The proposed system intends to address the existing problem using a cost effective solution towards addressing the defined problem statement. The next section outlines the adopted methodology.

4 Proposed Methodology

The proposed system is an extension of our prior model [32]. This part of the implementation will be carried out considering the analytical modeling approach where a novel concept of decentralized hashing operation is used for safeguarding the content. Figure 1 highlights the methodology to be implemented in this phase.

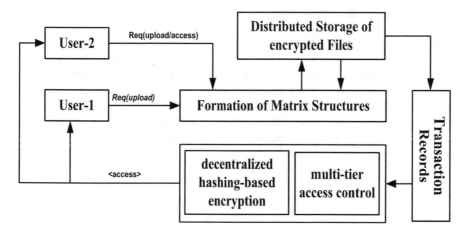

Fig. 1. Adopted methodology of phase-I

The implementation planning of this phase consists of constructing multiple numbers of processing. The first process will be responsible for developing a multi-tier access control policy that lets the user upload ciphered data along with an arbitrary security token. This process internally interacts with the second process that performs decentralized hashing-based encryption. This algorithm is the core implementation of this phase that will use a tree-structure to represents data as well as transaction. While designing this operation, the emphasis is given to make the process free from any form of social information associated with the transaction in order to facilitate the user's privacy. However, for facilitating decryption, a pointer-based addressing scheme is investigated so that data could be decrypted without any form of dependencies on the user's private data. The proposed system will deploy a matrix-structure that will reposit all the legal transaction that is subjected to hashing as well as encoding in order to formulate a secure tree-like structure. A process is designed to extract the security token from the prior matrix-structure and use it for generating new matrix structure with a link (or edge) between them. This process will be formed for ensuring the data integrity over a distributed and decentralized FIA with a prime target of not allowing any form of alteration of the data. One of the tentative benefits of this security planning is that it will enable the user to cross check and thereby audit the transaction without consuming an extra computational or network resources in FIA. Hence, it is expected to be highly cost effective.

5 System Implementation

The prime criteria of development of the proposed security algorithm are carried out on the basis that the new security approach should possess agility, scalability, supports integration, offers resiliency, and provides distributed security features. This section discusses the design components, methodology, and execution flow.

5.1 Design Attributes/Components

The proposed system consists of various design components as follows viz. (i) The first design component is, who is interested in forwarding their contents over FIA and is highly concern about the security of its data. (ii) The second design component *blocks* which are a distributed memory system to store the encrypted data of the user in a highly secured manner. (iii) The third design component is *Distributed storage* which maintains an underlying tree-schema to retain the encrypted data maintained under blocks. (iv) The fourth design component is the *pointer address* which retains the indexes of the blocks maintained in the form of the tree over the distributed data storage. (v) the fifth design component is an *auxiliary node* which is responsible for adding up blocks in the structure.

5.2 Design Methodology

The proposed system adopts the analytical research methodology in order to develop the security scheme where the case study considered is for a content-centric network system over FIA. The *first step* of the design is to develop a user who will perform ciphering of his data, followed by constructing a policy for performing authorization. The proposed system offers a mechanism that can control the authorization of the user (recipients only). After the encrypted data is obtained, the user will be required to generate security tokens (via hashing) and forwards it to the distributed storage system over FIA. The meta-data of this location (or address) details are retained separately. A tuple is constructed which retains the following information e.g., (i) the identity of the user that is attempting to access and (ii) discrete identity information of the owner of the content. The next operation is associated with the auxiliary node who appends the block over the tree structure in order to secure the data. However, the role of the auxiliary node is limited to this only. The next part of the implementation of the block-based access approach, which is a corruption-resistant entity in the network. The proposed system represents block-structure in the form of the tree. The study considers the leaf node of the tree to retain all the information associated with the transaction carried out by the devices. Interestingly, the proposed system doesn't use a specific form of pointers that bears the addresses of the data within the tree structure. The information that is retained in the transactional data is security tokens obtained by hashing, pointers to the location of data, signature, and hashing of the block data. It also bears the signature from the auxiliary node.

The first part of the implementation is focused on constructing a superior key obtained by hashing the private secret key and a random number, which is user-defined. The prime purpose of the superior secret key is to assist in computing crucial individual corresponding to the node. This superior key can also be recomputed if the user forgets about the secret key. The second part of the implementation is associated with performing the ciphering operation, followed by forwarding the data. The proposed system also calls for encrypting the data while forwarding the data by the user. The proposed system makes use of the symmetric encryption key for performing the ciphering process. The construction of the secret ciphering key is carried out by applying a hashing algorithm to the data that is obtained by concatenating private key and the frequency of the encryption operation. The proposed system performs hashing of the partial data that generates a unique hashing of the data as it will be nearly impossible for carrying out decryption of the such ciphered data. Therefore the ciphering of the secret key can be carried out by applying to hash over the value obtained by concatenation of the maximum value of the secret key with the index number. The next part of the implementation is associated with block processing in the master tree structure. Another essential factor in understanding is that auxiliary nodes are elected from the normal IoT nodes itself who will be responsible for authorizing the time-duration of the encryption process. A hashed-based block is used and is added to the total tree structure which is duly signed by the auxiliary node. Finally, the decryption process is initiated over the tree structure. This processing is carried out as follows:-as the data structure of hash values are maintained over the tree; therefore it becomes easier for the requester node to have access to the data that facilitates the user to obtain access to the data using the index value of the hash. It will mean that the system facilitates the requestor node to have access to the hash file of the original data, but it will not mean that it is given access to the direct data. The accessibility of the direct data is only given when the policy of the access rights are found satisfied by the proposed system. One of the unique parts of the implementation is that the extraction of hash value doesn't mean direct access to data chunk. A requestor node has to first seek permission from the auxiliary node for the access rights of the data. This operation is followed by cross-checking of identity information of the requester node, which is followed by obtaining the shared data from the auxiliary node. In case the data is missing from the auxiliary node, the query process is carried out over the main block structure in rode to obtain the location of the secret key that is followed by obtaining the encrypted message. Using a decryption key of requestor node followed by checking the access right, the access to data is offered to the requestor node.

5.3 Execution Flow

This paper dimension includes a new scheme of the context security for FIA for securely sharing the content by handling the trade-off of the degree of security level and resource overheads. The method adopts a decentralized hashing operation to secure the content by developing a multi-tier block-based access control policy where the user uploads a ciphered data along with an arbitrary security token. A tree-structure represents the data and the transactional details, where the algorithm does not consider any personal information into any of the transaction to provide the user's privacy. The deciphering process adopts a pointer based addressing scheme without having any dependencies on the user's profile data. The efficient matrix-based structure keeps records of all the legal transaction of hashed and encoded structure to form a secure tree like structure. The scheme extracts the security blocks from the secure tree structure and recursively generates a new set of matrix structure with a link of edge among them. Therefore, the content integrity over a distributed and decentralized FIA is achieved, where it is very hard to alter the data and the auditing process of the content integrity is achieved with optimal overheads to facilitate content security and integrity with authentication together is achieved only by hashing based operations.

Algorithm for content-centric encryption
Input: n (data)
Output: u_{data} (decrypted data received by user)
Start
1. **For** i=1:n
2. $C_{data}{\rightarrow}f(n)$
3. **For** j=1: m
4.   *construct* block$\rightarrow$gen_token(C_{data})
5.   **End**
6.   distribute block$\rightarrow$u
7. $u_{data}{\rightarrow}g(block)$
8. **End**
End

The algorithm takes the input of n (data) that after processing yields and outcome of u_{data} (decrypted data received by user). For all the data obtained in the forwarding process, the ciphered data is obtained by using hashing function $f(x)$ (Line-2). The next part of the study is about constructing block for the maximum value of hash m (Line-3). The blocks are constructed by applying a function *gen_token* that takes the input of ciphered data c_{data} (Line-4). After the blocks are obtained, the hash values of the blocks are now ready to be assigned to each user u (Line-6). The final stage of implementation is about applying a discrete encryption operation towards the finally obtained hash value of the blocks for facilitating the extraction of user data u_{data} (Line-7). This will complete the decryption process. Therefore, it can be seen that irrespective of any inclusion of any complex cryptography, the block-based access policy offers potential security towards the data forwarding process among the IoT devices in distributed networks. The next section discusses the results.

6   Results and Discussion

The implementation of the proposed system is carried out in MATLAB, considering the performance parameters of encryption time, latency, overhead, and processing time. As this part of the study introduces a very novel mechanism to boost up data security as well as data integrity with a proper authentication system; hence they are anticipated to offer both forward and backward secrecy with faster processing time. The computational and network overhead is anticipated to be much minimal owing to the usage of only hashing-based operation. Hence, this can be seen in Fig. 2.

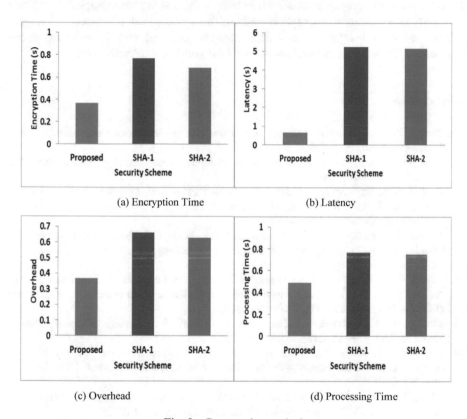

(a) Encryption Time   (b) Latency

(c) Overhead   (d) Processing Time

Fig. 2. Comparative analysis

The study outcome shows that the proposed system offers better performance in balancing the security demands as well as communication demands in IoT. Owing to the adoption of tree-based block policy, the proposed system never let any member have direct access to data which retains full resistance to any intruder while owing to hashing, the operation of encryption becomes very fast (Fig. 2(a)). This also affects the data transmission making lesser delay (Fig. 2(b)) as all the hash files are made in the form of a tree with connected blocks. As the complete authentication process is independent of data, therefore, overhead is comparatively lower in the proposed system (Fig. 2(c)). Apart from this, the proposed system also offers faster processing time exhibiting the lower computational complexity.

7   Conclusion

This paper discusses a novel framework that uses block-based access policy in order to offer secure transmission in FIA. The contribution of the proposed study are (i) cost effective usage of security protocol while using only hashing based approach, (ii) non-recursive mechanism of encryption making the system cost effective, (iii) a good balance between encryption and data transmission, and (iv) block-based access mechanism for faster authentication with robust resistance towards intruders.

References

1. Huang, T., Yu, F.R., Xie, G., Liu, Y.: Future internet architecture and testbeds. China Commun. **14**(10), iii–iv (2017)
2. Ding, W., Yan, Z., Deng, R.H.: A survey on future internet security architectures. IEEE Access **4**, 4374–4393 (2016)
3. Kent, S., Seo, K.: Security architecture for the internet protocol. Document RFC 4301 (2005)
4. Arends, R., Austein, R., Larson, M., Massey, D., Rose,S.: DNS security introduction and requirements. Document RFC 4033 (2005)
5. Yaqoob, I., Hashem, I.A., Ahmed, A., Kazmi, S.A., Hong, C.S.: Internet of things forensics: recent advances, taxonomy, requirements, and open challenges. Future Gener. Comput. Syst. **1**(92), 265–275 (2019)
6. Alberti, A.M., Scarpioni, G.D., Magalhães, V.J., Arismar Cerqueira, S., Rodrigues, J.J.P.C., da Rosa Righi, R.: Advancing NovaGenesis architecture towards future internet of things. IEEE Internet Things J. **6**(1), 215–229 (2019)
7. Mukherjee, B., Ferdousi, S.: The network user and its growing, influence. Comput. Commun. **131**, 43–45 (2018). ISSN 0140-3664
8. Aldossary, S., Allen, W.: Data security, privacy, availability and integrity in cloud computing: issues and current solutions. Int. J. Adv. Comput. Sci. Appl. (IJACSA) **7**(4), 485–498 (2016)
9. Cheng, L., Liu, F., Yao, D.D.: Enterprise data breach: causes, challenges, prevention, and future directions. Wiley Interdisc. Rev. Data Min. Knowl. Discov. **7**(5) (2017)
10. Zhang, J., Chen, B., Zhao, Y., Cheng, X., Hu, F.: Data security and privacy-preserving in edge computing paradigm: survey and open issues. IEEE Access **6**, 18209–18237 (2018)
11. Vidya, M.S., Patil, M.C.: Reviewing effectivity in security approaches towards strengthening internet architecture. Int. J. Electr. Comput. Eng. (IJECE) **9**(5), 3862–3871 (2019)
12. Granjal, J., Monteiro, E., Silva, J.S.: End-to-end transport-layer security for Internet-integrated sensing applications with mutual and delegated ECC public-key authentication. In: 2013 IFIP Networking Conference, Brooklyn, NY, pp. 1–9 (2013)
13. Han, Z., Li, X., Huang, K., Feng, Z.: A software defined network-based security assessment framework for cloud IoT. IEEE Internet Things J. **5**(3), 1424–1434 (2018)
14. Nizzi, F., Pecorella, T., Esposito, F., Pierucci, L., Fantacci, R.: IoT security via address shuffling: the easy way. IEEE Internet Things J. **6**(2), 3764–3774 (2019)
15. Liu, C., Zhang, Y., Li, Z., Zhang, J., Qin, H., Zeng, J.: Dynamic defense architecture for the security of the internet of things. In: 2015 11th International Conference on Computational Intelligence and Security (CIS), Shenzhen, pp. 390–393 (2015)
16. Singh, D., Tripathi, G., Jara, A.: Secure layers based architecture for Internet of Things. In: 2015 IEEE 2nd World Forum on Internet of Things (WF-IoT), Milan, pp. 321–326 (2015)

17. Ambrosin, M., Compagno, A., Conti, M., Ghali, C., Tsudik, G.: Security and privacy analysis of national science foundation future internet architectures. IEEE Commun. Surv. Tutor. **20**(2), 1418–1442 (2018)
18. Kim, H., Wasicek, A., Mehne, B., Lee, E.A.: A secure network architecture for the internet of things based on local authorization entities. In: 2016 IEEE 4th International Conference on Future Internet of Things and Cloud (FiCloud), Vienna, pp. 114–122 (2016)
19. Meng, Z., Chen, Z., Guan, Z.: Seamless service migration for human-centered computing in future Internet architecture. In: 2017 IEEE SmartWorld, Ubiquitous Intelligence and Computing, Advanced and Trusted Computed, Scalable Computing and Communications, Cloud and Big Data Computing, Internet of People and Smart City Innovation (SmartWorld/SCALCOM/UIC/ATC/CBDCom/IOP/SCI), San Francisco, CA, pp. 1–7 (2017)
20. Cha, S., Hsu, T., Xiang, Y., Yeh, K.: Privacy enhancing technologies in the internet of things: perspectives and challenges. IEEE Internet Things J. **6**(2), 2159–2187 (2019)
21. Yu, K., Arifuzzaman, M., Wen, Z., Zhang, D., Sato, T.: A key management scheme for secure communications of information centric advanced metering infrastructure in smart grid. IEEE Trans. Instrum. Meas. **64**(8), 2072–2085 (2015)
22. Zhang, J., Zhang, X., Imran, M.A., Evans, B., Zhang, Y., Wang, W.: Energy efficient hybrid satellite terrestrial 5G networks with software defined features. J. Commun. Netw. **19**(2), 147–161 (2017)
23. Prajapati, A., Sakadasariya, A., Patel, J.: Software defined network: future of networking. In: 2018 2nd International Conference on Inventive Systems and Control (ICISC), Coimbatore, pp. 1351–1354 (2018)
24. Ross, K.J., Hopkinson, K.M., Pachter, M.: Using a distributed agent-based communication enabled special protection system to enhance smart grid security. IEEE Trans. Smart Grid **4**(2), 1216–1224 (2013)
25. Kotenko, I., Saenko, I., Ageev, S.: Applying intelligent agents for anomaly detection of network traffic in internet of things networks. In: 2018 IEEE International Conference on Internet of Things and Intelligence System (IOTAIS), Bali, pp. 123–129 (2018)
26. Jing, X., Yan, Z., Pedrycz, W.: Security data collection and data analytics in the internet: a survey. IEEE Commun. Surv. Tutor. **21**(1), 586–618 (2019)
27. Oravec, J.A.: Emerging "cyber hygiene" practices for the Internet of Things (IoT): professional issues in consulting clients and educating users on IoT privacy and security. In: 2017 IEEE International Professional Communication Conference (ProComm), Madison, WI, pp. 1–5 (2017)
28. Rufus, R., Esterline, A.: Concatenating unprotected internet of things network event-driven data to obtain end-user information. In: 2017 26th International Conference on Computer Communication and Networks (ICCCN), Vancouver, BC, pp. 1–7 (2017)
29. Peng, Z., Kato, T., Takahashi, H., Kinoshita, T.: Intelligent home security system using agent-based IoT devices. In: 2015 IEEE 4th Global Conference on Consumer Electronics (GCCE), Osaka, pp. 313–314 (2015)
30. Nekrasov, H.A.: Development of the search robot for information security in web content. In: 2017 International Conference "Quality Management, Transport and Information Security, Information Technologies" (IT&QM&IS), St. Petersburg, pp. 79–81 (2017)
31. Simpson, S., Shirazi, S.N., Marnerides, A., Jouet, S., Pezaros, D., Hutchison, D.: An inter-domain collaboration scheme to remedy DDoS attacks in computer networks. IEEE Trans. Netw. Serv. Manag. **15**(3), 879–893 (2018)
32. Vidya, M.S., Patil, M.C.: Qualitative study of security resiliency towards threats in future internet architecture. In: Proceedings of the Computational Methods in Systems and Software, pp. 274–284. Springer, Cham (2018)

Computational Processes Management Methods and Models in Industrial Internet of Things

Semyon Potriasaev$^{(\boxtimes)}$, Viacheslav Zelentsov, and Ilya Pimanov

Laboratory of Information Technologies in System Analysis and Modeling,
St. Petersburg Institute for Informatics and Automation of the Russian Academy
of Sciences, 39, 14 Line V.O., 199178 Saint Petersburg, Russia
info@litsam.ru
https://litsam.ru

Abstract. The paper proposes a method and a multi-model complex for managing computations, which make it possible to increase the efficiency of production processes at existing and prospective industrial enterprises due to the optimal (rational) functioning of their information systems. The features of computational processes and architectures of information systems of modern enterprises based on the concept of Industry 4.0 are considered. A brief description of the software package is given, in which the task of structural-functional synthesis of the structure of an enterprise information system, as well as the task of building an operational schedule for its work, are simultaneously solved in an automated (automatic) mode.

Keywords: Industry 4.0 · Computation planning ·
Cyber-physical systems · Fog computing · Logical-dynamic models ·
Multicriteria optimization

1 Introduction

The methods to solve industrial management tasks have made a long way – from the Gantt Chart at the beginning of the 20th century and up to development of comprehensive systems for widespread planning and enterprise resources management [1–3]. The significant breakthrough took place in the field of software development for production processes scheduling after publication of works by Leonid Kantarovich (in 1939) and George Dantzig (in 1947), providing mathematical framework for solving tasks on manufacturing processes efficient planning [3,4]. The increasing complexity of production, development of hard- and software resulted in necessity to create automated control systems (ACS) for complex objects (CO) referred to as "human-machine" systems, providing efficient functioning of the above mentioned objects, ensuring collecting and evaluation of data, required to implement management functions by automation and computing facilities [6].

© Springer Nature Switzerland AG 2019
R. Silhavy et al. (Eds.): CoMeSySo 2019, AISC 1047, pp. 266–277, 2019.
https://doi.org/10.1007/978-3-030-31362-3_26

1.1   Stages of Production Management Automation

The directions of evolutionary development of ASCs for industrial enterprises have always been defined by tendencies in the field of corresponding ICT technologies and systems, which provide physical basis for implementation of the existing and perspective automated control technologies. Four major stages of the evolutionary development are usually outlined [7].

At the first stage of ASCs for industrial enterprises development, that took place in the mid 1960s – late 1970s in the USSR, only particular functions for information support and production operations control were automated. The first attempts were made to integrate various information technologies within the framework of relevant ASCs for technological processes (TP) and ACSs for production processes (PP).

Since the 1980s and till early 1990s the second stage of industrial enterprises management automation took place – firstly in the USSR and later in the Russian Federation – works were held to unify the processes of IT technologies utilization in the framework of broad integration of standard automation modules. Similar works on production processes automation were held abroad [8].

At the third stage (1990s–2010s) evolutionary complication and integration of information resources and technologies were performed in Russia, provided by both national and foreign companies.

The modern stage of ASCs for industrial enterprises development is characterized by large-scale integration of network-based microelectronic devices into production processes and is formally called the Fourth Industrial Revolution. It was named the project Industry 4.0 in the West. In practice its implementation currently allows to provide required level of production efficiency and additional income by utilizing digital technologies, supporting direct network interaction between involved suppliers and partners, utilizing brand-new production and financial business models.

The modern ASCs for industrial enterprises apply complex computing algorithms, while a big amount of indicators are taken into consideration – it results not only in new high requirements claimed to computing resources, but it also sets task to organize (plan) the computation process within the information system [7]. In the framework of the considered systems it is required to configure IS and relevant computational processes and to design the efficient work schedule for IS resources at the stage of production planning in order to, firstly, ensure regularity and sustainability of the production processes, and, secondly, provide balanced utilization of computing devices and telecommunication networks.

1.2   The Industrial Internet

The modern low-cost computing devices and mass-produced sensors are capable to interact with the physical world, implement computations and exchange information by standard web-communication protocols. Their broad utilization has changed the familiar image of IS with close integration of computing and physical processes, that formed the information technology concept for cyber-physical systems (CPS). One of the most quoted definitions for CPS defines the following. CPS is a system comprising a set of interacting physical and digital components, which may be centralized or distributed, that provides a combination of sensing, control, computation and networking functions to influence outcomes in the real world through physical processes [10].

Development of industrial information management systems is based on the concept of the Industrial Internet of Things (IIoT), comprising networked smart objects, cyber-physical systems, cloud or fog computing platforms, which enable real-time collection, analysis and data exchange in the industrial environment in order to boost productivity, reduce production costs and the resources consumption [11].

In addition to technical and technological changes the new concept of the Industrial Internet has facilitated the process of client-oriented, small-scale and flexible production development. The main feature of the above-mentioned productions distinguishing them from the large-scale production lines is the mobility (nonstationarity) of manufacturing process and characteristics of corresponding informational flows, computational processes. In order to support effective management of such production it is necessary to create new architecture, new models, methods and algorithms for CPSs management, unified into the Industrial Internet.

Unlike local data processing centers, traditionally used by former generation production enterprises, enterprises of the new type deploy computing resources in the framework of the Industrial Internet on the basis of cloud technologies, edge and fog computations [12].

CPSs are not only capable to transmit data flows from sensors to MES, but also to implement particular operations: aggregate them in order to receive a generalized state of a physical device, utilize them to make an immediate decision without transmitting data to the upper level, provide modeling and forecasting of a physical object further behavior, plan and implement proactive management influence on a physical object. Therefore, a new pool of computing resources appears within the process of CPS utilization, capable to implement some upper level (level of the cloud computing) management functions under particular conditions.

Thus, delegation of certain data processing functions to CPS allows to create architecture with edge and fog computations [13]. Their main aim is to reduce the amount of networking delays, reduce the load on communication channels and cloud computing resources (see Fig. 1).

Unlike the Internet of Things, on the Industrial Internet the sequence of computational processes as well as computational operations parameters are defined by the production technology. Within the cloud and fog computations there appears a big number of alternative methods to organize the computing process, which accompanies the production and provides required level of information. Each method and corresponding computation plan can be evaluated by criterial function: intensity of the implemented operations and works, amount of the consumed energy, indicators of load-balancing for the telecommunication channels, indicators of sustainability and robustness of the plan, etc.

Under conditions of flexible, small batch and client-oriented production – alongside regular production process planning – it is necessary to ensure proactive management of the accompanying computations, which means to provide reasonable selection of the ways to implement them and to plan computational process taking into consideration possible changes in the upcoming environment both for the CPSs and production processes.

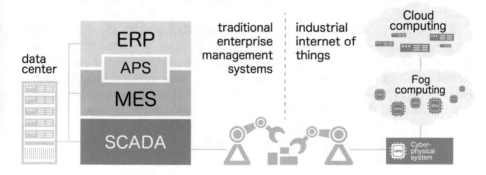

Fig. 1. ACS architecture in enterprises of different generations.

2   Computational Processes Management

Generally speaking about computational processes management on the Industrial Internet it is necessary to highlight the following major tasks:

1. To evaluate goals attainability to process data and information about physical objects by means of available computing capacities and within the set time limits.
2. To synthesize technologies capable to organize and implement computations out of the variety of alternative methods for their deployment.
3. To develop computational process plan.
4. To provide control over the plan implementation and computing process monitoring.
5. To ensure proactive management of the planned computational process implementation.

So far, quite an impressive number of models, methods and algorithms for operational planning and computational processes management have been developed in order to solve particular tasks on computational processes management [14]. Most of the applied approaches are focused on solving planning tasks in an enterprise separate computing elements and IS subsystems; they are not orchestrated and the factors of their mutual influence are not taken into consideration, as well as factors of uncertainty or factors related to setting and solving multi-objective tasks on computational processes efficient planning; coordinated account is not implemented for the Industrial Internet functioning main objective on an enterprise computing and production levels – that is to provide high-quality information support for production process synchronized in time and space [15]. These factors considerably reduce the amount of analyzed computation methods, lead to inefficient consumption of computing resources, make obstacles to receiving the best solutions, and result in peak loads on the Industrial Internet.

The defined disadvantages can be overcome with a broader view on the problem of a distributed SPC work planning, more specifically – considering it as an issue of organizing task-oriented and continued process of Industrial Internet various elements and subsystems functioning and cooperation, identification of their condition in each specific time point, their development forecasting for some period of time in the future, development (synthesis) and programming a set of their actions and plans, aimed to achieve the desired results [16, 17]. In this case the following major principles will define specific features of such computational processes planning [16, 18]:

- planning is the process of making preliminary decision about a hardware and software system architecture and functions, that provides achievement of the tasks set within specified time frame;
- planning result in a system of related solutions, distributed both in space and time, influencing each other;
- the planning process is permanently approaching its final point, but never reaches it due to two reasons: firstly, there is a possibility to endlessly reconsider the previously-taken decisions, secondly, internal and external environment can change within the planning process, thus the formed plans have to be permanently corrected;
- planning is aimed to prevent incorrect actions and reduce the number of untapped opportunities.

That being said, the above described tasks on complex planning of computational processes of a modern enterprise, affiliated to the Industrial Internet, can be formed as follows [16–18]: (1) to define goals and tasks assigned to an information system (IS), including identification of corresponding computational process desirable conditions, and to define time frame for the goals and tasks on data and information processing, ensuring production efficiency (productivity) increase; (2) to define resources required to reach these goals and tasks; (3) to define resources required for plans implementation, and to develop principles, methods and methodologies, required to distribute the resources within an enterprise information system elements and subsystems; (4) to develop (synthesize) an enterprise information system image (and, first of all, its basic structures) and its functioning mechanisms (algorithms), which ensure consistency, good timing and quality of the planning processes and the planned solutions implementation.

So far, most of the evaluated tasks in the field of scheduling theories, that in most cases include all the above mentioned tasks on computation planning in CPS, are referred to NP-difficult [14]. Traditional polynomial heuristic scheduling algorithms can not be applied within practical utilization to solve tasks of the computational processes optimization in production due to extremely large dimension and a number of such tasks peculiarities. Various metaheuristic approaches such as genetic and ant algorithms [19,20], simulated annealing method [21], based on scheduling theory and combinatorial mathematics, allow to receive a possible calculation schedule, but do not provide information about assessed inclination indicators of a particular plan quality in comparison to the aimed indicators [14]. The gaining in popularity constraint-based scheduling method also faces the problem of computing complexity, that is difficult to overcome, and this method inherits other disadvantages of heuristic algorithms [22].

2.1 Logical-Dynamic Models for Computation Planning

In order to solve the described tasks it is suggested to use a mathematical tool, developed in the framework of the modern theory on complex dynamic objects management with reconfigurable architecture [23–25]. Within this approach a dynamic alternative system multigraph can be used to describe a lot of methods to conduct required computation, which provide corresponding production processes with required level of data and informational support. In such case the multigraph peaks match computing operations and its arcs set restrictions on their implementation sequence. Alternative methods for computing process implementation are in this case described with polymodel complex, including [23–26]: logical-dynamic models for management of operations, flows, resources, operations parameters, structures, additional dynamic models to record requirements related to restrictions to interrupt particular operations. On the whole the generalized logical-dynamic model for computational processes management on the Industrial Internet represents a large-scale terminated finite-dimensional non-stationary non-linear differential dynamic system with variable range of the acceptable controlling actions, with partly-fixed border-line conditions in the initial and final time points. In the works [18,20] it was shown that in the framework of this formal description the task to synthesize a way (technology) to implement computations and develop computing process plan can be formed as a multi-objective problem on searching the preferable program management with the mentioned generalized dynamic model.

The formal problem definition is as follows: it is required to develop principles, approaches, models, methods, algorithms allowing to find such $<U_*^t, U_\delta^{*t_f}>$, whereby the following conditions are met:

$$J_\theta(X_\chi^t, \Gamma_\chi^t, Z_\chi^t, F_{<\chi,\chi'>}^t, \Pi_{<\widetilde{\delta},\widetilde{\widetilde{\delta}}>}^t, t \in (t_0, t_f)) \xrightarrow[<U^t, S_\delta^{*t_f}>\in\Delta_g]{} extr,$$

$$\Delta_g = \{<U^t, S_\delta^{t_f}>|R_\beta(X_\chi^t, \Gamma_\chi^t, Z_\chi^t, F_{<\chi,\chi'>}^t, \Pi_{(\widetilde{\delta},\widetilde{\widetilde{\delta}})}^t) \leq \widetilde{R}_g; \quad (1)$$

$$U^t = \Pi_{<\delta_1,\delta_2>}^{t_1} \circ \Pi_{<\delta_2,\delta_3>}^{t_2} \Pi_{<\delta,\delta>}^{t_2}; \chi \in B\}$$

where χ is an index, characterizing various types of industrial enterprise information systems structures, $\chi \in \{Top, Func, Tech, SW\}$ is an indexing set relevant for the topological, functional, technical and mathematical software structures, $t \in T$ – set of time points; $X_\chi^t = \{x_{\chi^l}, l \in L_\chi\}$ – set of elements, composing the structure of dynamic alternative multigraph G_χ^t; $\Gamma_\chi^t = \{\gamma_{<\chi,l,l'>}^t, l, l' \in L_\chi\}$ – set of multigraph edges of the type G_χ^t, reflecting its elements interconnections at a time point t; G_χ^t; $Z_\chi^t = \{z_{<\chi,l,l'>}^t, l, l' \in L_\chi\}$ — set of parameters values, quantitatively characterizing connections of the multigraph relevant elements. $F_{<\chi,\chi'>}^t$ – reflection of an enterprise information system various structures on each other at a time point t, $\Pi_{<\tilde{\delta},\tilde{\tilde{\delta}}>}^t$ – composition operation of the multistructural macrostates with the numbers $\tilde{\delta}, \tilde{\tilde{\delta}}$ at a time point t; U^t – management influence, allowing to synthesize both information system structures, and its functioning processes; J_Θ – value, time and resource indicators, characterizing the quality of an enterprise information system functioning, $q \in Q = \{1, ..., l\}$ – set of indicators numbers; Δ_q – a set of dynamic alternatives (a set of enterprise information system structures and parameters, a set of their functioning programs); $\mathbf{B}$ – a set of numbers for spatiotemporal, technical and technological constraints, defining the computational processes implementation; $\tilde{R}_g$ – set values; $T = (t_0, t_f]$ – time frame for technologies and computational processes synthesis [27].

The problem presented by formula (1) on simultaneous multi-criteria choice for both technologies to implement computational processes, and the relevant plans for the indicated computations management, is suggested to be solved simultaneously in the framework of a multiple-stage procedure. At the first step the acceptable variants of multi-structural macrostates for an enterprise information system should be formed, or, in other words, structural-functional synthesis of its new image should be implemented, corresponding to the forecast situation. At this stage a set of nonterminal decisions is formed (a set of non-dominated, effective alternatives, Pareto set) based on multidimensional orthogonal projecting on attainability set, formed as the result of complex modeling of alternative scenarios on the Industrial Internet functioning, a set, which helps to form requirements, applied to local indicators on computational processes efficiency implemented in IS. Thus, we solve the task to calculate achievability of the set goals for data and information processing using available computation capacities of an IS in the framework of the set time limits. The final selection of effective system engineering solutions must be made out of the defined set. At the second stage the particular version of the finite IS multistructural macro-state must be selected (and corresponding versions to organize computational processes) accompanied by simultaneous synthesis (construction) of adaptive plans (schedules) for management of an enterprise information system transition from the current state into the required multistructural macrostate of IS. Furthermore, the considered plans must ensure implementation of programs for sustainable management of the computational processes within intermediate macrostates.

So, the suggested approach allows to simultaneously solve the tasks on both synthesis of technology for implementing computation out of a set of available alternative ways, and the tasks to construct computation process plan using the unified poly-model logical-dynamic description within the considered subject field. The original combined methods and algorithms to synthesize technologies of an enterprise IS management – by far developed in the framework of this approach – and the corresponding work schedules allow to receive the appropriate (model) solutions based on the goals of a production enterprise functioning and the relevant quality criteria of accompanying computational processes management in the dynamically changing environment. The tasks on computing process monitoring and control, as well as its implementation management, are solved outside the framework of the suggested approach, but if a discrepancy is identified at the stage of computing plan implementation with the help of suggested models, there is also a possibility to solve the rescheduling tasks (plan correction, returning to the plan pathway).

3   Way to Implement the Suggested Approach

Practical implementation of the suggested approach on the Industrial Internet does not require to create new hardware facilities or considerably complicate architecture of an enterprise information system. Within the next scheduled time period – during production process planning – it is required to form and solve the tasks on calculating the set goals attainability, the tasks on selecting way for computation implementation and plan development for organizing computing process on the available resources. After that some software dispatcher provides commands for the computational processes launch and controls their implementation according to the developed plan. As mentioned above, consideration of ISs (which include CSP) functioning is outside the framework of this paper in case of abnormal situations.

The software package prototype developed by the author can serve as an example of architecture for subsystem on computational processes planning on the Industrial Internet.

It is based on a service-oriented architecture and business-processes performer, described with WS-BPEL language. All the algorithms required to implement the computation were implemented as web-services and distributed both in the fog layer and in the cloud. Meanwhile, the same services can be disposed both in CPS – in order to provide edge computations – and in the cloud, or on specialized computational nodes of an enterprise network, i.e. in micro data centers (MDC). Every service is provided with metadata about its operations duration and about energy requirements for their implementation. Due to such a structure-functional redundancy there appears a possibility to diversify the ways to implement computations, accompanying the production process, which means to flexibly reallocate functions among the Industrial Internet nodes of the cloud and fog layers on the basis of formally specified objective.

In order to formulate tasks on computing reachability of goals specified for the Industrial Internet and tasks on computation technologies synthesis and

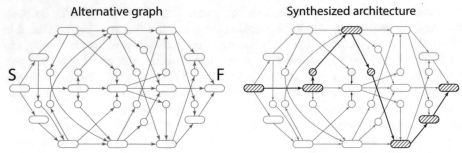

a) initial alternative graphic chart and example of synthesized architecture

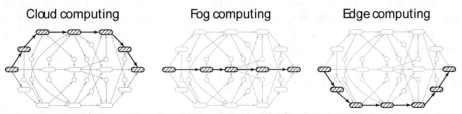

b) degenerated variants of the synthesized architectures

Fig. 2. Enterprise information system image synthesis results based on the example of one computing process.

corresponding plan development – it is enough to use Business Process Model and Notation (BPMN). It is thoroughly described in [28] why it is reasonable to apply it, as well as peculiarities of alternative or graph setting and separate parameters of the computing process operations. The initial data recorded in this way can be directly used in the developed software complex. As a result the hardware complex provides solution for the following tasks:

- analysis of possibility to implement computing process in the specified conditions;
- synthesis of certain computing technology out of the set of alternatives presented in the BPMN file;
- construction of scheduling for computing resources utilization;
- development of an executed scenario that is used at the stage of production process implementation in order to manage accompanying computing process.

The scenario is created in a form of BPEL file, that can be deployed in the business-processes implementation environment, like Apache ODE. Thus, if an alternative dynamic multigraph is set in the form of BPMN diagram for each production process, it enables the hardware complex to independently in the automatic mode reallocate the computing operations among the nodes of the inner network or transmit data for processing into the "cloud". Herewith, there would be utilized the best possible technology and calculations plan in terms of specified preference pattern on a set of particular quality indicators i.e. from the point of cyber-physical systems energy consumption minimization, time reduction for plans computation and managerial actions development in general, but at the same time high pressure on the cloud services would be allowed. An example of technologies selection out of a set of alternative options is provided in Fig. 2. A dynamic alternative multigraph is set, describing all the possible architectures and options for the computing process implementation, starting from receiving information from the sensors (peak S) and up to to control activities delivery (peak F). Degenerated options of the synthesized architectures are shown in Fig. 2b. An example of synthesized architecture in picture Fig. 2a corresponds to some hybrid choice of an architecture and computing process, which involves all levels of the Industrial Internet.

4   Conclusion

The approaches and methods considered in this paper allow to solve a number of important tasks related to the modern hi-tech production in the framework of the developed mathematical tool. First of all – in order to practically connect the common objectives of an industrial enterprise functioning to the objectives, which are implemented within the computing process management in its information-control system. And besides – to reasonably define and select the sequences of tasks solving and operations implementation, in other words, to synthesize the technology of computational processes management on the Industrial Internet by means of redundancy and heterogeneity of its computing devices and their relations. And finally – to deliberately find compromise solutions within distribution of the limited computing resources of an enterprise. The indicated results were reached by using dynamic decomposition of the initial task on complex computation planning, as well as using dynamic limitations, which allow to consider only relevant computational processes at every step of the process of solving the overall task to generate schedules, based on the limitations of their implementation sequence. In this case the workload of solving the initial task on planning the computations according to preliminary assessment takes the intermediate position between the classes P and NP of tasks and is defined by the level of partial order of the computing operations planned for implementation.

Acknowledgements. The researches implemented on this theme were partly financed with the grant support of the Russian Foundation for Basic Research (#17-29-07073-ofi-m, 17-06-00108, 18-07-01272, 18-08-01505, 19-08-00989), in the framework of budget theme #0073-2019-0004.

References

1. Zaytsev, N.G.: Information and software support for ECSP. Kiev (1974)
2. Yakobson, B.M., Rozinkin, A.E.: Automated Production Management Systems. Soviet Radio, Moscow (1971)
3. Piatkowski, O.I.: Automated Production Management System: Study Guide. Altai State Technical University, Barnaul (2010)

4. Kantorovich, L.V.: Works in Mathematical Economics. Nauka, Novosibirsk (2011)
5. Dantzig, G.B.: Maximization of a linear function of variables subject to linear inequalities. In: Koopmans, T.C. (ed.) Activity Analysis of Production and Allocation, Cowles Commission Monograph, vol. 13. Wiley, New York (1951)
6. Sovetov, B.Y.: Theoretical Framework for Automated Management: Textbook for Higher Educational Istitutions. Vysshaya Shkola, Moscow (2006)
7. Sokolov, B.V., Tsivirko, E.G., Yusupov, R.M.: Influence analysis of informatics and computer science on development of theory and systems of control by complex objects. In: SPIIRAS Proceedings, vol. 1, no. 11, pp. 11–51. SPIIRAS, St. Petersburg (2009)
8. Meyer, H.: Manufacturing Execution Systems: Optimal Design, Planning, and Deployment. McGraw-Hill, New York (2009)
9. Business Portal TAdviser: The Fourth Industrial Revolution. Populary on the main technological trend of the 21st century. http://www.tadviser.ru/a/371579
10. Boyes, H.: A Security Framework for Cyber-Physical Systems. University of Warwick, Coventry (2017)
11. Boyes, H., Hallaq, B., Cunningham, J., Watson, T.: The industrial internet of things (IIoT): an analysis framework. Comput. Ind. **101**, 1–12 (2018)
12. Lee, J., Bagheri, B., Kao, H.: A cyber-physical systems architecture for industry 4.0-based manufacturing systems. Manuf. Lett. **3**, 18–23 (2015)
13. Chiang, M., Zhang, T.: Fog and IoT: an overview of research opportunities. IEEE Internet Things J. **3**(6), 854–864 (2016)
14. Lazarev, A.A., Gafarov, E.R.: Scheduling Theory. Tasks and Algorithms. Lomonosov Moscow State University (MSU), Moscow (2011)
15. Sokolov, B.V.: Dynamic models and algorithms of comprehensive scheduling for ground-based facilities communication with navigation spacecrafts. In: SPIIRAS Proceedings, vol. 13, pp. 7–44. SPIIRAS, St. Petersburg (2010)
16. Ackoff, R.L.: The Art of Problem Solving. Wiley, New York (1978)
17. Klir, G.J.: Architecture of Systems Problem Solving. Plenum Press, New York (1985)
18. Gupta, M.M., Sinka, N.K.: Intelligent Control Systems: Theory and Applications. IEEE Press, New York (1996)
19. Vikhar, P.A.: Evolutionary algorithms: a critical review and its future prospects. In: Proceedings of the 2016 International Conference on Global Trends in Signal Processing, Information Computing and Communication (ICGTSPICC), Jalgaon, pp. 261–265 (2016)
20. Dorigo, M., Caro, G., Gambardella, L.: Ant algorithms for discrete optimization. Artif. Life **5**(2), 137–172 (1999)
21. Kirkpatrick, S., Gelatt, C., Vecchi, M.: Optimization by simulated annealing. Science **220**(4598), 671–680 (1983)
22. Baptiste, P., Le Pape, C., Nuijten, W.: Constraint-Based Scheduling: Applying Constraint Programming to Scheduling Problems. Kluwer Academic Publishers, Netherlands (2001)
23. Burakov, V.V., Zelentsov, V.A., Potryasaev, S.A., Sokolov, V.B., Kalinin, V.N.: Methodological and methodical basis of evaluation and choice of automatic control technology for active moving objects on the basis of integrated modeling. HES Res. J. **8**(3), 6–12 (2016)
24. Kalinin, V.N., Sokolov, B.V.: Multi-model description of control processes for airspace crafts. J. Comput. Syst. Sci. Int. **1**, 149–156 (1996)

25. Ohtilev, M.Y., Zelentsov, V.A., Potryasaev, S.A., Sokolov, B.V.: Complex technical objects proactive control conception and its implementation technologies. J. Instrum. Eng. **55**(12), 73–75 (2012)
26. Potryasaev, S.A.: Synthesis of structural dynamics modeling scenarios for automated control systems of active moving objects. J. Instrum. Eng. **57**(11), 46–52 (2012)
27. Ohtilev, M.Y., Sokolov, B.V., Yusupov, R.M.: Intellectual Technologies for Monitoring and Control of Structure-Dynamics of Complex Technical Objects. Nauka, Moscow (2006)
28. Potryasaev, S.A.: Integrated modelling of complex processes based on BPMN notation. J. Instrum. Eng. **59**(11), 913–920 (2016)

Applied Quadratic Programming with Principles of Statistical Paired Tests

Tomas Barot[1($\boxtimes$)], Radek Krpec[1], and Marek Kubalcik[2]

[1] Department of Mathematics with Didactics, Faculty of Education,
University of Ostrava, Fr. Sramka 3, 709 00 Ostrava, Czech Republic
{Tomas.Barot,Radek.Krpec}@osu.cz
[2] Department of Process Control, Faculty of Applied Informatics,
Tomas Bata University in Zlin, Nad Stranemi 4511, 760 05 Zlin, Czech Republic
Kubalcik@fai.utb.cz

Abstract. In applied research areas, various types of mathematical disciplines have been advantageously connected together with wide corresponding applications. As an applied proposal of this connection, a numerical optimization method of the quadratic programming particularly modified by a principle of statistical hypothesis testing can be seen in this paper. With regards to a computational complexity, algorithms of multivariable Model Predictive Control (MPC) can be considered as procedures with a higher computational complexity caused by the multi-variability, higher horizons and included constraints conditions. A wide spectrum of modifications has been proposed in the optimization subsystem of MPC controller yet; however, approaches based on including the hypotheses testing have not been widely considered in applied optimization method. A number of operations should be decreased; however, a control quality may be slightly influenced with regards to this aim. Therefore, the proposed modification is advantageous in an applied form of the quadratic programming technique where necessary information for following steps of a process control are provided. Achieved results are discussed in order to the incorporating of the principle of hypotheses testing in the modified numerical method of the applied quadratic programming.

Keywords: Hypothesis testing · Optimization · Quadratic programming · Multivariable MPC · Control quality

1 Introduction

In the process control [1], a various spectrum of proposed approaches can be seen, e.g. in [2–7]. A higher algorithmically complexity of a control algorithm reduced by modified methods may have slightly influenced control quality. The control quality can be monitored by various types of control quality criterions. According to this principle, a control algorithm of Model Predictive Control (MPC) [8–10] is modified in this paper. Particularly, the MPC is an appropriate example of an applied control where the control of a multivariable process considering higher horizons with regards to constraints conditions is bound with a high number of floating-point operations [11].

© Springer Nature Switzerland AG 2019
R. Silhavy et al. (Eds.): CoMeSySo 2019, AISC 1047, pp. 278–287, 2019.
https://doi.org/10.1007/978-3-030-31362-3_27

In applied cases of the MPC, modifications are usually based on an optimization subsystem which corresponds with a predictor block. These parts of the MPC controller cooperate on the receding horizon strategy [10]. According to this strategy, a trajectory of future manipulated variables [8] is given by solving an optimization problem incorporating a suitable cost function and constraints. In the further considered discrete control case, optimization problem is repeated in each sampling period. Input data for the optimization part of the MPC are based on information from previous values of control variables and predicted values determined in an incorporated model of a controlled process. Using the model, predictions of future outputs of the control process are expressed. According to these future outputs equations, optimization method is able to return a required final vector of increments of future manipulated variables [8].

Optimization techniques for achievement of these vectors have been modified by many interesting strategies, e.g. [12] in which is proposed as an external optimization approach with offline optimization precomputations. As can be seen e.g. in [10], approaches can be also aimed on a preparation of own optimization problem considering the opposite online optimization strategy without precomputations. In the optimization part of the MPC controller, the incorporating of modifications of the numerical algorithm can be realized, e.g. in [13] where the numerical technique bounded with modification of the Hildreth method was proposed. The Hildreth method is a widely recommended method for solving optimization problems in MPC [13]. This method is based on a dual solution of the quadratic programming [14] problem with a quadratic cost function and considered constraints [14], frequently appeared in MPC. The Hildreth method belongs to approaches solved in the dual variable area [14] further transposed able to the primary variable area.

In the situation, when a numerical optimization algorithm terminates, a computed optimum in the form of a multidimensional vector of future manipulated variables belongs to all declared conditions in the quadratic programming. If a decreasing of a computational complexity is necessary in case of an applied MPC, it could be advantageous to stop an optimization algorithm in step in which an inter-result will belong to considered constraints; however, a global optimum could not be achieved. Therefore, a final modified solution can be slightly different in comparison to the global optimum with regards to applied purposes. Corresponding to this modification of the optimization algorithm, the control quality can be slightly influenced; although, the complexity of performing the optimization task could decrease. This particular case is realized by extending the optimization [8] using a principle of hypothesis testing [15], e.g. in applied forms [16, 17]. Concretely, there is tested whether two last local extremes in the Hildreth method, which belong to all declared constraints, are significantly different. For this purpose, a significance level is obviously defined as 0.001 in order to technical applications. In each loop of numerical cycle, this hypothesis testing is being performed in the numerical optimization. If two vectors are not significantly different, the cycle of the searching the dual variable is terminated and transformed as usual into primary form of variable.

In this paper, the hypothesis testing can be seen as a mathematical method which builds a part of a numerical applied optimization method. A significance level 0.001 can help to keep accuracy of given solutions; however, corresponding inaccuracies can slightly appear in applied MPC. A global advantage can be brought by a potential decreasing of a number of whole operations in the iterative cycles. These assumptions are further detailed analyzed in this paper with regards to achieved control quality indices. These criterions can be complemented by information based on determination of a number of floating-point operations in an optimization part of the MPC. Achieved results are discussed in order to the incorporating of the principle of hypothesis testing in the modified numerical method of the applied quadratic programming.

2 Multivariable Model Predictive Control

Model Predictive Control (MPC) [8] is one of the most effective approaches for a control of multivariable systems. An advantage of the MPC is that the multivariable systems can be controlled considering constraints conditions. In MPC controller, optimization and prediction parts [9] are incorporated. These subsystems cooperate on the receding horizon strategy [10]. According to this strategy, a trajectory of future manipulated variables is given by solving an optimization problem including a cost function and constraints [8]. Previous signal values of control variables, predictions computed from predictor are interconnected with optimization problem of a quadratic programming [14]. In the MPC research, a recommendation of an implementation provided by the Hildreth method has been appeared e.g. in [13]. In order to including the mathematical model of a controlled process in the predictor, a following mathematical model of multivariable process [18] (1) with two inputs and two outputs is further considered in this paper. This matrix fraction is a multiplication of two particular matrices $A(z^{-1})$ and $B(z^{-1})$, as can be seen in (2) with extending description of polynomials in (3)-(4).

$$G(z^{-1}) = A^{-1}(z^{-1})B(z^{-1}) \tag{1}$$

$$A(z^{-1}) = \begin{bmatrix} \alpha_{11}(z^{-1}) & \alpha_{12}(z^{-1}) \\ \alpha_{21}(z^{-1}) & \alpha_{22}(z^{-1}) \end{bmatrix}, B(z^{-1}) = \begin{bmatrix} \beta_{11}(z^{-1}) & \beta_{12}(z^{-1}) \\ \beta_{21}(z^{-1}) & \beta_{22}(z^{-1}) \end{bmatrix} \tag{2}$$

$$\left. \begin{array}{l} \alpha_{11}(z^{-1}) = 1 + \alpha_{111}z^{-1} + \alpha_{112}z^{-2}, \ \alpha_{12}(z^{-1}) = \alpha_{121}z^{-1} + \alpha_{122}z^{-2}, \\ \alpha_{21}(z^{-1}) = \alpha_{211}z^{-1} + \alpha_{212}z^{-2}, \ \alpha_{22}(z^{-1}) = 1 + \alpha_{221}z^{-1} + \alpha_{222}z^{-2} \end{array} \right\} \tag{3}$$

$$\left. \begin{array}{l} \beta_{11}(z^{-1}) = \beta_{111}z^{-1} + \beta_{112}z^{-2}, \ \beta_{12}(z^{-1}) = \beta_{121}z^{-1} + \beta_{122}z^{-2}; \\ \beta_{21}(z^{-1}) = \beta_{211}z^{-1} + \beta_{212}z^{-2}, \ \beta_{22}(z^{-1}) = \beta_{221}z^{-1} + \beta_{222}z^{-2} \end{array} \right\} \tag{4}$$

Using CARIMA model (5) (Controlled AutoRegressive Integrated Moving Average) [8], calculations of predictions (6) corresponding to a maximum prediction horizon N_2 are computed in each sampling period of the discrete control. The manipulated variable is $u(k)$. The vector of future increments of this manipulated variable has N_u elements. Output signal is denoted as $y(k)$. Variable $e(k)$ is a control error. Variable $w(k)$ is a reference signal. Each variable is two-dimensional.

$$A(z^{-1})y(k)2 = B(z^{-1})u(k) + \Delta^{-1}(z^{-1})C(z^{-1})e_s(k), \Delta(z^{-1}) = \begin{bmatrix} 1 - z^{-1} & 0 \\ 0 & 1 - z^{-1} \end{bmatrix}$$

$$(5)$$

$$\left.\begin{array}{c} y(k) = A_1 y(k-1) + A_2 y(k-2) + A_3 y(k-3) + B_1 \Delta u(k-1) + B_2 \Delta u(k-2), \\ A_1 = \begin{bmatrix} (1 - \alpha_{111}) & -\alpha_{121} \\ -\alpha_{211} & (1 - \alpha_{221}) \end{bmatrix}, \ A_2 = \begin{bmatrix} (\alpha_{111} - \alpha_{112}) & (\alpha_{121} - \alpha_{122}) \\ (\alpha_{211} - a_{212}) & (\alpha_{221} - \alpha_{222}) \end{bmatrix}, \\ A_3 = \begin{bmatrix} \alpha_{112} & \alpha_{122} \\ \alpha_{212} & \alpha_{222} \end{bmatrix}; \ B_1 = \begin{bmatrix} \beta_{111} & \beta_{121} \\ \beta_{211} & \beta_{221} \end{bmatrix}, B_2 = \begin{bmatrix} \beta_{112} & \beta_{122} \\ \beta_{212} & \beta_{222} \end{bmatrix} \end{array}\right\}$$

$$(6)$$

The predictive equations with utilized CARIMA model (6) and complemented by matrices P and G, can be seen in Eq. (7) with vectors without consideration of the noise signal $e_s(k)$. N_1 and N_2 are minimum and maximum prediction horizons defined in (8)–(9) where Z is a zero matrix of a given dimension.

$$\left.\begin{array}{c} \underbrace{\begin{bmatrix} y(k+N_1) \\ \vdots \\ y(k+N_2) \end{bmatrix}}_{y} = P \begin{bmatrix} y(k) \\ y(k-1) \\ y(k-2) \\ \Delta u(k-1) \end{bmatrix} + G \underbrace{\begin{bmatrix} \Delta u(k) \\ \Delta u(k+1) \\ \vdots \\ \Delta u(k+N_u-1) \end{bmatrix}}_{\Delta u} \\ \\ P = \begin{bmatrix} P_{11} & P_{12} & \cdots & P_{14} \\ P_{21} & P_{22} & \cdots & P_{24} \\ \vdots & & \ddots & \vdots \\ P_{i1} & P_{i2} & \cdots & P_{i4} \end{bmatrix}, G = \begin{bmatrix} G_{11} & G_{12} & \cdots & G_{1j} \\ G_{21} & G_{22} & \cdots & G_{2j} \\ \vdots & & \ddots & \vdots \\ G_{i1} & G_{i2} & \cdots & G_{ij} \end{bmatrix} \end{array}\right\}$$

$$(7)$$

$$\left.\begin{array}{c} P \in \mathcal{R}^{2N_2.8}; \\ P_{11} = A_1; P_{12} = A_2; P_{13} = A_3; P_{14} = B_2; \\ P_{21} = A_1^2 + A_2; P_{22} = A_1 A_2 + A_3; \\ P_{23} = A_1 A_3; P_{24} = A_1 B_2; \\ P_{31} = A_1^3 + A_1 A_2 + A_3 + A_1 A_2; \\ P_{32} = A_1^2 A_2 + A_1 A_3 + A_2^2; \\ P_{33} = A_1^2 A_3 + A_3 A_2; P_{34} = A_1^2 B_2 + A_2 B_2; \\ \left(P_{ij} = A_1 P_{(i-1)j} + A_2 P_{(i-2)j} + A_3 P_{(i-3)j}\right), \\ i = 4,..., N_2; j = 1,..., i \end{array}\right\}$$

$$(8)$$

$$
\left.\begin{array}{c}
\mathcal{G} = Z; Z \in \mathcal{R}^{2N_2 - 2N_1 + 2, 2N_2}; \\
\mathcal{G}_{11} = \mathcal{G}_{22} = \mathcal{G}_{33} = B_1; \\
\mathcal{G}_{21} = \mathcal{G}_{32} = (A_1 B_1 + B_2); \\
\mathcal{G}_{31} = (A_1^2 B_1 + A_1 B_2 + A_2 B_1); \\
\left(\begin{array}{c}
\mathcal{G}_{i1} = A_1 \mathcal{G}_{(i-1)1} + A_2 \mathcal{G}_{(i-2)1} + A_3 \mathcal{G}_{(i-3)1} \\
\mathcal{G}_{i(j-1)} = A_1 \mathcal{G}_{(i-1)(j-1)} + A_2 \mathcal{G}_{(i-2)(j-1)} + \\
+ A_3 \mathcal{G}_{(i-3)(j-1)} + B_2 \\
\mathcal{G}_{ij} = B_1
\end{array}\right), \\
i = 4, \ldots, N_2; \ j = 1, \ldots, i
\end{array}\right\}
\tag{9}
$$

In the form of a minimization problem by the quadratic programming, the vector of future increments of manipulated variable is determined by solving Eq. (10). Particularly, requirements on the MPC are projected into a minimization (11) of a quadratic cost function J [14] in order to the minimization of powers of future increments of the manipulated variable and powers of predicted control errors. This definition of the quadratic programming problem is bound with a declaration of constraints included in matrices M and in vector γ. Matrix I is an identity matrix of a required dimension.

$$
J = (y - w)^T (y - w) + \Delta u^T \Delta u
\tag{10}
$$

$$
\left.\begin{array}{c}
\min\left\{ J = \dfrac{1}{2} \Delta u^T H \Delta u + b^T \Delta u \ \middle| \ M \Delta u \le \gamma \right\}, \\
H = \mathcal{G}^T \mathcal{G} + I, b = \mathcal{G}^T \left(P \begin{bmatrix} y(k) \\ y(k-1) \\ y(k-2) \\ \Delta u(k-1) \end{bmatrix} - w \right)
\end{array}\right\}
\tag{11}
$$

3   Applied Quadratic Programming Technique

Widely recommended implementation of an applied quadratic programming problem (11) in MPC can be read in [13]. In this approach, a numerical method is solved in iterative cycles, which are performed in each sampling period of MPC and are terminated if last computed local optimums are similar with consideration constraints.

Particular inter-results (12) are gradually improving into the form of a global minimum d; however, this solution is taken into account with regards to the dual variables solving the dual optimization problem (13) [14]. A connection between the vector of future increments Δu of this manipulated variable can be expressed from the dual form d using rule (14). In Eq. (14), a transformation to a primary variable [14] is provided. Algorithmically steps of this described principle is concretely described by a code for MATLAB in [13].

$$d(i) = [\, d_1(i) \quad \cdots \quad d_{Nu}(i) \,]^T \tag{12}$$

$$\left.\begin{array}{l} d = \begin{bmatrix} d_1 & \cdots & d_{Nu} \end{bmatrix}^T = \arg\min\left\{\dfrac{1}{2}d^T N d + o^T d \mid d \geq 0\right\}; d \in \mathcal{R}^{Nu,1} \\[2mm] N = MH^{-1}M^T; N_{ij} \in N \in \mathcal{R}^{Nu,Nu}; o = MH^{-1}b^T + \gamma; o_{ij} \in o \in \mathcal{R}^{Nu,1} \end{array}\right\} \tag{13}$$

$$\Delta u(k) = -H^{-1}(M^T d + b^T) \tag{14}$$

4 Paired Tests Applied in Quadratic Programming in MPC

With regards to an applied realization of multivariable Model Predictive Control (MPC), a modification of an optimization method is proposed in this paper. In order to this assumed aim, some slightly influenced control quality indices can be analyzed; however, simplifications can be advantageous in favour of a decreasing of a computational complexity of the solved optimization part in MPC. This part is significantly time-demanding in case of multi-variability of a controlled process, higher horizons and constraints [10].

Modification of the optimization subsystem of the MPC controller consists of a proposal based on including the hypothesis testing into a numerical optimization algorithm of the quadratic programming [14] as a termination procedure in its cycles. On the significance level 0.001, the proposal considers an exit condition where last computed two vectors will not be significantly different. In favour of a reduction of a number of operations, both vectors may be only local extremes; however, both vectors have to belong to all declared constraints in a quadratic programming assumptions.

Statistical methods corresponding to these purposes can be based on strategies of paired tests. For case of occurring of normality of data [19, 20], Paired T-test should be used. In opposite case, the Wilcoxon test should be applied. However, in order to situation, that data of the last vectors have a low number of values expressing a non-normal probability distribution, the Wilcoxon test is further selected [15].

Results obtained in hypothesis testing in each cycle of the numerical optimization algorithm have a form of p-values. If p is greater or is equal to the significance level, then significant differences between two vectors are not statistically identified. In opposite case, significant differences appear. Final forms of hypotheses occurred in the modified Hildreth method follow rules, as can be seen in Table 1. Each k-th hypothesis in a corresponding k-th step of an optimization procedure can be divided into two partial hypotheses – a zero hypothesis kH_0 and an alternative hypothesis kH_1. Situation, whether a zero hypothesis is rejected or failed to reject, depends on p value compared with the significance level.

Table 1. Hypotheses for purposes of realized termination in the Hildreth method

Hypothesis	Assumption
k-th H_0	There are not statistically significant differences between paired values of vectors $d(i-1)$ and $d(i)$
k-th H_1	There are statistically significant differences between paired values of vectors $d(i-1)$ and $d(i)$

5   Results

Using described proposal based on incorporated hypothesis testing into the quadratic programming, a complexity of the modified approaches in multivariable Model Predictive Control has been programmed and analyzed in MATLAB. A grade of accuracy is given by the determined significance level 0.001. In general, modifications proposed in process control can be appropriate e.g. for decreasing a computational complexity of algorithm; however, a control quality can be slightly influenced. For these purposes, the modified optimization Hildreth method is further considered.

In the simulation, multivariable process (15) with two inputs and two outputs was used. This model was considered also in [11] for purposes of further possible comparisons of authors' proposals. The minimum horizon N_1 was set as value of 1. Both control N_u and maximum horizons N_2 were same in this simulation; however, these horizons were being steeply increased in a complexity analysis of a number of floating point operations and values of control quality criterions.

$$
\left.
\begin{aligned}
A(z^{-1}) &= \begin{bmatrix} 1 - 1.32640z^{-1} + 0.3271z^{-2} & 0.0240z^{-1} - 0.0029z^{-2} \\ -0.0711z^{-1} + 0.0759z^{-2} & 1 - 1.0911z^{-1} + 0.1340z^{-2} \end{bmatrix} \\
B(z^{-1}) &= \begin{bmatrix} 0.2983z^{-1} - 0.0970z^{-2} & 0.0930z^{-1} + 0.0682z^{-2} \\ 0.1755z^{-1} + 0.0688z^{-2} & 0.1779z^{-1} + 0.1065z^{-2} \end{bmatrix}
\end{aligned}
\right\}
\tag{15}
$$

According to the process model (14), for both dimensions of variables in the multivariable MPC, the constraints were defined as follows: $u_{min} = -2$, $u_{max} = 2$. Definition of these constraints was considered in the form of matrix M and in vector γ by rule (16). T is a lower triangular matrix.

$$
M = \begin{bmatrix} -T \\ T \end{bmatrix}, \gamma = \begin{bmatrix} [u(k-1) - u_{min} & \cdots & u(k-1) - u_{min}]^T \\ [u_{max} - u(k-1) & \cdots & u_{max} - u(k-1)]^T \end{bmatrix}
\tag{16}
$$

A rule for determination of the numbers of floating-point-operations (flops) was include in the code for each program-command according to [21]. Then flops were analyzed due to various horizons setting of both control and maximum horizons which were being steeply increased from value of 5 to 35 by step 5. The control quality of proposals of modifications was analyzed using control quality criterions based on sums of powers of control increments and on sums of powers of control errors [22] (17). Achieved results of both types of a complexity analyses can be seen in Table 2 complementing information about a number of solved constraints conditions in the quadratic programming problem.

Table 2. Achieved results of flops analysis and control quality criterions

Horizons	Non-modified MPC			MPC with proposed modification		
$N_2 = N_u$	Flops	J_1	J_2	Flops	J_1	J_2
5	2406341	62.775	207.152	**104173**	62.855	208.334
10	20487679	58.321	195.404	**5924731**	**57.688**	195.987
15	70638817	56.026	196.084	**36456801**	56.303	200.614
20	169254155	55.090	197.662	**117430831**	**55.054**	201.642
25	332728093	55.628	196.733	**204548737**	**54.334**	203.061
30	577455031	58.191	193.095	**155330795**	58.867	**191.520**
35	919829369	58.270	192.947	**265571771**	59.034	**192.692**

$$
\left.
\begin{aligned}
J_1 &= \sum_k [\Delta u_1(k)]^2 + \sum_k [\Delta u_2(k)]^2, \\
J_2 &= \sum_k [w_1(k) - y_1(k)]^2 + \sum_k [w_2(k) - y_2(k)]^2
\end{aligned}
\right\}
\tag{17}
$$

As can be seen in Table 2, the number of floating-point operations significantly decreased using proposed modification of the applied quadratic programming. However, in appeared cases influence of the proposal on the control quality is obvious according to the results of control quality criterions. An advantageous improving of these analyzed indicators can be seen using highlighted values by bold font in Table 2.

6   Conclusion

With slightly influenced control quality, a decreased number of operations in algorithm of multivariable Model Predictive Control was achieved. Due to the applied modification of the quadratic programming problem, this numerical method was improved for the applied purposes of MPC with higher computational requirements. Concretely, the optimization procedure was extended by the principle of the hypotheses testing based on paired comparison of inter-results in steps of this optimization method with regards to a fulfilling the declared constraints. The hypothesis testing was performed on the significance level 0.001 frequently appeared in technical applications. Therefore, achieved results contained only slightly different control quality criterions in comparison to the case of non-modified control algorithm, particularly in the applied case of MPC.

References

1. Corriou, J.P.: Process Control: Theory and Applications. Springer, Heidelberg (2004)
2. Xu, S., Ni, D., Lu, S., et al.: A novel digital multi-mode control strategy with PSM for primary-side flyback converter. Int. J. Electron. **104**(5), 840–854 (2017). https://doi.org/10.1080/00207217.2016.1253783. ISSN 0020-7217
3. Li, C., Mao, Y., Yang, J., et al.: A nonlinear generalized predictive control for pumped storage unit. Renew. Energy **114**, 945–959 (2017). https://doi.org/10.1016/j.renene.2017.07.055. ISSN 0960-1481
4. Sun, D., Xu, S., Sun, W., et al.: A new digital predictive control strategy for boost PFC converter. IEICE Electron. Exp. **12**(23), 9 (2015). https://doi.org/10.1587/elex.12.20150726. ISSN 1349-2543
5. Zheng, Y., Zhou, J., Zhu, W., et al.: Design of a multi-mode intelligent model predictive control strategy for hydroelectric generating unit. Neurocomputing **207**, 287–299 (2016). https://doi.org/10.1016/j.neucom.2016.05.007. ISSN 0925-2312
6. Abraham, A., Pappa, N., Honc, D., et al.: Reduced order modelling and predictive control of multivariable nonlinear process. Sadhana – Acad. Proc. Eng. Sci. **43**(3) (2018). https://doi.org/10.1007/s12046-018-0798-x. ISSN 0256-2499
7. Navratil, P., Pekar, L., Klapka, J.: Load distribution of heat source in production of heat and electricity. Int. Energy J. **17**(3), 99–111 (2017). ISSN 1513-718X
8. Camacho, E.F., Bordons, C.: Model Predictive Control. Springer, Heidelberg (2004)
9. Rossiter, J.A.: Model Based Predictive Control: A Practical Approach. CRC Press, Boca Raton (2003)
10. Kwon, W.H.: Receding Horizon Control: Model Predictive Control for State Models. Springer, Heidelberg (2005)
11. Kubalcik, M., Bobal, V., Barot, T.: Modified Hildreth's method applied in multivariable model predictive control. In: Innovation, Engineering and Entrepreneurship. Lecture Notes in Electrical Engineering, vol. 505, pp. 75–81. Springer (2019). https://doi.org/10.1007/978-3-319-91334-6_11. ISBN 978-3-319-91333-9
12. Ingole, D., Holaza, J., Takacs, B., et al.: FPGA-based explicit model predictive control for closed loop control of intravenous anesthesia. In: 20th International Conference on Process Control (PC), pp. 42–47. IEEE (2015). https://doi.org/10.1109/pc.2015.7169936
13. Wang, L.: Model Predictive Control System Design and Implementation Using MATLAB. Springer, Heidelberg (2009)
14. Dostal, Z.: Optimal Quadratic Programming Algorithms: With Applications to Variational Inequalities. Springer, Heidelberg (2009)
15. Kitchenham, B., Madeyski, L., Budgen, D., et al.: Robust statistical methods for empirical software engineering. Empir. Softw. Eng. **22**, 1–52 (2016)
16. Sulovska, K., Belaskova, S., Adamek, M.: Gait patterns for crime fighting: statistical evaluation. In: Proceedings of SPIE - The International Society for Optical Engineering, vol. 8901. SPIE (2013). https://doi.org/10.1117/12.2033323. ISBN 978-081949770-3
17. Pivarc, J.: Ideas of Czech primary school pupils about intellectual disability. Educ. Stud. Taylor & Francis (2018, in press). https://doi.org/10.1080/03055698.2018.1509784. ISSN 0305-5698
18. Navratil, P., Balate, J.: One of possible approaches to control of multivariable control loop. IFAC Proc. **40**(5), 207–212 (2007). https://doi.org/10.3182/20070606-3-mx-2915.00033. ISSN 1474-6670
19. Alizadeh Noughabi, H.: Two powerful tests for normality. Ann. Data Sci. **3**(2), 225–234 (2016). ISSN 2198-5812

20. Vaclavik, M., Sikorova, Z., Barot, T.: Particular analysis of normality of data in applied quantitative research. In: Computational and Statistical Methods in Intelligent Systems. Advances in Intelligent Systems and Computing, vol. 859, pp. 353–365. Springer (2019). https://doi.org/10.1007/978-3-030-00211-4_31. ISBN 978-3-319-91333-9
21. Hunger, R.: Floating point operations in matrix-vector calculus. (Version 1.3), Technical Report. Technische Universität München, Associate Institute for Signal Processing (2007)
22. Kubalcik, M., Bobal, V.: Adaptive control of coupled drives apparatus based on polynomial theory. Proc. IMechE Part I: J. Syst. Control Eng. **220**(I7), 641–654 (2006). https://doi.org/10.1109/cca.2002.1040252

Migration and Self-selection: An Agent-Based Model

Samira Boulahbel-Bachari[1]([⊠]) and Nadjia El Saadi[2]

[1] National Center of Research in Applied Economics for Development (CREAD), Bouzareah, Algiers, Algeria
boulahbelsamira@gmail.com
[2] Higher National School of Statistics and Applied Economics Kolea (ENSSEA), Tipaza, Algeria
enadjia@gmail.com

Abstract. We construct an agent-based economic model that is capable of reproducing the mechanisms of internal migration of workers by focusing on the auto-selective characteristic of the subject-matter. Our model considers the potential of people to migrate as well as those who do migrate. Thus, the model introduces a pre-selection stage into the act of migrating. This is mainly performed using a hypothesis, which was previously discussed in other theoretical works covering the dynamics of migration, but rarely tested. To migrate, an individual must have a certain profile and a real pecuniary gain primarily sought through highly probable hiring strategies.

We use an agent-based model to simulate individuals' migration. Two types of agents are considered: Individuals and Regions. We describe each type of agents by a set of equations. We also consider many variables to introduce Individuals' and Regions' properties and heterogeneity.

The use of agent-based model allows for the consideration of a high heterogeneity of agents while at the same time permitting us to test the effect of individuals 'self-selection'. The obtained results show the importance to take self-selection aspect and heterogeneity in migration analysis.

Keywords: Agent-based model · Self-selection · Labor migration · Heterogeneity

1 Introduction

The study of migration has long been at the heart of research in various disciplines like sociology, geography, mathematics or economics. Analyzing such a phenomenon implies real consideration to several parameters that relate to its economic, sociological and demographical aspects. However, the consideration of all these aspects is difficult to realize using the usual methods. The use of agent-based simulation allows the use of several parameters simultaneously and offers the opportunity to study systems at a microeconomic level and simulate the emerging macroeconomic behavior. Agent-based models also allow the testing of multiple configurations of the studied processes quickly and without experiments.

© Springer Nature Switzerland AG 2019
R. Silhavy et al. (Eds.): CoMeSySo 2019, AISC 1047, pp. 288–303, 2019.
https://doi.org/10.1007/978-3-030-31362-3_28

During the last decade, several works studying migration with an agent-based model have been realized. Depending on the area of research and the desired objective of the study, the model's assumptions are addressed. From a sociological point of view, the social impact of migration on the migrants and the region of destination has been examined [1]. For demographers, analyses of migration effect focus on the projection of population growth [2].

In economics, research on modelling migration considers principally the monetary aspect of migration [3]. For example, the work of [4] that discussed agent-based modelling of rural-urban migration considering a statistical mechanic model. The translation into ABM of the work of [5] made by [6] or the work of [7] that studied the relationship between labour migration and rice production to improve management practices are other examples of agent-based economic modelling of migration.

In this work, we construct an agent-based economic model that considers the effect of socioeconomic characteristics of individuals in the propensity and decision of migration. The pecuniary gain of migration and the probability to find a job are also considered. The model describes also the decisions that an individual can take during his lifecycle.

Individuals' decisions are influenced by external events related to employment policies. Variables influencing the propensities of migration are determined from the empirical literature on internal migration [8–10].

The agents' behaviour is controlled by a set of rules of evolution and constraints that are described by mathematical equations. The use of agent-based model allows the consideration of a high heterogeneity of agents and testing the effect of individuals 'self-selection'.

The structure of this paper is as follows. In Sect. 2, the global structure of the proposed migration model is presented. Section 3 presents the main features of the migration agent-based model. Section 4 describes the tested scenarios and the main results obtained followed by a conclusion that summarizes the work presented.

2   Global Structure of the Model

While the economic factor is decisive for internal migration, it is usually found to be highly dependent on the individual characteristics of migrants. In fact, not all individuals react in the same way to pecuniary incentive and chances of finding a job. Therefore, it is important to consider the profile of the individual, his skills and his experiences that constitute his human capital.

This is why, we consider in our model the human capital of an individual by taking into account her\his socioeconomic and demographic characteristics and the pecuniary incentive. Thus, individuals with identical wage differentials and distinct sociodemographic characteristics will have different propensities to migrate. That is we call self-selective aspect of migration. This aspect is introduced in our migration model to determine potential migrants. Several sociodemographic variables are considered to describe this aspect. These variables are age [11], gender [12], qualification [13], marital status [14], employment status [15] and social network [16].

We assume that young and single individuals are more mobile than their elders or those married with children. Also, by gender, we attribute a higher propensity to migrate to men. In terms of educational attainment, we give qualified individuals greater propensities for migration.

The pecuniary incentive of migration is described by the expected gains from migration. These gains increase with individuals' qualifications.

We assume that workers migrate from an Origin Region with a relatively low decent job creation rate to a Destination Region with greater economic potentiality and job offers. The distribution of wages in these two regions is different: the region of origin has lower wages and is widely dispersed, this is mainly due to the preponderance of precarious jobs, unlike the Destination Region that has higher wages and less important wage dispersion. The Destination Region also has a larger share of skilled jobs.

At first, we consider job seekers of the Origin Region. We select those with high migration propensity that will be defined as potential migrants. Once all job seekers in the Origin Region are considered, we calculate for the potential migrants, the expected gains of migration. If these gains are found to be higher than expected gains in the Origin Region, then potential migrants move to the Destination Region. When all potential migrants decide to migrate, or not, the job search begins in the two regions. Individuals with higher hiring probabilities will be hired at first (Fig. 1).

2.1   Conceptual Model

We define two types of agents: Individual Agents (IA) and Region Agents (RA). The Individual Agents seek to maximize their income by choosing migration. These agents are confronted with several choices throughout their lives, while the decisions they take, emerge from a comparison of the different opportunities they meet.

The Region Agents represent the firms and therefore the creation and destruction of jobs.

Individual Agents are defined by the following system of equations:

$$
\begin{cases}
F_j(t) = \sum_{i=1}^{N} C_{ij} P_i \\
W_{i,j} = \bar{W} \times Q_i + \sigma_j \times r(0,1) \\
\pi_i^j(t) = \pi_i^j(t-1) + \left(1 - \pi_i^j(t-1)\right) \times P_i^j(t)
\end{cases}
\tag{1}
$$

Whereby:

- $F_j(t)$ the score's function of the individual j at time t;
- P variable preferences related to each variable;
- C_{ij} the scores attributed to the different modalities of the variables;
- N the number of variables;

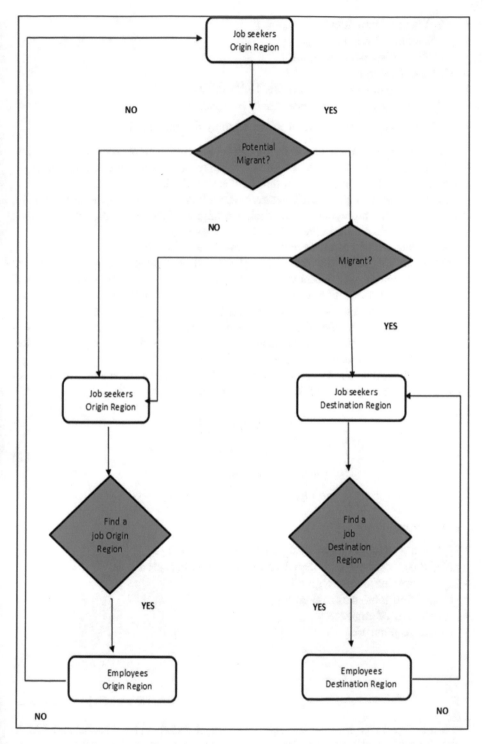

Fig. 1. Individuals' decisions in both regions

- $W_{i,j}$ Wage of individual i in region j;
- $\bar{W}$ National Minimum Wage;
- σ_j Wages dispersion in region j;
- Q_i Qualification of individual i;
- r $(0, 1)$ a random realization of the Uniform law;
- $W^e_{i,j}$ expected wage of the individual i in region j;
- P^j_i Probability to find a formal job in region j for individual i.

Score functions are calculated for IA in a situation of unemployment. This function is then weighted by the tension of the urban labor[1] market and the realization of the uniform law, and is compared to a threshold set by the user

IAs with scores higher than this threshold will have the choice of either migrating or staying to seek employment. They decide to migrate if the wage differential between the Destination Region and the Origin Region is positive.

An update of the economic and demographic variables is done.

Once the migration process is completed, all unemployed people in the two Regions start looking for decent work (formal sector). The probabilities of recruitment are then calculated and IAs with the highest probabilities will be recruited first. An update of the economic and demographic variables of the regions is made.

The last stage of the model concerns the economic and demographic transitions of the two regions as well as a last update of the economic and demographic variables. The simulator then undertakes a new cycle.

The Region Agents represent the firms and therefore the creation and destruction of jobs. Each RA is defined by this system of equations:

$$\begin{cases} QP^j_i(t) = \frac{Pos_j(t)}{A_j} \\ NqP^j_i(t) = \frac{NqPos_j(t)}{A_j} \\ QPos_j(t) = C \times \alpha_j \times Empl_j(t-1) \\ NqPos_j(t) = (1-C)\alpha_j \times Empl_j(t-1) \end{cases} \quad (2)$$

Whereby:

- $QPos_j$ the number of qualified jobs created in region j;
- $NqQPos_j$ the number of unqualified jobs created in region j
- α_j job creation rate in region j;
- C qualified jobs' part in region j;
- $Empl_j$ total of employees;
- A_j Active population.

---

[1] Labour market tension measures the ability of the labour market to absorb new entrants into the labour market. It equals to the ratio between vacancies and the number of new unemployed.

In each region, employment positions are created with different job rates relatively to the jobs existing in the previous period.

3   Simulation and Results

3.1   Simulations

The characteristics of the artificial population used in this work are inspired by those relating to two Algerian cities (Blida and Medea), which both know a significant migratory flow. These two cities correspond respectively to the Destination and the Origin Regions of our model, respectively. Important characteristics of these two regions are summarized in this table (Table 1):

Table 1. Characteristics of region agents

Region	Origin	Destination
Part of skilled employment in total employment	40%	20%
Minimum National Salary Guaranteed	18 000 DZ	
Qualified salaries dispersion	51 000 DZ	94 000 DZ
Unqualified salaries dispersion	25 000 DZ	37 000 DZ

For Individual Agents, the following parameters are used (Table 2):

Table 2. Scores of some selected variables

Variables	Variables' modalities and scores					
Age	16–24	25–34	35–44	45–54	55–59	
Score	6	8	5	2	0	
Education	Without instruction	Primary	College	High school	University	
Score	2	3	4	4	7	
Employment status	1	2	3	4	5	6
Score	5	3	8	0	0	0

The agent-based model contains five main procedures:

Setup Procedure to define the characteristics of the Agents Individuals and Agents Regions:

```
to Setup
      Initiate Population
      Generate Global variables
         Population, ;     Employmentj ;    QEmploymentj ;
UqEmploymentj ;
         Unemployment j;        QUnemploymentj ;      UqUn-
employmentj ;
         Positionsj ;      QPsitionsj ;     UqPsitionsj ;
MarketTensionj;
         NewUnemploymentj ;    MigrationFlow ;    QMi-
grationFlow ;
         UqMigrationFlow.
end
```

Migration Procedure: relating to the region of origin only:

```
to Migration
      Agent Score
      Potential Migrant
          Score x MarketTension x u > = Threshold true
          Then Agent.IsAPotentialMigrant = True
      Expected gains
          If WageDifferential> = 0,
          Then, Agent.IsAMigrant = True
          Move agent to Destination agents' collection
          Else, do nothing
All agents with PotentialMigrant = True were treated?
    Update global variables
end
```

Job Searching Procedure: concerns both regions:

```
to Job Searching
     Calculate HiringProba
HiringProbaj = HiringProbaj + (1 - HiringProbaj) * Selec-
tionProbaj
All unemployed agents were treated?
     Selection of hired agents
      Update global variables
end
```

Updates Procedure: includes all demographic and socio-economic updates for Agents Individuals and Agents Regions.

```
to Updates
t = t + 1
Deaths:
Set Age = Age + 1;
Birth, Marriages, Divorces, New students, School drop-
outs, Educational promotions, Retirement, Entrepreneur-
ship, Inactivity, New labour market entry
end
```

3.2   Scenarios Tested

We consider that an individual decides to migrate if his chances of finding a better-paid formal job in the Destination Region are greater than his chances in the Origin Region. We assume that young and single individuals are the most mobile and can derive profits from migration over a longer period than their elders or those married. Also, according to gender, we assume that a higher proportion of migrants are men. Skilled individuals can move quickly and have a higher probability of finding a job through probably more efficient use of information. We assume that the Destination Region has more potential for hiring and better wages.

Two scenarios relating to migratory flows were selected:

Scenario 1: Massive Migration
The job creation rates of both regions are set so that a massive migration occurs. These job creation rates correspond to 10% for the Destination Region and 2% for the Origin Region.

Scenario 2: Moderate Migration
Moderate migration is obtained by applying a job creation rate of 10% in both regions.

Regarding the number of cumulative migrants in the two situations considered, it is relatively close during the first years, but the gap widens after about twenty years. At the end of the considered period, the number of migrants is close to 8,000 when the migration is massive and 5,000 when it is moderate (Fig. 2).

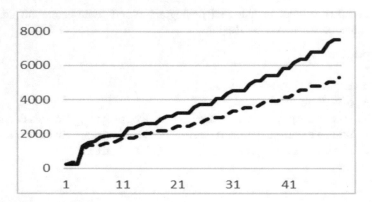

Fig. 2. Cumulative migration evolution in case of massive migration (bold line) and in case of moderate migration (dashed line)

3.3   Results

Migration and Selectivity

Looking at the evolution of migrants number in the case of massive migration, it appears that the number of potential migrants is almost identical to the number of migrants. All potential migrants decide to migrate.

When migration is moderate, (Fig. 3) the potential migrants curve is dissociated from migrants' areas. The number of potential migrants is higher than the number of migrants. Since the chances of finding a job in the Origin Region are greater than that in the case of massive migration, jobseekers stay in their home city to do their prospecting. The number of migrants and potential migrants is increasing over time as the population grows.

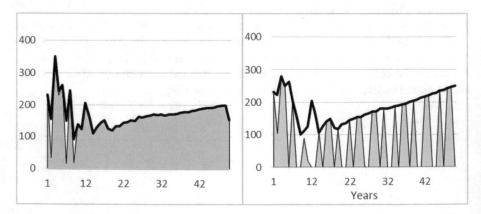

Fig. 3. Evolution of migrants (gray surfaces) and potential migrants (curve) in the presence of massive migration (left) and moderate migration (right).

The comparison between the number of potential migrants and the number of migrants in both cases (massive and moderate migration), shows a selectivity in the act of migrating in the case of moderate migration: not all potential migrants decide to migrate. The probability of finding a job, particularly in the Origin Region, plays an important role in the decision to migrate and deters jobseekers from moving.

This probability is rarely considered in the analysis of migration because it is considered that an individual can work at any time (even if the work remains very precarious).

By looking at the number of potential migrants and the number of migrants according to the qualifications of the individuals, the previous conclusions are confirmed.

Indeed, in the case of massive migration, the number of qualified potential migrants and the number of skilled and unskilled migrants are almost identical (Fig. 4).

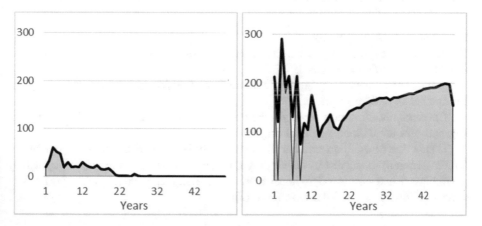

Fig. 4. Evolution of migrants (Gray Surfaces) and potential migrants (Curve) in the presence of massive migration for qualified individuals (Left) and unqualified individuals (Right).

In the case of a moderate migration, the consideration of individuals' qualifications generally gives the same conclusions as those made for all potential migrants and those migrating. However, for skilled individuals, the gap between the number of potential migrants and that of migrants is greater than that found amongst unskilled individuals (Fig. 5)

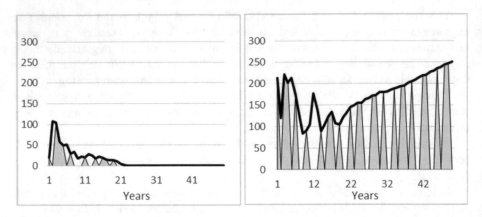

Fig. 5. Evolution of migrants (gray surfaces) and potential migrants (curve) in presence of moderate migration for qualified individuals (left) and unqualified individuals (right).

Migration and Heterogeneity

The aspect of heterogeneity is relative to the composition of migration flows between skilled and unskilled migrants

In the case of massive migration, the number of unskilled migrants is higher than the number of skilled migrants.

The same observation is made for the evolution of the number of qualified potential migrants and unqualified potential migrants.

This is due to the high number of unskilled individuals in the studied population. This is generally observed in developing countries. In Algeria, the active population (unemployed and employed) with a degree (university or vocational training) represents only 18.5% of the total workforce (Fig. 6).

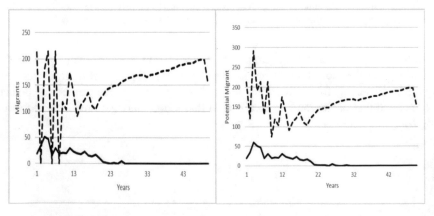

Fig. 6. Evolution of the number of migrants (left) and potential migrants (right) by qualification in the case of a massive migration (qualified individuals in solid line, unqualified individuals in dashed line)

In the case of a moderate migration, we find similar differences between unskilled and skilled migrants (qualified/unqualified potential migrants) in a situation of a massive migration.

This confirms the heterogeneity of migration's flow composition and the necessity to consider it.

Even if the propensities of migration were important for skilled individuals, migration flows remain largely composed of unskilled migrants because of their large numbers in the original population (Fig. 7).

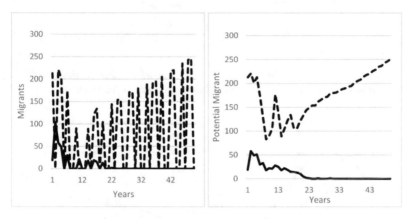

Fig. 7. Evolution of the number of migrants (left) and potential migrants (right) by qualification in the case of a moderate migration (qualified individuals in solid line, unqualified individuals in dashed line).

Heterogeneity and Employment

The evolution of employment is different by region. For the Origin Region, the employment decreases when migration is massive (low job creation rate) and increase when the migration is moderate (high job creation). For the Destination Region, we applied the same high job creation rate (10%). Thus, the trend of employment's evolution in this region has not changed; the only difference is in the volume. That is why, in the next section we will focus in the Origin Region's evolutions.

For the evolution of rural employment in both migration situations, we find that global employment increases in a situation of moderate migration (high job creation rate) and decreases in a situation of mass migration (low job creation rate, significant migration). However, changes in employment by qualification show differences between the evolution of skilled and unskilled employment.

The evolution of unskilled employment is similar to the evolution of total employment, whereas the evolution of skilled employment is different: it increases first and then decreases in the case of massive and moderate migration (Figs. 8, 9).

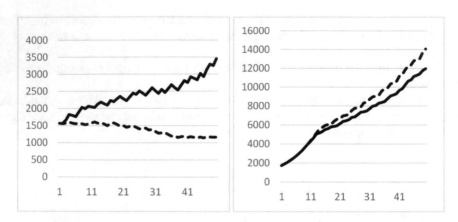

Fig. 8. Employment evolution in case of massive migration (dashed line) and in case of moderate migration (solid line) in the Origin Region (left) and in the Destination Region (right)

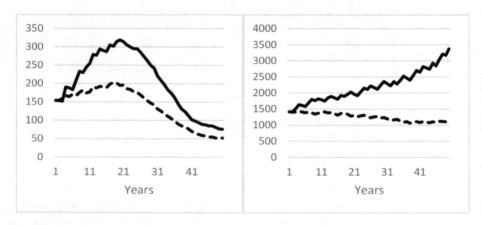

Fig. 9. Employment evolution in the Origin Region by qualification (qualified employment left, unqualified employment right) in case of massive migration (dashed line) and moderate migration (solid line).

Heterogeneity and Unemployment

The same observations are made for the evolution of the unemployment rate. Similarities between the evolution of the global unemployment rate and the unskilled unemployment rate (increase when the migration is massive, decay and stagnation at a low level for the moderate migration), and different evolution for the qualified ones (decrease up to cancellation in case of moderate migration, decrease and then vacillations between increase and cancellation in case of massive migration) (Figs. 9, 10).

These results show the need to take into account the heterogeneity of migrants and workers in the analysis of employment dynamics in order to better target public employment policies (Fig. 11).

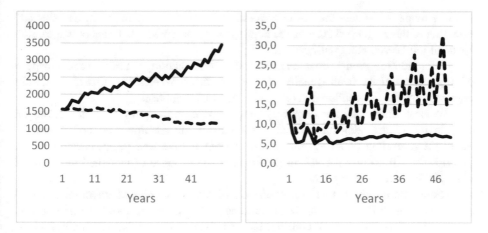

Fig. 10. Unemployment rate evolution (right) and Employment evolution in the Origin Region (left) in case of massive migration (dashed line) and moderate migration (solid line).

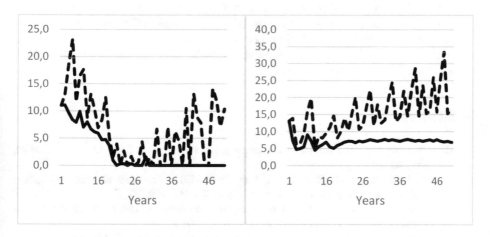

Fig. 11. Unemployment rate evolution in the Origin Region by qualification (qualified unemployment left, unqualified unemployment right) in case of massive migration (dashed line) and moderate migration (solid line).

4   Conclusion

The main aim of our proposed model is to present an economic model that is capable of reproducing the migration process and its implications on the evolution of labor market evolution by considering the self-selection aspect and the heterogeneity characteristics of migration. This has been accomplished through an agent-based approach, with two types of agents: individuals and regions.

The model considers the socioeconomic characteristics of migrants and the self-selection of migration and ensures the heterogeneity of workers in terms of wages, time spent in job search and qualification.

The simulation results show the importance of considering the self-selection aspects of migration and the heterogeneity of workers, most are elements that are generally neglected by the conventional economic models of migration.

We have found that the migratory process as described in our model shows the complexity of analyzing the results: a low rate of job creation does not necessarily incite the migration of all the diasporas, at least for the first years, because skilled individuals will occupy the vacant positions at the expense of unskilled unemployed who will be forced to migrate.

Results found show that the evolution of employment and unemployment by qualification is different in both considered cases (massive and moderate migration).

It is therefore important to consider the specification of each type of unemployment in order to emit the right conclusions about the analysis of labor migrations.

Nevertheless, the model presents some limitations. We are working on a more elaborate version considering the informal employment and the effect of the social network on the decision of migration and on the probability of finding a job.

The aim of this work was to show the possibilities and facilities offered by agent-based models for economic modelling. In the long term, the ambition of this work is to provide a tool that will allow the evaluation and prediction of economic policies for the labor market in developing countries.

References

1. Haug, S.: Migration networks and migration decision-making. J. Ethn. Migr. Stud. **34**(4), 585–605 (2008)
2. Wu, B.M., Birkin, M.H., Rees, P.H.: A dynamic MSM with agent elements for spatial demographic forecasting. Soc. Sci. Comput. Rev. **29**, 145–160 (2011)
3. Klabunde, A., Willekens, F.: Decision-making in agent-based models of migration: state of the art and challenges. Eur. J. Popul. **32**(1), 73–97 (2016)
4. Espındola, A.L., Silveira, J.J., Penna, T.J.P.: A Harris-Todaro agent-based model to rural-urban migration. Braz. J. Phys. **36**(3A), 603 (2006)
5. Harris, J.R., Todaro, M.P.: Migration, unemployment and development: a two-sector analysis. Am. Econ. Rev. **60**(1), 126–142 (1970)
6. El Saadi, N., Bah, A., Belarbi, Y.: An agent-based implementation of the Todaro model. In: Advances in Practical Multi-Agent Systems, pp. 251–265 (2010)
7. Naivinit, W., Le Page, C., Trebuil, G., Gajaseni, N.: Participatory agent-based modelling and simulation of rice production and labour migrations in Northeast Thailand. Environ. Modell. Softw. **25**, 1345–1358 (2010)
8. Stillwell, J., Hussain, S., Norman, P.: The internal migration propensities and net migration patterns of ethnic groups in Britain. Migr. Lett. **5**(2), 135 (2008)
9. Eppley, S., Rosenstiel, T., Graves, C., García, E.: Limits to sexual reproduction in geothermal bryophytes. Econ. Dev. Cult. Change **58**(2), 323–344 (2011)

10. Foulkes, M., Newbold, K.B.: Migration propensities, patterns, and the role of human capital: comparing Mexican, Cuban, and Puerto Rican interstate migration, 1985–1990. Prof. Geogr. **52**(1), 133–145 (2000)
11. Duboz, P., Macia, E., Gueye, L., Boëtsch, G., Chapuis-Lucciani, N.: Migrations internes au Sénégal. Caractéristiques socioéconomiques, démographiques et migratoires des Dakarois. Diversité urbaine, **11**(2), 113–135 (2011)
12. Caldwell, J.C.: African rural-urban migration: the movement to Ghana's towns (1969)
13. Diagne, B., Sakrya, Y.: De La Migration Interne Au Sénégal : Déterminants Et Impact Sur La Pauvreté. Document D'étude N°31 (2015)
14. Graves, P.E., Linneman, P.D.: Household migration: theoretical and empirical results. J. Urban Econ. **6**(3), 383–404 (1979)
15. Da Vanzo, J.: Does unemployment affect migration? Evidence from micro data. Rev. Econ. Stat. **60**(4), 504–514 (1978)
16. Massey, D.S.: Economic development and international migration in comparative perspective. Popul. Dev. Rev. **14**(3), 383–413 (1988)

Analysis of Independences of Normality on Sample Size with Regards to Reliability

Marek Vaclavik[✉], Zuzana Sikorova, and Iva Cervenkova

Department of Education and Adult Education, Faculty of Education,
University of Ostrava, Fr. Sramka 3, 709 00 Ostrava, Czech Republic
{Marek.Vaclavik,Zuzana.Sikorova,
Iva.Cervenkova}@osu.cz

Abstract. In the previous contribution, a guarantee of normality property was proved in case of applied pedagogical research. However, an analysis of this confirmed behaviour can be interestingly identified in case of a research with a large scale sample of data. Therefore, an extended analysis is provided in this paper. According to added of reliability indicators, this analysis is performed on dynamical changed population of the whole file of data with random principle of selection of sub-parts of this file for purposes of their analysis. With regards to this variable sample size, the guarantee of the normality and corresponding conclusions about testing considered hypotheses are discussed in this paper with aim to confirm assumed independences of the normality on the sample size.

Keywords: Quantitative research · Testing normality · Testing hypotheses · Reliability · Applied research

1 Introduction

Behaviour of data properties can be considered as one of an important assumption in the quantitative research [1]. Especially, a property of normality [2] of data is a guarantee for selection of an appropriate statistical method in the testing hypotheses [3]. In this paper, previous research analysis [4] is extended by reliability [5, 6] measurement in case of a large scale sample of research data.

The quantitative research has a significant role, especially in the field of social science; concretely, interesting examples can be seen in widely researched area of the educational science, e.g. in modified e-learning approaches [7, 8], applied pedagogical principles [9–11] as well as proposals and interconnections between process modeling and the pedagogical cybernetics [12–14]. In applied quantitative educational research, testing normality has not been frequently so widely seen and appeared as should be necessary. Due to this underestimation, an incomplete research cannot have then a meaningful value. Therefore, the normality of data [2] should be considered as a primarily based rule before testing the hypotheses on the determined significant level α [3].

The significance level is a carrier assumption in the considered applied research area. In the social science, value of 0.05 can be frequently seen. In the medical science, value of 0.001 appears. Each research hypothesis is expressed by a zero and by an

© Springer Nature Switzerland AG 2019
R. Silhavy et al. (Eds.): CoMeSySo 2019, AISC 1047, pp. 304–311, 2019.
https://doi.org/10.1007/978-3-030-31362-3_29

alternative hypothesis. Results of testing normality and testing hypotheses are achieved in the form of p value [3]. This decision indicator is important for rejecting or fail to rejecting the zero hypothesis. Whether p value is greater or equal to the significance level, then the zero hypothesis is failed to reject. In the opposite case, the zero hypothesis is rejected in favour of the alternative hypothesis [3].

Especially, in an educational based research, the reliability [5] is considered as an important criterion for a determination whether analyzed problems are accurately captured using questionnaires. The achieved reliability greater or equal to 0.7 expresses a suitable designed collecting phase and reproducibility in the quantitative research. Beside the p values of testing normality and testing the hypotheses, reliability coefficients are determined as a support of an accuracy of realized measurements [5].

In this paper, the previous research analysis [4] is extended by the testing the reliability in case of a large scale sample of research data of realized research [15] of Z. Sikorova and her colleagues. In this applied educational research, consequences of an influence of the changing the sample size is analyzed with regards to the measured normality property of data. Achieved results, supported by reliability property of data, can be helpful in favour of the obtainment of a guarantee in the general statistical processing in the research.

2 Testing Normality of Data in Context of Testing Hypotheses

For purposes of testing hypotheses in the quantitative research [1], a foregoing procedure of testing the normality of data [2] is an important introductory step. The testing the normality of n cardinal variables X_i can be performed by various established methods. As frequently recommended methods, methods Anderson-Darling, Shapiro-Wilk or Kolmogorov-Smirnov have been frequently used in the practice. In particular, the first pair of these methods is appeared in freeware software PAST Statistics [16]. Due to utilization of the SPSS version 25 in this paper, Shapiro-Wilk and Kolmogorov-Smirnov tests are applied. In all denoted tests, matrix X (1) can be proved on the normality property. According to the comparison of achieved p values on significance level α, obtained result of testing the i-th hypothesis can be classified into two cases: in (2) the normality is confirmed and in (3) the normality is not proved.

$$i = 1, \ldots, n : X = [X_1 \quad \cdots \quad X_i \quad \cdots \quad X_n]^T \tag{1}$$

$$p \geq \alpha \Rightarrow iH_0^N : X_i \sim N(\mu = 0; \sigma^2 = 1) \tag{2}$$

$$p < \alpha \Rightarrow iH_1^N : X_i \nsim N(\mu = 0; \sigma^2 = 1) \tag{3}$$

With regards to the normality property of data X, appropriate method for testing the research hypotheses can be determined. In order to non-normality property of data, non-parametrical tests are used. In the opposite case, the parametrical tests are utilized. Wide practically utilizations can be seen e.g. in [3, 20–23].

In the category of non-parametrical tests [17], Mann-Whitney test (2 categorical items) or Kruskal-Wallis (3 and more categorical items) are incorporated. T-test [18] (2 categorical items) or ANOVA [19] (3 and more categorical items) belong to the parametrical tests. All denoted tests consider following zero and alternative hypotheses (Table 1) in the frame of testing the existence of statistically significant dependences of the cardinal variable on categorical m items [3].

Table 1. Hypotheses in testing existence of statistically significant dependences

Hypothesis		Definition
iH	$iH_0 : \mu_1 = \cdots = \mu_m$	There are not statistically significant dependences of cardinal variable X_i on categorical variable with m items
	$iH_1 : \mu_1 \neq \cdots \neq \mu_m$	There are statistically significant dependences of cardinal variable X_i on categorical variable with m items

Achieved results of testing the existence of the statistically significant dependences (Table 1) are divided into two cases (4)–(5) according to the returned p value in the statistical software.

$$p \geq \alpha \Rightarrow iH_0 : \mu_1 = \cdots = \mu_m \tag{4}$$

$$p < \alpha \Rightarrow iH_1 : \mu_1 \neq \cdots \neq \mu_m \tag{5}$$

3   Consequences of Data Normality Supported by Reliability

In this paper, a main part of the performed analysis is based on an identification of a stability behavior of results of the normality. Concretely, dependence or independence of the normality property on a size of data samples. The previous research [4] has been published focusing on the applied normality tests; however, case of a large sample of population is considered in this paper.

Large population of respondents can bring a stable background for the testing properties of data in the frame of the quantitative research [1]. In case of changing the sample size, an appropriate behaviour of considered property can be widely analyzed.

For a guarantee of the normality behavior of cardinal data X should obtain the aimed conclusion (6), that the normality is consisted across various sizes of samples of data. As various samples of data, random parts of matrix X are taken into account. In expression (6), a sharp disjunction is considered.

$$\forall p : (p \geq \alpha) \underline{\vee} (p < \alpha) \tag{6}$$

Beside the p values of testing the normality and testing the hypotheses, reliability coefficients are determined as a support of an accuracy of realized measurements. Also in these analyses, random parts of matrix X are considered. The achieved reliability greater or equal to 0.7 express a suitable designed collecting phase and reproducibility in the quantitative research. Following reliability criterions are used in SPSS version 25, as can be seen in Table 2.

Table 2. Utilized reliability coefficients [6] supporting analysis of normality

Symbol	Coefficient of reliability
r_α	Cronbach Alpha
r_P	Parallel Reliability Coefficient
r_G	Guttman's Reliability Coefficient
r_{SP}	Strict Parallel Reliability Coefficient

4   Results

In this paper, the influence of a sample size of data is analyzed in case of the applied educational research. In comparison with previous research [4], a large size of samples is appeared in the practically investigated quantitative research. Two tests of Shapiro-Wilk and Kolmogorov-Smirnov achieved conclusions about a normality of data and are discussed after their processing in the software IBM SPSS 25.

Behaviour of p values was analyzed in both cases of testing the normality of data and testing the hypotheses. There was declared the significance level 0.05.

In the educational research of Z. Sikorova and her colleagues in 2019 [15], data was collected from October to December 2018. This research was focused on a wide spectrum of research hypotheses based on analyses of using the educational materials by university students. There was also performed the confirmation factor analysis. Questionnaire items corresponded to the Likert scales and there were identified three approaches to learning: deep, surface and strategic. Respondents had the structure of a whole population of 2671 respondents from different faculties of University of Ostrava.

In this paper, 83 cardinal variables in matrix X were declared and following hypotheses (Table 3) were selected for purposes of the proposed statistical analysis.

According to the changed sample size of data, following results of the normality and reliability tests were achieved as can be seen in Table 4. Applied reliability coefficients are also displayed in Fig. 1.

Table 3. Hypotheses in testing existence of statistically significant dependences

Hypothesis		Definition
$1H$	$1H_0$	There are not statistically significant dependences of using materials of the current teacher on gender
	$1H_1$	There are statistically significant dependences of using materials of the current teacher on gender
$2H$	$2H_0$	There are not statistically significant dependences of the appearance of learning in semester on the faculty (Faculty of Education/Faculty of Medicine)
	$2H_1$	There are statistically significant dependences of the appearance of learning in semester on the faculty (Faculty of Education/Faculty of Medicine)
$3H$	$3H_0$	There are not statistically significant dependences of the frequency of using webinars or e-courses on the faculty (Faculty of Education/Faculty of Medicine)
	$3H_1$	There are statistically significant dependences of the frequency of using webinars or e-courses on the faculty (Faculty of Education/Faculty of Medicine)
$4H$	$4H_0$	There are not statistically significant dependences of studies without visiting a library on the year of study ($1^{st}/2^{nd}/3^{rd}$)
	$4H_1$	There are statistically significant dependences of studies without visiting a library on the year of study ($1^{st}/2^{nd}/3^{rd}$)
$5H$	$5H_0$	There are not statistically significant dependences of using textbooks on the form of study (full-time/part-time)
	$5H_1$	There are not statistically significant dependences of using textbooks on the form of study (full-time/part-time)

Table 4. Achieved reliability coefficients for purposes of supporting normality

Size [%] of X	r_α	r_P	r_G	r_{SP}	Normality of data X
100	0.777	0.777	0.768	0.706	$p = 0.000$ Rejected
95	0.774	0.780	0.765	0.706	$p = 0.000$ Rejected
90	0.774	0.779	0.770	0.708	$p = 0.000$ Rejected
85	0.776	0.784	0.767	0.705	$p = 0.000$ Rejected
80	0.771	0.773	0.770	0.712	$p = 0.000$ Rejected
75	0.773	0.775	0.778	0.707	$p = 0.000$ Rejected
70	0.774	0.778	0.763	0.714	$p = 0.000$ Rejected
65	0.775	0.772	0.765	0.704	$p = 0.000$ Rejected
60	0.777	0.770	0.763	0.718	$p = 0.000$ Rejected
55	0.785	0.765	0.764	0.712	$p = 0.000$ Rejected
50	0.782	0.780	0.774	0.707	$p = 0.000$ Rejected
45	0.773	0.777	0.774	0.716	$p = 0.000$ Rejected
40	0.783	0.770	0.774	0.716	$p = 0.000$ Rejected
35	0.767	0.766	0.794	0.707	$p = 0.000$ Rejected
30	0.767	0.771	0.770	0.706	$p = 0.000$ Rejected
25	0.773	0.761	0.756	0.720	$p = 0.000$ Rejected
20	0.790	0.790	0.778	0.713	$p = 0.000$ Rejected
15	0.779	0.765	0.776	0.709	$p = 0.000$ Rejected
10	0.781	0.768	0.725	0.710	$p = 0.000$ Rejected

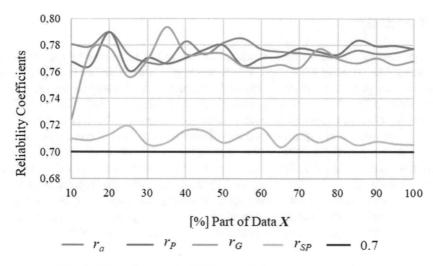

Fig. 1. Dependences of reliability coefficients on size of samples

Due to the changed sample size of data, same conclusions of testing hypotheses 1*H*–5*H* were achieved, as can be seen in Table 5. For testing the hypotheses 1*H*, 2*H*, 3*H* and 5*H*, Mann-Whitney test was applied. In hypothesis 4*H*, Kruskal-Wallis test was used. Results of *p* values, which correspond to failing to reject the zero hypotheses are written by bold font in this table.

Table 5. Achieved results of testing considered hypotheses

Size [%] of X	p (1H)	p (2H)	p (3H)	p (4H)	p (5H)
100	0.000	0.000	0.000	**0.318**	**0.378**
95	0.000	0.000	0.000	**0.457**	**0.345**
90	0.000	0.000	0.000	**0.514**	**0.453**
85	0.000	0.000	0.000	**0.372**	**0.655**
80	0.000	0.000	0.000	**0.179**	**0.437**
75	0.000	0.000	0.000	**0.830**	**0.144**
70	0.000	0.000	0.000	**0.710**	**0.151**
65	0.000	0.000	0.000	**0.430**	**0.093**
60	0.000	0.000	0.000	**0.455**	**0.591**
55	0.000	0.000	0.000	**0.219**	**0.200**
50	0.000	0.000	0.000	**0.707**	**0.787**
45	0.000	0.000	0.000	**0.840**	**0.566**
40	0.010	0.000	0.000	**0.884**	**0.898**
35	0.030	0.000	0.000	**0.537**	**0.268**
30	0.013	0.001	0.000	**0.911**	**0.927**
25	0.049	0.000	0.000	**0.312**	**0.278**
20	0.000	0.000	0.002	**0.118**	**0.923**
15	0.004	0.002	0.000	**0.119**	**0.975**
10	0.010	0.000	0.013	**0.781**	**0.450**

5   Conclusion

On a practical example of the educational research with large size of population, the normality data property was tested with supporting by the reliability coefficients. This analysis was performed using dynamical changed size of samples by the random principle in SPSS 25. The normality was tested using two methods of Shapiro-Wilk and Kolmogorov-Smirnov. Achieved results confirmed the stable behaviour of the analyzed normality property during the changed size of samples. Data and its considered parts had not the normal probability distribution. Reliability coefficients of data and of its tested parts were greater than 0.7. This conclusion is advantageous with respect to the guarantee for the obtainment of the stable results even in the testing the hypotheses. Also, the similar stable behaviour of the conclusion of testing the considered hypotheses was proved. Therefore, the normality can be considered as guaranteed property on the applied significance level 0.05 in the research with large sample of data.

References

1. Stockemer, D.: Quantitative Methods for the Social Sciences. Springer, Heidelberg (2019). https://doi.org/10.1007/978-3-319-99118-4
2. Alizadeh Noughabi, H.: Two powerful tests for normality. Ann. Data Sci. **3**(2), 225–234 (2016). https://doi.org/10.1007/s40745-016-0083-y. ISSN 2198-5812
3. Kitchenham, B., Madeyski, L., Budgen, D., et al.: Robust statistical methods for empirical software engineering. Empir. Softw. Eng. **22**, 1–52 (2016). https://doi.org/10.1007/s10664-016-9437-5. ISSN 1573-7616
4. Vaclavik, M., Sikorova, Z., Barot, T.: Particular analysis of normality of data in applied quantitative research. In: Computational and Statistical Methods in Intelligent Systems. Advances in Intelligent Systems and Computing, vol. 859, pp. 353–365. Springer (2019). https://doi.org/10.1007/978-3-030-00211-4_31. ISBN 978-3-030-00210-7
5. Zacks, S.: Introduction to Reliability Analysis: Probability Models and Statistical Methods. Springer, Heidelberg (1992)
6. IBM Knowledge Center: Reliability Analysis. https://www.ibm.com/support/knowledge center/el/SSLVMB_23.0.0/spss/base/idh_reli.html
7. Kostolanyova, K.: Adaptation of personalized education in e-learning environment. In: 1st International Symposium on Emerging Technologies for Education, pp. 433–442. Springer (2017). https://doi.org/10.1007/978-3-319-52836-6_46. ISBN 978-3-319-52835-9
8. Schoftner, T., Traxler, P., Prieschl, W., et al.: E-learning introduction for students of the first semester in the form of an online seminar. In: Pre-Conference Workshop of the 14th E-Learning Conference for Computer Science, pp. 125–129. CEUR-WS (2016). ISSN 1613-0073
9. Simbartl, P., Honzikova, J.: Using demonstration to design creative products focused on folk traditions. In: 8th International Conference on Education and New Learning Technologies, pp. 2832–2837. IATED (2016). https://doi.org/10.21125/edulearn.2016.1613
10. Krammer, G., Vogl, H., Linhofer, S., et al.: Learning opportunities provided by erasmus as mobility programme in teacher education. In: Holz, O. (ed.) Current Trends in Higher Education in Europe, pp. 279–288. Lit Veralg (2016). ISBN 978-3643134196
11. Vicherkova, D.: Communication strategy in the area of readership of 15-year-old pupils in the Moravian-Silesian Region. Grant J. **7**(1), 107–111 (2018). ISSN 1805-0638

12. Barot, T.: Possibilities of process modeling in pedagogical cybernetics based on control-system-theory approaches. In: Cybernetics and Mathematics Applications in Intelligent Systems. Advances in Intelligent Systems and Computing, vol. 574, pp. 110−119. Springer (2017). https://doi.org/10.1007/978-3-319-57264-2_11. ISBN 978-3-319-57263-5

13. Vaclavik, M., Sikorova, Z., Barot, T.: Approach of process modeling applied in particular pedagogical research. In: Cybernetics and Automation Control Theory Methods in Intelligent Algorithms. In: Advances in Intelligent Systems and Computing, vol. 986, pp. 97−106. Springer (2019). https://doi.org/10.1007/978-3-030-19813-8_11. ISBN 978-3-030-19812-1

14. Barot, T.: Adaptive control strategy in context with pedagogical cybernetics. Int. J. Inf. Commun. Technol. Educ. 6(2), 5–11 (2017). ISSN 1805-3726

15. Sikorova, Z., Barot, T., Vaclavik, M., Cervenkova, I.: Czech university students' use of study resources in relation to the approaches to learning. New Educ. Rev. Adam Marszalek Publishing House (Article in Press)

16. Hammer, O., Harper, D.A.T., Ryan, P.D.: PAST: paleontological statistics software package for education and data analysis. Palaeontologia Electronica. 4(1), 9 (2001). http://palaeo-electronica.org/2001_1/past/issue1_01.htm

17. Fischer, D., Oja, H.: Mann-Whitney type tests for microarray experiments: the R package gMWT. J. Stat. Softw. 65(1), 1–19 (2015). https://doi.org/10.18637/jss.v065.i09. ISSN 1548-7660

18. Lazic, S.E.: Why we should use simpler models if the data allow this: relevance for ANOVA designs in experimental biology. BMC Physiol. 8(18), 16 (2008). https://doi.org/10.1186/1472-6793-8-16. ISSN 1472-6793

19. Rasch, D.: The two-sample t test: pre-testing its assumptions does not pay off. Stat. Pap. 52(1), 219–231 (2011). https://doi.org/10.1007/s00362-009-0224-x. ISSN 1613-9798

20. Sulovska, K., Belaskova, S., Adamek, M.: Gait patterns for crime fighting: statistical evaluation. In: Proceedings of SPIE - The International Society for Optical Engineering, vol. 8901. SPIE (2013). https://doi.org/10.1117/12.2033323. ISBN 978-081949770-3

21. Pivarc, J.: Ideas of Czech primary school pupils about intellectual disability. Educ. Stud. Taylor & Francis (2018, in press). https://doi.org/10.1080/03055698.2018.1509784. ISSN 0305-5698

22. Gruzenkin, V.D., Mikhalev, A.S., Grishina, G.V., et al.: Using blockchain technology to improve N-version software dependability. In: Computational and Statistical Methods in Intelligent Systems. Advances in Intelligent Systems and Computing, vol. 859, pp. 132−137. Springer (2019). https://doi.org/10.1007/978-3-030-00211-4_14. ISBN 978-3-030-00210-7

23. Petlak, E., Tistanova, K., Juszczyk, S.: Undesirable behaviour of pupils towards teachers in Slovak schools. New Educ. Rev. 55(1), 170–184 (2019). https://doi.org/10.15804/tner.2019.55.1.14. ISSN 1732-6729

Novel Analytical Model for Resource Allocation Over Cognitive Radio in 5G Networks

B. P. Vani[(⊠)] and R. Sundaraguru

Department of Electronics and Communication Engineering,
Sir M. Visvesvaraya Institute of Technology, Bengaluru, India
vani.smvit@gmail.com

Abstract. The framework of 5G with cognitive radio network faces a challenge of heterogeneity and non-synchronization. Existing approaches towards utilizing cognitive radio over 5G network for the purpose of resource allocation is strictly defined on smaller subsets of the nodes as well as non-consideration of practical parameters that is affected by the presence of artifacts like interference. Therefore, the proposed manuscript introduces a novel evaluation model that investigates the impact of mathematical optimization over resource allocation over 5G networks. The study uses an analytical research methodology where disciplined convex optimization is used as a part of improving the process of resource allocation along with novelty of consideration of practical parameters. The simulated outcome of study shows that proposed system offers better communication performance in contrast to existing resource allocation schemes.

Keywords: Resource allocation · Optimization · Spectral efficiency ·
Cognitive radio network · 5G network

1 Introduction

The evolution of wireless mobile communication system is being evolved from 2G family to 3G, 4G and now 5G with evolving technologies like GSM, CDMA, TDMA, EDGE, W-CDMA, HSUPA, WIMAX, LTE and LTE-A for different data rate requirement [1]. The 5G aims for optimal data rate, bandwidth and delay to have a seamless operation of internet of things (IoT), robotics and Industry 4.0 based connected control system etc. with a futuristic application paradigm that operates globally irrespective of the device type and location [2]. The targeted applications expect to design a suitable carrier to meet a latency less than 1 ms as well the connectivity should be many folds faster as compared to the existing 4G-LTE to have higher scale network among billions of devices from IoT. It is an evidential fact of growing subscription of LTE demands a better quality of user experience at minimal cost. It can be achieved by means of advancement into terminals and collaboration of many cells so that various communication and mobile data-based services is conceptualized which is the vision of upcoming mobile technologies and 5G standard. The transition from 4G to 5G for its associated devices, network service providers and infrastructure builders are on the fast

© Springer Nature Switzerland AG 2019
R. Silhavy et al. (Eds.): CoMeSySo 2019, AISC 1047, pp. 312–321, 2019.
https://doi.org/10.1007/978-3-030-31362-3_30

pace of development, where small cell and multi-cell coordinated test beds are already into existence. The 5G based applications will exploit many radio access networks for both short- and long-range communication in highly coordinated manner in every layer of communication. This paper discusses one such solution of optimal resource allocation in 5G network. Section 2 discusses about the existing research work followed by problem identification in Sect. 3. Section 4 discusses about proposed methodology followed by elaborated discussion of algorithm implementation in Sect. 5. Comparative analysis of accomplished result is discussed under Sect. 6 followed by conclusion in Sect. 7.

2 Related Work

This section brief of review in approaches used in existing system towards allocation of resources [3]. Optimization scheme considering the cognitive network was found to be effective toward resource allocation scheme as seen in work of Hong et al. [4]. The work of Ejaz et al. [5] has addressed the resource allocation problem considering the case study of device-to-device communication over cognitive network in 5G. Liu et al. [6] have targeted to enhance the cumulative capacity of secondary user for maximized energy conservation as well as efficiency of spectrum. Rattarao et al. [7] have discussed about a preemptive mechanism of allocation of multiple resources using admission control theory. Allocation of subcarrier towards improving the accessibility of spectrum is carried out by Wang et al. [8] with pure focus on power allocation. Power efficiency was also improved by using cognitive radio as witnessed in model of Zappone et al. [9]. Study towards radio resource management towards upcoming network was discussed by Lien et al. [10]. The work of Hu et al. [11] have used cognitive radio considering the constraint of outage possibility of primary users. Adoption of cognitive radio over content-centric network was seen in the work of Gur and Kafiloglu [12]. Park and Hwang [13] have proved that cognitive radio over the femto-cells could offer energy efficiency in 5G networks. He et al. [14] have used Markov principle for an effective formulation of decision towards specific allocation of resources. Zhang et al. [15] have carried out research work towards allocation of spectrum using swarm-based optimization principle. Sboui et al. [16] have proved that adoption of cognitive radio significantly contributes towards obtaining demanded transmission rate over unmanned vehicles. Literature has also presented discussion where it was claimed that pairing concept of user over the 5G network is studied (Ding et al. [17]). Kibria et al. [18] have presented a study towards improving the data delivery performance considering multicarrier system with cognitive radio. Diamantoulakis et al. [19] have presented a discussion of using convex optimization tool for improving the data aggregation property of relay node in cooperative network with cognitive radio. Apart from this, there are various other works being carried out towards resource allocation in cognitive network viz [20–34]. The next section outlines the problems.

3   Problem Description

Review of existing resource allocation approaches shows that majority of the schemes have never found to consider the multi-cellular structure of 5G network. This non-consideration of structure causes the modeling to limit to only single link which is practically associated with limited resource allocation for data transmission. Apart from this, the consideration of cognitive radio has been found without much association with cost effective beam forming which significantly affects the antenna configuration when the downlink transmission takes place. There is also a lack of a standardized and benchmarked modeling practice which leads to less exploration of an effective test-bed. Therefore, problem statement is "to develop a computational model that is capable of an effective resource allocation over 5G network equipped with cognitive radio".

4   Proposed Methodology

The proposed study aims for developing a scheme that can perform an effective analysis of the resource allocation when carried out over multi-cellular 5G networks. The complete resource allocation strategy and algorithm is formulated on the basis of the users (primary/secondary) operation over cognitive radio network. Figure 1 highlights the scheme.

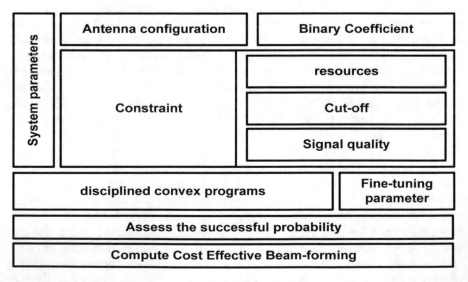

Fig. 1. Proposed schematic diagram

Figure 1 highlights the considering of novel system parameters for proposed system that consists of parameters derived from cognitive radio (antenna configuration, signal quality) mainly that offers better spectral utilization over multi-cellular communication in 5G. The model considers practical constraint factors followed by applying convex optimization scheme in order to obtain probability of successful allocation of resources over multicell 5G network and to compute the beam forming.

5 System Implementation

The core goal of the proposed study is to ensure that inclusion of cognitive radio over multi-cell 5G networks offers an effective resource allocation process. The main idea is also to offer an evaluation platform where the analysis of resource allocation can be effectively carried out for a given set of constraints connected with practical resource consumption while propagating data over multi-cell 5G networks. This section offers a discussion of strategy adopted for implementation followed by flow adopted for execution of proposed algorithm.

5.1 Implementation Strategy

The primary strategy of the proposed system is to consider the presence of cognitive radio in 5G network in order to offer more scalability towards spectral presence over multiple wireless devices over 5G network. Therefore, main idea by inclusion of cognitive radio will be to address the issue associated with the scarcity of the spectrum and enhance the efficiency of the spectrum. The complete implementation strategy of the proposed system is carried out on the basis of the fact that – in order to carry out allocation of dynamic resources while attempting to transmit data over multi-cells, it is essential to maximize the utility function. This operation can be carried out by assigning a specific resource required for transmitting data towards all the connected users considering diversified direction over the multi-cells in 5G networks. For the purpose of proper operation, it is required that the process of such resource allocation should cater up the practical constraints of resources exclusively for the downlink data transmission as well as allocation of specific resources. By doing this, the problems associated with the interference vector between the secondary and primary users in cognitive radio are addressed. The proposed system introduces a scheme that is basically responsible for exploring the degree of probability towards obtaining better quality of service while performing peak data transmission over 5G network on multi-cells. The study considers impairment caused over the communication device responsible for rendering data forwarding and receiving signals over the cognitive radio device. The formulation of the model of proposed study also considers artifact value in order to represent this extent of impairment from the hardware aspect of the transmitting and receiving device. It should be noted that artifact value is basically a constant parameter that is dependent upon power factor associated with distortion and signal. According to this strategy, the proposed system renders a spontaneous communication between the primary and secondary base station which leads to generation of a highly connected multi-cells in 5G network. The core impact of this implementation strategy is basically a seamless communication between the primary and secondary user where dynamic traffic has no much adverse effect over the communication quality over the 5G network. Therefore, the proposed system advocates that scheme of considering essential constraint factors connected with quality of service of cognitive radio, which directly influences the conservation of essential resources of network. This scheme can also offer practical implementation scenario as well as evaluation test-bed for assessing performance of communication over multi-cell 5G network.

5.2   Execution Flow

In order to carry out implementation, the proposed system constructs an algorithm that is responsible for primarily explore if there is a better way to utilize the concept of resource allocation using cognitive radio network over multi-cell network like 5G. An algorithm is constructed and it takes the input of α (antenna configuration), β (binary coefficient), θ (resource constraint), γ (cut-off resource constraint), ψ (signal quality constraint), and τ (artifact value). The algorithm after processing yields and outcome of ϕ (probability), χ (cost effective beam forming). Following are the details of the input files:

- α (Antenna Configuration): It is basically a matrix constructed for respositing the index values of the antenna connected with transmitter and receiver of cognitive radio. The dimension of this matrix is (u x u x n).
- β (Binary Coefficient): This variable is another matrix constructed diagonally with its each element 1 if they are part of transmit cognitive radio and 0 if it is part of receiving cognitive radio.
- θ (resource constraint): This matrix reposits all the values of the c resource constraints with a dimension of n x n x c.
- γ (cut-off resource constraint): It acts as threshold for c number of resource constraint while communication takes place over multiple cell.
- ψ (signal quality constraint): This variable represents a communication vector over multi-cell in 5G with abstract specification of signal quality over presence of noise applicable for all users.

In the initialization step (Line-1), the algorithm initializes number of users u, quantity of total transmit cognitive radio n, and quantity of resource constraints c. The proposed system utilizes *disciplined convex programs* [35] that offers wider range of supportability for various mathematical problems. This scheme is used for ensuring minimization of resource consumption considering various demands of quality of service while communicating under multiple-cells over 5G. The study constructs a 3 discrete variables viz. i) a complex variable W mapping with matrix of beamforming, ii) a fine-tuning parameter f_p associated with resource constraint, and iii) $\arg_{min}(f_p)$ for reducing the resource consumption by fine-tuning resource constraints. The algorithm considers all the value of mobile user u (Line-2 and Line-3) and compute allocated channel system with respect to resources available (Line-4). The computation is carried out by scalar product of antenna coefficient α and binary coefficient β (Line-4). The next part of the algorithm is associated with extraction of imaginary ig and real re values obtained from allocated channel Z (Line-6 and Line-7). The next part of the implementation is associated with the normalization of the resource constraints for all the value of it i.e. c (Line-9 and Line-10). According to this conditional check, the proposed system assess if the normalized value of resource constraint θ is less than fine-tuning parameter for resource constraint i.e. f_p (Line-10). The computation of the proposed *disciplined convex programs* is executed till this stage to reposit all the values of f_p to be positive integers.

The next part of the algorithm is all about analysis of the outcome in order to ensure better computation of probability of the resource constraint allocation system over multi-cells of cognitive radio in 5G. In this, the algorithm checks if both problem of reducing resources as well as probability are practical or not (Line-12). The value of the ϕ (probability) variable is assigned false index or the value of the fine-tuning parameter is found to be more than 1 than the proposed system finds the status of *disciplined convex programs* to be impractical or else it allocates the truth index to the ϕ (probability) variable while χ (cost effective beam forming) is computed from W (Line-18).

Algorithm forProbability Calculation for Resource Allocation

Input: α (antenna configuration), β (binary coefficient), θ (resource constraint), γ (cut-off resource constraint), ψ (signal quality constraint), τ (artifact value)

Output: ϕ (probability), χ (cost effective beam forming)

Start

1. init $\alpha, \beta, \gamma, \theta, \psi, \tau, \phi, \chi$

2. **For** i=1: u

3.   **For** j=1: u

4. $Z \rightarrow \alpha * \beta$

5.   **End**

6. ig(Z)$\rightarrow$0

7. ......re(Z)$\rightarrow \sqrt{} \psi(i).norm(Z)$.

8. **End**

9. **For** m=1: c

10.   norm(θ)$\leq$fp. ($\sqrt{\gamma}$)

11. **End**

12. **If** (cond=u_{solved})

13. $\phi \rightarrow$false

14. **elseif** fp>1

15. $\phi \rightarrow$false

16. **Else**

17. $\phi \rightarrow$true

18. $\chi \rightarrow$W

End

6   Results Discussion

The implementation of the proposed work is carried out in MATLAB considering 500 nodes spread across the multi-cell in 5G network. For an effective analysis, the outcome of the proposed study was compared with frequently existing approaches i.e. scheduling using MIMO [36–38], energy harvesting [39], and bio-inspired approach [40, 41] as showcased in Table 1.

Table 1. Comparative analysis

Resource allocation schemes		Supports peak traffic	Higher device connectivity	Higher applicability
Existing system	Scheduling using MIMO [36–38]	Yes	Yes	No
	Energy harvesting [39]	No	No	Yes
	Bio-inspired [40, 41]	Yes	No	Yes
Proposed		Yes	Yes	Yes

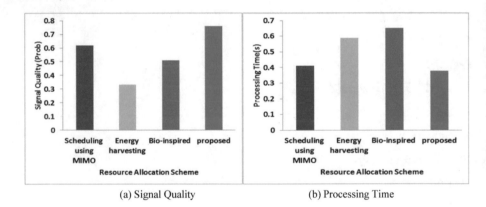

(a) Signal Quality   (b) Processing Time

Fig. 2. Comparative analysis outcomes

The outcome shown in Fig. 2 and Table 1 highlights that proposed system is quite capable of introducing a cost effective modeling of resource allocation. The outcome also shows better signal quality (Fig. 2(a)) and reduced processing time (Fig. 2(b)) in contrast to existing system. The prime reason behind this is –proposed system offers highly reduced dependencies on apriori computation of resources and does all the computation dynamically that reduces the processing time and reduced dependencies of network parameters. However, existing system introduces quite a complex mechanism of computing resources and exploring the elite outcome with inclusion of massive parameters. Apart from this, the practicality of the proposed system is quite more as the entire modeling is carried out considering real-time constraints associated with cognitive radio operation over 5G environment shrouded with artifacts. Hence, signal quality also proves better spectral efficiency computed in probability score to claim that better communication standards are achieved by proposed system.

7 Conclusion

There is no doubt that introduction of cognitive radio networks offers extensive spectral efficiency of the wireless network especially in large scale cellular network. However, existing system has less evidence to prove the optimality of the solution using certain benchmarked approach. Therefore, the proposed system introduces a computational mechanism to investigate the effect of proposed resource allocation concept over multi-cellular structured in 5G networks. The study also incorporates cognitive radio for better decision making of users by the system followed by added advantage of improved spectral efficiency. Apart from this, the consideration of real-time constraints also offers preemptive and practical applicability of the proposed system. The study outcome shows that proposed system offers cost effective allocation of resource from the existing resource allocation practices.

References

1. Ahmadi, S.: Mobile WiMAX: A Systems Approach to Understanding IEEE 802.16m Radio Access Technology, p. 784. Academic Press, Technology & Engineering, Cambridge (2010)
2. Casagrande, L., Gruber, V., Marcelino, R.: IoT and the industry 4.0: principles and educational applications. In: SPS, Educational Technology, p. 76 (2016)
3. Ramchandran, K., Zeien, E., Andreasen, N.C.: Distributed neural efficiency: intelligence and age modulate adaptive allocation of resources in the brain. Trends Neurosci. Educ. **15**, 48–61 (2019)
4. Hong, X., Zheng, C., Wang, J., Shi, J., Wang, C.: Optimal resource allocation and EE-SE trade-off in hybrid cognitive Gaussian relay channels. IEEE Trans. Wirel. Commun. **14**(8), 4170–4181 (2015)
5. Ejaz, W., Ibnkahla, M.: Multiband spectrum sensing and resource allocation for IoT in cognitive 5G networks. IEEE Internet Things J. **5**(1), 150–163 (2018)
6. Liu, M., Song, T., Hu, J., Sari, H., Gui, G.: Anti-shadowing resource allocation for general mobile cognitive radio networks. IEEE Access **6**, 5618–5632 (2018)
7. Rattaro, C., Bermolen, P., Belzarena, P.: Multi-resource allocation: analysis of a paid spectrum sharing approach based on fluid models. IEEE Trans. Cogn. Commun. Netw. **4**(3), 607–617 (2018)
8. Wang, D., Ren, P., Du, Q., Sun, L.: Reciprocally benefited spectrum access scheme with joint power and subcarrier allocation in a software-defined network. IEEE Access **3**, 1248–1259 (2015)
9. Zappone, A., Matthiesen, B., Jorswieck, E.A.: Energy efficiency in MIMO underlay and overlay device-to-device communications and cognitive radio systems. IEEE Trans. Sig. Process. **65**(4), 1026–1041 (2017)
10. Lien, S., Chen, K., Liang, Y., Lin, Y.: Cognitive radio resource management for future cellular networks. IEEE Wirel. Commun. **21**(1), 70–79 (2014)

11. Hu, H., Zhang, H., Li, N.: Location-information-assisted joint spectrum sensing and power allocation for cognitive radio networks with primary-user outage constraint. IEEE Trans. Veh. Technol. **65**(2), 658–672 (2016)
12. Gür, G., Kafiloğlu, S.: Layered content delivery over satellite integrated cognitive radio networks. IEEE Wirel. Commun. Lett. **6**(3), 390–393 (2017)
13. Park, H., Hwang, T.: Energy-efficient power control of cognitive femto users for 5G communications. IEEE J. Sel. Areas Commun. **34**(4), 772–785 (2016)
14. He, H., Shan, H., Huang, A., Sun, L.: Resource allocation for video streaming in heterogeneous cognitive vehicular networks. IEEE Trans. Veh. Technol. **65**(10), 7917–7930 (2016)
15. Zhang, X., Zhang, X., Han, L., Xing, R.: Utilization-oriented spectrum allocation in an underlay cognitive radio network. IEEE Access **6**, 12905–12912 (2018)
16. Sboui, L., Ghazzai, H., Rezki, Z., Alouini, M.: Achievable rates of UAV-relayed cooperative cognitive radio MIMO systems. IEEE Access **5**, 5190–5204 (2017)
17. Ding, Z., Fan, P., Poor, H.V.: Impact of user pairing on 5G nonorthogonal multiple-access downlink transmissions. IEEE Trans. Veh. Technol. **65**(8), 6010–6023 (2016)
18. Kibria, M.G., Villardi, G.P., Ishizu, K., Kojima, F.: Throughput enhancement of multicarrier cognitive M2M networks: universal-filtered OFDM systems. IEEE Internet Things J. **3**(5), 830–838 (2016)
19. Diamantoulakis, P.D., Pappi, K.N., Muhaidat, S., Karagiannidis, G.K., Khattab, T.: Carrier aggregation for cooperative cognitive radio networks. IEEE Trans. Veh. Technol. **66**(7), 5904–5918 (2017)
20. Al-Dulaimi, A., Al-Rubaye, S., Cosmas, J., Anpalagan, A.: Planning of ultra-dense wireless networks. IEEE Netw. **31**(2), 90–96 (2017)
21. Hao, Y., Tian, D., Fortino, G., Zhang, J., Humar, I.: Network slicing technology in a 5G wearable network. IEEE Commun. Stand. Mag. **2**(1), 66–71 (2018)
22. Pan, C., Yin, C., Beaulieu, N.C., Yu, J.: Distributed resource allocation in SDCN-based heterogeneous networks utilizing licensed and unlicensed bands. IEEE Trans. Wirel. Commun. **17**(2), 711–721 (2018)
23. Hu, F., Chen, B., Zhu, K.: Full spectrum sharing in cognitive radio networks toward 5G: a survey. IEEE Access **6**, 15754–15776 (2018)
24. Li, B., Fei, Z., Xu, X., Chu, Z.: Resource allocations for secure cognitive satellite-terrestrial networks. IEEE Wirel. Commun. Lett. **7**(1), 78–81 (2018)
25. Xu, L., Cai, L., Gao, Y., Xia, J., Yang, Y., Chai, T.: Security-aware proportional fairness resource allocation for cognitive heterogeneous networks. IEEE Trans. Veh. Technol. **67**(12), 11694–11704 (2018)
26. Jiang, C., Wang, B., Han, Y., Wu, Z., Liu, K.J.R.: Exploring spatial focusing effect for spectrum sharing and network association. IEEE Trans. Wirel. Commun. **16**(7), 4216–4231 (2017)
27. Alabbasi, A., Rezki, Z., Shihada, B.: Outage analysis of spectrum sharing Over M-block fading with sensing information. IEEE Trans. Veh. Technol. **66**(4), 3071–3087 (2017)
28. Tehrani, R.H., Vahid, S., Triantafyllopoulou, D., Lee, H., Moessner, K.: Licensed spectrum sharing schemes for mobile operators: a survey and outlook. IEEE Commun. Surv. Tutor. **18**(4), 2591–2623 (2016)
29. Li, B., Fei, Z., Chu, Z., Zhou, F., Wong, K., Xiao, P.: Robust chance-constrained secure transmission for cognitive satellite–terrestrial networks. IEEE Trans. Veh. Technol. **67**(5), 4208–4219 (2018)
30. Khwandah, S., Cosmas, J., Glover, I.A., Lazaridis, P.I., Araniti, G., Zaharis, Z.D.: An enhanced cognitive femtocell approach for co-channel downlink interference avoidance. IEEE Wirel. Commun. **23**(6), 132–139 (2016)

31. Alnakhli, M., Anand, S., Chandramouli, R.: Joint spectrum and energy efficiency in device to device communication enabled wireless networks. IEEE Trans. Cogn. Commun. Netw. **3**(2), 217–225 (2017)

32. Chen, H., Liu, L., Novlan, T., Matyjas, J.D., Ng, B.L., Zhang, J.: Spatial spectrum sensing-based device-to-device cellular networks. IEEE Trans. Wirel. Commun. **15**(11), 7299–7313 (2016)

33. Zhao, F., Tang, Q.: A KNN learning algorithm for collusion-resistant spectrum auction in small cell networks. IEEE Access **6**, 45796–45803 (2018)

34. Lorenzo, B., Gonzalez-Castano, F.J., Fang, Y.: A novel collaborative cognitive dynamic network architecture. IEEE Wirel. Commun. **24**(1), 74–81 (2017)

35. Liberti, L., Maculan, N.: Global optimization: from theory to implementation, p. 427. Springer, Heidelberg (2006)

36. Benmimoune, M., Driouch, E., Ajib, W., Massicotte, D.: Joint transmit antenna selection and user scheduling for massive MIMO systems. In: Proceedings of IEEE Wireless Communication Network Conference (WCNC 2015), New Orleans, LA, USA, 9–12 March, pp. 381–386 (2015)

37. Barayan, Y., Kostanic, I., Rukieh, K.: Performance with MIMO for the downlink 3GPP LTE cellular systems. Univers. J. Commun. Netw. **2**(2), 32–39 (2014)

38. Rusek, F., et al.: Scaling up MIMO: opportunities and challenges with very large arrays. IEEE Signal Process. Mag. **30**(1), 40–60 (2013)

39. Liu, G., Sheng, M., Wang, X., Jiao, W., Li, Y., Li, J.: Interference alignment for partially connected downlink MIMO heterogeneous networks. IEEE Trans. Commun. **63**(2), 551–564 (2015)

40. Olwal, T.O., Masonta, M.T., Mekuria, F.: Bio-inspired energy and channel management in distributed wireless multi-radio networks. IET Sci. Meas. Technol. **8**(6), 380–390 (2014)

41. Olwal, T.O., van Wyk, B.J., Kogeda, P.O., Mekuria, F.: FIREMAN: foraging-inspired radio communication energy management in green multi-radio networks. In: Khan, S., Mauri, J.L. (eds.) Green Networking and Communications: ICT for Sustainability, pp. 29–47. Taylor & Francis, CRC Press, Boca Raton (2013)

Numerical Analysis of Discrete L^p-norm Error for Rayleigh-Ritz Variational Technique and Its Application to Groundwater Flows Model

P. H. Gunawan$^{(\boxtimes)}$ ⓘ and A. Aditsania ⓘ

School of Computing, Telkom University, Jl. Telekomunikasi no 1,
Terusan Buah Batu, Bandung 40275, Indonesia
{phgunawan,aaditsania}@telkomuniversity.ac.id

Abstract. The investigation of computational finite element method based on variational technique for approximating groundwater flow model is elaborated in this paper. Here, the Rayleigh-Ritz scheme will be used in order to obtain the numerical solution for steady state form of groundwater flow model. The groundwater model is given in one-dimensional parabolic type partial differential equation (PDE). To obtain steady state form, the finite difference approach is used for discretizing time differential equation in PDE form. Two numerical tests are elaborated to see the robustness of the numerical scheme to approximate the continuous model. The convergence rates of discrete L^p-norm error with $p = 1$, $p = 2$ and $p = \infty$ are shown satisfying for two numerical tests. The result of Rayleigh-Ritz scheme is obtained in second order approximation. Moreover, the comparison of current result and result from another literature for groundwater flow simulation is shown in a good agreement.

Keywords: Groundwater · Finite element method · Rayleigh-Ritz · Variational approach · Convergence

1 Introduction

Groundwater flow is the water flow which is flowing beneath the ground surface. Here, the origin of water could be from the surface and then stream down into the ground trough porous soil [2]. The mathematical model of groundwater flow is needed to investigate the behaviour of flow which depends on type of soils. Therefore, groundwater flow model in horizontal level under the soil will be elaborated in this paper.

The horizontal profile of groundwater flows is governed in the mass equation form such as

$$\frac{\partial \Psi}{\partial t} = -\frac{\partial (Hu)}{\partial x} + \frac{\omega}{p}, \tag{1}$$

© Springer Nature Switzerland AG 2019
R. Silhavy et al. (Eds.): CoMeSySo 2019, AISC 1047, pp. 322–331, 2019.
https://doi.org/10.1007/978-3-030-31362-3_31

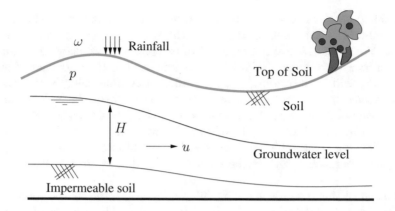

Fig. 1. The groundwater level under the horizontal slice view.

where $\Psi(x,t)$ is groundwater level, $H(x,t)$ is water depth, $u(x,t)$ is horizontal average of velocity, ω is rain fall in volume per area and p is soil porosity. Moreover, time and space are denoted by t and x respectively (Fig. 1).

According [11], the velocity of water in porous medium can be defined as

$$u(x,t) = -\kappa \frac{\partial \Psi(x,t)}{\partial x}, \tag{2}$$

where κ is the soil permeability coefficient. Thus, combining mass equation (1) and (2), the diffusion equation is obtained such as,

$$\frac{\partial \Psi}{\partial t} = -\frac{\partial}{\partial x}\left(-\kappa H \frac{\partial \Psi}{\partial x}\right) + \frac{\omega}{p}. \tag{3}$$

If the water depth H is constant along the x direction, thus the equation above can be transformed into the ordinary diffusion equation such as heat transfer

$$\frac{\partial \Psi}{\partial t} = D \frac{\partial^2 \Psi}{\partial x^2} + \frac{\omega}{p} \tag{4}$$

where $D = \kappa H$ is diffusion coefficient. Note that, (4) is the non-homogeneus parabolic type of partial differential equations.

Some numerical methods to approximate the solution of groundwater flow are available in many literatures [2,11]. For instance, in [11], the model is given in parabolic partial differential equations (PDE) and discretized using finite different method (FDM) in straightforward approach. Some results regarding the choice of numerical schemes with FDM have interesting issue. The explicit scheme of FDM leads to the unstable results if the stability conditions is not considered. Whereas, the implicit scheme of FDM shows the unconditionally stable can be obtained. The detail about those methods can be found in some literatures [5,6,8] or [9].

In this paper, another numerical method will be considered. The finite element method (FEM) based on Rayleigh-Ritz variational technique will be elaborated. This method is the combination of two methods which are FDM and the Rayleigh-Ritz variational method for approximate the PDE solution. First, the PDE model will be discretized only in differential time using FDM, and then it transformed into boundary-value problem (BVP) form. Therefore, Rayleigh-Ritz variational method can be used to solve BVP. The similar procedure can be found in FEM with FreeFem++ software [4] or in the paper of [7]. However in the FreeFem++ [4], the Galerkin approach is used instead of the Rayleigh-Ritz variational approach, which is the solution is computed directly from the weak form.

The rest of this paper is given as follows, in Sect. 2, the semi-discrete form of one-dimensional model of groundwater flows is given. The Rayleigh-Ritz method to approximate the solution of BVP and its algorithm are given in Sect. 3. The robustness of numerical scheme using analytical solution in two numerical tests is shown in Sect. 4. Moreover, the numerical solution in groundwater flow model is also elaborated in Sect. 4. Finally, in Sect. 5, the conclusion of this work is drawn.

2   Semi-discrete Model

Here discretization of time dependent of problem (4) is considered. Denote the final time of domain is T on finite time domain $[0, T]$. Then, time domain is discretized into T_n grid points, such that $t^n = n \times \Delta t$ with $n \in \mathcal{T} = \{0, 1, 2 \cdots T_n\}$ and time step is given as $\Delta t = T/T_n$. Thus we have $\Psi(\cdot, t^n) \approx \Psi^n$ which presents the function Ψ at present time and $\Psi(\cdot, t^{n+1}) \approx \Psi^{n+1}$ at next time level. Finally, the first-order finite different method in time can be written as

$$\frac{\partial \Psi}{\partial t} \approx \frac{\Psi^{n+1} - \Psi^n}{\Delta t} + \mathcal{O}(\Delta t). \tag{5}$$

The semi-discrete model of (4) over one-dimensional domain Ω is given as

$$\frac{\Psi^{n+1} - \Psi^n}{\Delta t} = D\frac{\partial^2 \Psi}{\partial x^2} + \frac{\omega}{p}. \tag{6}$$

Since the proposed scheme is addressed in the second derivation in space by the new time step Ψ^{n+1}, thus the equation can be rewritten as follow

$$-D\frac{d^2\Psi^{n+1}}{dx^2} + \frac{1}{\Delta t}\Psi^{n+1} = \frac{1}{\Delta t}\Psi^n + \frac{\omega}{p}. \tag{7}$$

Therefore the boundary value problem (BVP) is obtained as shown in (7). The variational formulation now is in $L^2(0, T; H^1(\Omega))$; indeed the goal is to seek $\Psi^{n+1}(x)$ satisfying for all $\varphi \in H_0^1(\Omega)$.

$$-\int_\Omega \left(D\frac{d^2\Psi^{n+1}}{dx^2} + \frac{1}{\Delta t}\Psi^{n+1}\right)\varphi \, dx = \int_\Omega \left(\frac{1}{\Delta t}\Psi^n + \frac{\omega}{p}\right)\varphi \, dx, \tag{8}$$

$$\int_\Omega \left(D\frac{d\Psi^{n+1}}{dx}\frac{d\varphi}{dx} + \frac{1}{\Delta t}\Psi^{n+1}\varphi\right) dx = \int_\Omega \left(\frac{1}{\Delta t}\Psi^n + \frac{\omega}{p}\right)\varphi \, dx. \tag{9}$$

Above equations are also known as the weak form of (7). Unfortunately, the Rayleigh-Ritz method does not work in the weak form. However, the Rayleigh-Ritz method approximates the solution by minimizing an integral over all functions in $H_0^1(\Omega)$, such as

$$I(\varphi) = \min_{\varphi \in H_0^1(\Omega)} \left\{ \int_\Omega \left[D[\varphi']^2 + \frac{1}{\Delta t}[\varphi]^2 - 2 \left(\frac{1}{\Delta t}\Psi^n + \frac{\omega}{p} \right) \varphi \right] dx \right\} \qquad (10)$$

In order to solve this equation, we would like to recall the solution of BVP by Rayleight-Ritz method in the next section.

3   Rayleigh-Ritz for Boundary Value Problem

In order to find the solution of (7), first the following boundary value problem (BVP) is considered.

$$-\frac{d}{dx}\left(p(x)\frac{d\Psi}{dx} \right) + q(x)\Psi = f(x), \qquad \text{for } x \in \Omega \qquad (11)$$

with the boundary conditions

$$\Psi(x) = 0, \qquad \forall x \in \partial\Omega. \qquad (12)$$

Moreover, assume that $p(x) \in C^1(\Omega)$, and $q(x), f(x) \in C(\Omega)$, and further assume there exists $\delta > 0$, a constant such that,

$$p(x) \geq \delta, \text{ and that } q(x) \geq 0, \forall x \in \Omega. \qquad (13)$$

From [1,3], previous assumptions are sufficient to guarantee that the boundary-value problem (11–12) has a unique solution. In order to solve (11) or in this case to solve (7), then the piecewise-linear basis function can be used as shown in [3]. Given $y(x)$ is an approximation function for $\Psi(x)$, then piecewise linear basis function can be given as follows,

$$\Psi(x) = \sum_{i=1}^N \gamma_i y_i(x),$$

where γ_i is the i-th coefficient of function $y_i(x)$ and $N \in \mathbb{Z}^+$. Moreover the basis function is given as follows,

$$y_i(x) = \begin{cases} 0, & \text{if } 0 \leq x \leq x_{i-1}, \\ \dfrac{x - x_{i-1}}{x_{i+1} - x_i}, & \text{else if } x_{i-1} < x \leq x_i, \\ \dfrac{x_{i+1} - x}{x_{i+1} - x_i}, & \text{else if } x_i < x \leq x_{i+1}, \\ 0, & \text{otherwise.} \end{cases} \qquad (14)$$

Note that, the final problem is to find the coefficients γ_i for $i \in \{1, \cdots, N\}$. Detail of piecewise-linear Rayleigh-Ritz method can be seen in [3]. Then, the algorithm to solve (7) can be seen in Algorithm 1.

Algorithm 1. Rayleigh-Ritz scheme for heat equation

1: Start
2: **Input** $N \geq 1 \in \mathbb{N}$; points $x_0 = 0 < x_1 < ... < x_N < x_{N+1} = 1.$; $n = 1$
3: **Define** initial condition $\Psi_i^0 \; \forall i \in \{0, \cdots, N+1\}$
4: **While** $n < FinalIter$ **do**
5:     **Solve** (7) using basis function (14) in order to obtain $\gamma_i, \quad i \in \{1, \cdots, N\}$
6:     **Set** $\Psi_i^n = \gamma_i \quad i \in \{1, \cdots, N\}$ and $n = n + 1$
7: **End**

4   Numerical Computation

Several numerical simulations will be given in this section. Two numerical tests will be elaborated to show the robustness of the proposed numerical scheme by showing the comparison of numerical and analytical solutions. Moreover, the proposed scheme will be used to approximate the solution for groundwater flow model in the next section.

4.1   Accuracy

In order to see the accuracy of Rayleigh-Ritz numerical scheme, two numerical tests will be given with the availability of analytical solutions for each tests. First, the homogeneous heat diffusion problem with smooth initial condition over domain $\Omega = [0, 1]$ is given as follows:

$$\frac{\partial \Psi}{\partial t} = \frac{\partial^2 \Psi}{\partial x^2}, \quad x \subset (0, 1), \quad t > 0, \tag{15}$$

$$\Psi(0, t) = 0, \quad \Psi(1, t) = 0, \quad t \geq 0, \tag{16}$$

$$\Psi(x, 0) = 3\sin(\pi x) + 5\sin(4\pi x), \quad x \in [0, 1] \tag{17}$$

Assume that, $u(x, t)$ is the analytical solution of (15). The analytical solution of this problem can be computed by separation variables technique as,

$$u(x, t) = 3e^{-\pi^2 t}\sin(\pi x) + 5e^{-16\pi^2 t}\sin(4\pi x). \tag{18}$$

The detail of separation variables method can be found in [10].

The result of this test can be seen in Fig. 2. Here, numerical and analytical solution are shown satisfied and close enough. In order to obtain the detail accuracy of Rayleigh-Ritz numerical scheme, the L^1, L^2 and L^∞ norm errors are computed using the references solutions. Here the definition of L^p norm error is given as follows,

$$||error||_{L^1} := \left(\int_\Omega |u - \Psi|^p \; \mathrm{dx} \right)^{1/p} \approx \left(\sum_i^N |u(x_i) - \Psi(x_i)|^p \Delta x_i \right)^{1/p} \tag{19}$$

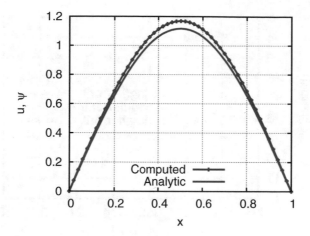

Fig. 2. The comparison of numerical (Ψ) and analytical (u) solutions at fix time $t = 0.1$ first test

The table of error in discrete L^1, L^2 and L^∞ norm errors at fixed final time $t = 0.1$ is given in Table 1. Here, discrete errors are collected from several grid numbers, 50, 100, 200, 400 and 800 partitions. Table 1 shows the decreasing of error number as the increasing of grids number of discrete partitions as expected.

Table 1. The discrete L^1, L^2 and L^∞ norm error for first test.

N	$\|error\|_{L^1}$	$\|error\|_{L^2}$	$\|error\|_{L^\infty}$
50	2.17E-04	6.05E-08	3.50E-04
100	5.47E-05	3.79E-09	9.00E-05
200	1.37E-05	2.41E-10	3.00E-05
400	3.20E-06	2.14E-11	1.00E-05
800	7.43E-07	4.45E-12	1.00E-05

The convergence rate from table error (Table 1) is shown in Fig. 3. The convergence rate for each simulations of the numerical results are shown satisfied. The log of error in L^2 is smaller than the log of error in L^1 and L^∞. In the log of error L^1 the graph shows decreasing linear along the increasing the log of grid numbers N. Please see [12] for more detail about this convergence rate.

The second example will be given similar as the first problem (the homogeneous heat diffusion problem), however the discrepancy is in initial condition, where here a constant initial condition will be given,

$$u(x,0) = 1, \qquad x \in (0,1), \tag{20}$$

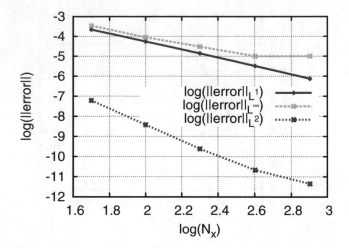

Fig. 3. The convergence rate error for first test.

with the analytical solution is given as

$$u(x,t) = \frac{4}{\pi} \sum_{k=1}^{N} \frac{1}{2k-1} e^{-((2k-1)\pi)^2 t} \sin((2k-1)\pi x), \qquad (21)$$

where $(N \geq 1) \in \mathbb{Z}^+$.

The comparison of numerical and analytical solution in second test is shown in Fig. 4. Again, the numerical solution is shown in a good agreement with the analytical solution. The analytical solution is obtained by the separation variables method with the coefficient Fourier. Moreover, the table error of discrete L^1, L^2 and L^∞ norm error at final time $t = 0.1$ is given in Table 2.

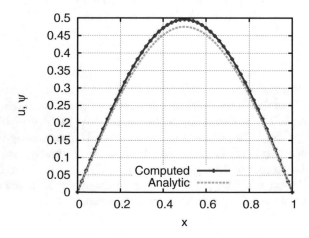

Fig. 4. The comparison of numerical (Ψ) and analytical (u) solutions at fix time $t = 0.1$ second test

The convergence rates τ for each discrete L norm errors are also shown in Table 2. From Table 2, clearly that in L^1 norm error, the convergence rate of Rayleigh-Ritz numerical scheme converges to more than two, thus Rayleigh-Ritz numerical scheme is obviously in second order scheme. In the next section, Rayleigh-Ritz numerical scheme will be used to approximate the solution of groundwater flow model.

Table 2. The discrete L^1, L^2 and L^∞ norm error for second test.

N	$\|error\|_{L^1}$	τ	$\|error\|_{L^2}$	τ	$\|error\|_{L^\infty}$	τ
50	1.96E-04	/	4.81E-08	/	2.97E-04	/
100	4.94E-05	1.99E+00	3.00E-09	4.01E+00	7.50E-05	1.99E+00
200	1.23E-05	2.01E+00	1.83E-10	4.03E+00	1.90E-05	1.98E+00
400	2.92E-06	2.07E+00	1.04E-11	4.14E+00	5.00E-06	1.93E+00
800	6.10E-07	2.26E+00	6.00E-13	4.12E+00	1.00E-06	2.32E+00

4.2 Groundwater Flow

Assume that groundwater level is located near a river. At the initial time, the water level of river and groundwater at equilibrium state. Moreover, no rainfall and a horizontal water level are assumed in this simulation. At a particular time, suddenly the water level of river is going down and remains constant. This causes water flowing under the ground, where water under soil flowing due to the presence of tangential water level between river and groundwater. This problem can be illustrated in Fig. 5.

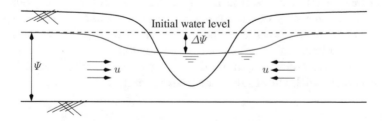

Fig. 5. The groundwater flow due to the sudden down of water level in the river.

Considering the length of groundwater domain in the right side of river is $L = 100$ m. Initially, the water level of river is 10 m, then suddenly going down to 0 m and remains constant. In this simulation, diffusion coefficient is a constant $D = 10^{-3}$ and domain is partitioned into 20 grids. The simulation result by the water level profile is shown in Fig. 6.

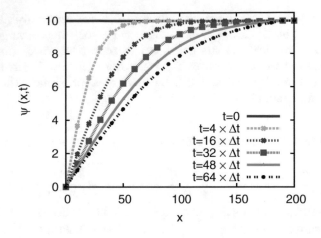

Fig. 6. The numerical simulation of groundwater flow.

In Fig. 6, water level is shown at various final time simulations $t = 0$, $t = 4 \times \Delta t$, $t = 8 \times \Delta t$, $t = 16 \times \Delta t$, $t = 32 \times \Delta t$ and $t = 64 \times \Delta t$. It can be seen that, after water level of river going down to 0, then groundwater flows to the right which is shown by the decreasing of groundwater level. This result is shown in good agreement with the result in the manuscript of [11].

5   Conclusion

The Rayleigh-Ritz method to approximate the solution of the parabolic type of partial differential equations is presented in this paper. The approximation solution is shown in a good agreement with the analytical solution using two numerical tests. Here, the discrete L^1, L^2 and L^∞ norm errors are computed in order to see the accuracy of the numerical scheme. Moreover, the convergence rate of two simulations shows that Rayleigh-Ritz method produces second order approximation. Meanwhile, in this paper the simulation of groundwater flow is also presented. The numerical simulation using Rayleigh-Ritz method shows in a good agreement with the result in another reference.

References

1. Bailey, P.B., Shampine, L.F., Waltman, P.E.: Nonlinear Two Point Boundary Value Problems. Academic Press, Cambridge (1968)
2. Bear, J., Cheng, A.H.D.: Modeling Groundwater Flow and Contaminant Transport, vol. 23. Springer, Heidelberg (2010)
3. Burden, R.L., Faires, J.: Numerical Analysis. Thomson Learning Inc., Stamford (2001)
4. Hecht, F.: New development in freefem++. J. Numer. Math. **20**(3–4), 251–266 (2012)

5. Hoffman, J.D., Frankel, S.: Numerical Methods for Engineers and Scientists. CRC Press, Boca Raton (2001)
6. Mattheij, R.M., Rienstra, S.W., ten Thije Boonkkamp, J.H.: Partial Differential Equations: Modeling, Analysis, Computation. SIAM, University City (2005)
7. Salamah, U., Aditsania, A., Gunawan, P.H.: Simulasi numerik pada aliran air tanah menggunakan collocation finite element method. E-Jurnal Matematika **7**(1), 5–10 (2018)
8. Strauss, W.A.: Partial Differential Equations, vol. 92. Wiley, New York (1992)
9. Thomas, J.W.: Numerical Partial Differential Equations: Finite Difference Methods, vol. 22. Springer, Heidelberg (2013)
10. Tveito, A., Winther, R.: Introduction to Partial Differential Equations: A Computational Approach, vol. 29. Springer, Heidelberg (2004)
11. Vreugdenhil, C.B.: Computational Hydraulics: An Introduction. Springer, Heidelberg (2012)
12. Zain, F.M., Khadafi, M.G., Gunawan, P.H.: Analisis konvergensi metode beda hingga dalam menghampiri persamaan difusi. E-Jurnal Matematika **7**(1), 1–4 (2018)

Parallel Computing Application for Testing of Parameters of Active Phased Antenna Array Noise Modules

M. V. Orda-Zhigulina[(✉)] and D. V. Orda-Zhigulina

Southern Scientific Center of the Russian Academy of Sciences, st. Chehova, 41,
Rostov-on-Don 344006, Russia
jigulina@mail.ru

Abstract. The paper is suggested a new theoretical method of calculating of chaotic and deterministic modes of an optical microwave modulator to optimize manufacturing of the noise modules of the radar systems. An optical microwave modulator is important part of a noise module of an active phased-array antenna which is significant part of a radar system. The parameters of functional modes of oscillating system of injection semiconductor laser are investigated. Calculating of the deterministic and chaotic modes produces the large data volumes. So, parallel computing increases speed of calculating with the same meaning of accuracy of calculations. Also it leads to reducing of the time and expenses of manufacturing of the optical microwave modulators as an active part of AFAR.

Keywords: Processing systems · Poincare maps ·
Injection semiconductor laser · Microwave optical modulator ·
Hardware and software parallelism

1 Introduction

Microwave element base is a set of complex integrated devices, which are implemented at the base of complex composite materials in industrial production usually [1, 2]. There is a need for complete control of the electrical parameters of each manufactured microwave device as it is the lack of repeatability of the initial parameters of a such device. It takes considerable time. Therefore, this may not always be carried out as the high cost of the measuring equipment and the lack of work-time of a design engineer. So, an urgent technical task is to replace laboratory measurements with theoretical modeling of the parameters of manufactured microwave devices.

An optical microwave modulator which is based on an injection semiconductor laser (IPL) is an important component of the noise modules of the active phased antenna arrays (AFAR). This is an urgent and complex technical task of manufacturing and designing of microwave optical modulator. Therefore, the analysis of the stability of the modulated IPL and the searching for areas of the electrophysical parameters of the IPL should be calculated before measurements. The chaotic mode of optical modulator is the basis for creating a noise module of AFAR.

© Springer Nature Switzerland AG 2019
R. Silhavy et al. (Eds.): CoMeSySo 2019, AISC 1047, pp. 332–338, 2019.
https://doi.org/10.1007/978-3-030-31362-3_32

It was studying the states of a complex oscillatory system of IPL, which is modulated by a microwave signal. A calculation method and a mathematical model have been proposed for the analysis of deterministic and chaotic modes of operation of optical microwave modulators which are based on IPL. Numerical computing of the operating modes of the IPL were made to find the values of the modulation coefficient, the amplitude of natural oscillations of the IPL, and the supply voltage when a stable or chaotic operating mode of an optical microwave modulator was realized.

A large amount of raw data for further processing was generated as a result of numerical computing. Originally, a sequential implementation of the algorithm for computing was used to process the obtained data. A method for parallel computing was proposed to reduce the time of numerical computing and the time of debugging of the AFAR noise module. So, it is advisable to optimize the computing time of the electrical parameters of each debugging AFAR noise module and its microwave devices which increasing productivity and reducing time of debugging.

2   Researching of the Scenario of the Shift of Microwave Modulators from Deterministic to Chaotic Mode and the Model for Calculating of Modulator's Parameters

IPL is a nonlinear oscillatory system when its laser beam is modulated with a microwave signal. There are various generation modes in the system, for instance, the noise (or chaotic) oscillations in the optical range. Stable and unstable modes of the oscillatory system were studied by help of classical mathematical methods of the theory of oscillations. Various systems of the IPL rate equations were analyzed [3–9]. For these systems of rate equations, it was estimated the influence of the modulating microwave signal parameters on the stability of oscillations of IPL. It was computing the parameters of IPL deterministic and chaotic oscillations for the base part of noise modules of active phased antenna arrays in the radar systems.

The states of the dynamical oscillatory system were studied by the Poincare maps. The Poincare maps lets to visualize the phase portrait of the oscillatory system "from the inside". The mapping method is flexible, as it can be modified to various classes of oscillatory systems.

The system of rate equations of Hanin-Pikovsky [3] for solid-state lasers with periodic loss modulation was taken:

$$
\begin{cases}
\dfrac{dM}{dt} = B \cdot M \cdot N - (1 + \beta \cdot \cos(\omega t)) \cdot \dfrac{M}{T_c}, \\
\dfrac{dN}{dt} = -B \cdot M \cdot N + \dfrac{N_0 - N}{T_1};
\end{cases}
\tag{1}
$$

Where M is the number of photons, N is the population difference, B is the Einstein coefficient, β is the modulation depth, ω is the oscillation frequency, N_0 is the saturated value of the population difference, T_c is the photon lifetime, T_1 is the relaxation time.

The analysis was realized at the base of numerical solutions of the system of equations for various values of the frequency and modulation depth of IPL.

The results are presented in Fig. 1. Phase portraits are shown in Fig. 1b, c. This figure shown the tracks twist and this behavior it counts in favor of quasiperiodic system mode. A similar picture is observed in the phase portrait for the population difference. The figure shows that dots look like closed circular arc and one more enclosed in it another circular arc. Location of dots shows quasiperiodic oscillation system's mode because of conglomerate of dots represents a complex geometric structure dispersed in a circle.

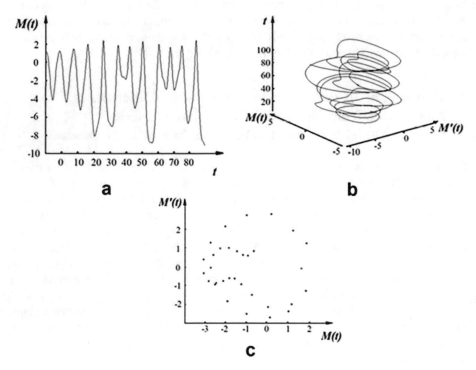

Fig. 1. Temporary response (a), Phase portrait (b), Poincare map (c) for the population difference at the modulation frequency and modulation depth.

If it is varying of parameters we can achieve the instability of the system, as the numerical method gives solutions for very short time periods. This method can be used to fix the behavior of a dynamic system in terms of deterministic chaos [6, 9].

A general analysis of the dynamics of the Hanin-Pikovsky laser shows the sensitivity of the system to changes of parameters and the unpredictability behavior of the oscillatory system. The obtained results let to argue there is such a set of parameters that corresponds to the chaos modes of IPL.

So, it is necessary to realize numerical computing to identify areas of the initial parameters which corresponding to chaotic or deterministic optical IPL modes during designing and debugging of microwave optical modulators.

The theoretical base for such modeling is analysis of the rate equations system of Hanin-Pikovsky. The analysis compare Poincare maps, temporary response, phase portraits and strange attractors at the same time. The presented mathematical model and the method of analysis can be applied at the phase of designing and debugging of microwave electro-optical devices, in particular, the AFAR noise modules.

3   The Parallelization of Computing of Parameters of Optical Microwave Modulators and the Application of Distributed Computing Systems

As mentioned above, the theoretical basis for computing of the parameters of optical microwave modulators is the system of differential Eqs. (1) by Hanin-Pikovsky [4–8]. Algorithms of the numerical computing are used to find temporal solutions, the construction of strange attractors, phase portraits, bifurcation diagrams and Lyapunov values. The comparison of all this results leads to analyze it is the chaotic or deterministic state of the oscillatory system of IPL.

At present, the existing algorithms for computing of temporary response, Poincare maps and phase portraits are implemented sequentially for theoretical analysis of the parameters of microwave optical modulators at their designing. So, there is a need for extra computing resources to realize numerical computing with high accuracy in a short time.

So, the sequential algorithm for analyzing deterministic and chaotic modes for the system of differential Eqs. (1) consists of the following tasks:

to find the temporary response – task 1,

to find the Poincare map – task 2,

to find the phase portrait – task 3,

to find the strange attractor – task 4,

to analyze tasks 1–4 by a design engineer, to answer question it is chaos or deterministic mode of optical microwave modulator. The sequential algorithm is shown in Fig. 2.

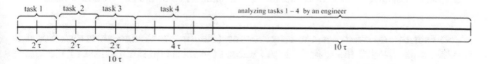

Fig. 2. Sequential method of computing of the parameters of the AFAR noise modules.

The solution of each task 1–4 gives information whether deterministic or chaotic mode will arise with the fixed values of the initial electrophysical parameters of IPL. As a result, the design engineer will correct the documentation and changes to the circuits at the pre-manufacturing of the microwave modulators. When it is sequential computing all the data become available for analysis only after sequential realizing of each tasks 1–4 as shown in Fig. 2.

As mentioned above, the theoretical analysis could replace some experimental measurements of the IPL parameters at the step of debugging of microwave optical modulators during their manufacture.

It was found the longest computing time of calculating the parameters of microwave optical modulators is the computing of strange attractors (approximately 40% of the time of all calculations). It is faster to find the Poincare map, which is a system of differential equations on a phase space of a lower dimension with discrete time. The times of finding temporal responses, Lyapunov values and bifurcation diagrams are comparable to the time for computing points on the Poincare map (approximately 20% of computing time for the whole task).

Therefore, it is advisable to parallelize the computing by each task, as shown in Fig. 3. This lets to speed up the theoretical modeling of the parameters of the microwave modulator.

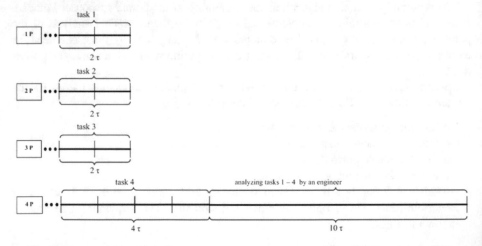

Fig. 3. Sequential method of computing of the parameters of the AFAR noise modules.

Solving the task 4 takes the longest time (building a strange attractor). So, it is possible to parallelize the task 4. It will reduce the total computing time of the whole task, as shown in Fig. 4.

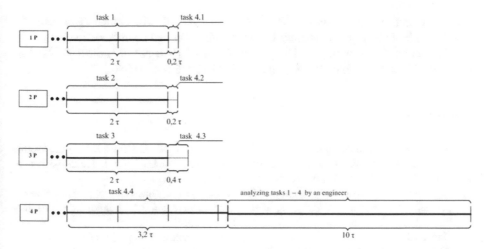

Fig. 4. Sequential method of computing of the parameters of the AFAR noise modules.

In this paper, it is proposed to use the parallel computing [10] when analyzing the parameters of a microwave optical modulator. It allows to calculate the region of values of the modulation factor, the supply voltage and the amplitude of the optical oscillations of IPL. The parallel computing leads to decreasing time of computing the whole task, but the computing accuracy is the same. So, the time of manufacturing of the noise modules will decrease too.

The time of theoretical analysis of switching of the microwave optical modulator to the chaotic mode can be reduced to 2.8–3.5 working days (for parallel computing) instead of 4–5 working days (for sequential computing). The computing was realized for tasks 1–4 for each IPL with different initial electrophysical parameters when debugging 100 AFAR.

4   Conclusions

The proposed model for analyzing the chaotic modes of IPL lets to find the initially parameters of the oscillating system of IPL when the deterministic of chaotic modes of the microwave optical modulators are realized. The obtained results can be applied for theoretical analysis of deterministic and chaotic modes of the microwave optical modulators as a part of the AFAR noise modules at the debugging during their manufacturing. Replacing experimental measurements with theoretical modeling will reduce the cost of manufacturing the AFAR noise modules as it is reducing the costs of the debugging of the optical microwave modulators.

The proposed method of parallel computing will increase the speed of finding of the parameters of microwave optical modulators with the same meaning of the accuracy of the calculations. This method of parallel computing will reduce the time of control parameters of microwave optical modulators when debugging. At the result the time of the debugging and the costs of manufacturing of AFAR will reduce. The achieved results can be used in the designing of radar with improved characteristics.

The publication was prepared as part of GZ SSC RAS N GR project № AAAA-A19-119011190173-6 and RFBR projects 18-29-22086 и 18-05-80092.

References

1. Blohin, Je.E., Pashhenko, A.S., Lunin, L.S., Chebotarev, S.N., Alfimova, D.L.: Issledovanie geterostruktur InAs/GaAs s potencial'nymi bar'erami AlGaAs. Vestnik Juzhnogo nauchnogo centra RAN, vol. 13, №. 1, pp. 11–17 (2017)
2. Arzhanov, S.N., Barov, A.A, Gjunter, V.Ja.: Razrabotka i organizacija proizvodstva priemo-peredajushhih modulej AFAR s ispol'zovaniem sobstvennoj jelektronnoj komponentnoj bazy SVCh. Fazotron, vol. 15, №. 1–2, pp. 14–18 (2011)
3. Hanin, Ja.I.: Osnovy dinamiki lazerov. M.: Nauka, Fizmatlit, 368 p. (1999)
4. Alekseev, Ju.I., Orda-Zhigulina, M.V., Desjatnikova, E.V., Cherednikova, S.A.: Analiz stohasticheskoj dinamiki opticheskih moduljatorov na osnove IPLD. Izvestija VUZov. Fizika, vol. 56, №. 8–2, pp. 344–346 (2013)
5. Alekseev, Y., Demyanenko, A.V., Orda-Zhigulina, M.V., Semernik, I.V.: An experimental study of the development dynamics of the chaotic oscillation mode in a deterministic self-oscillatory microwave system. Instrum. Exp. Tech. **57**(3), 307–310 (2014)
6. Bahrah, L.D., Bliskavickij, A.A.: Primenenie lazerov i volokonno-opticheskih sistem dlja upravlenija formirovaniem SVCh – signalov i ih raspredelenija v antennyh reshjotkah. Kvantovaja jelektronika, vol. 15, №. 5, pp. 879–914 (1988)
7. Naim, M., Tailozin, A.: Enhanced chaos in optically injected semiconductor lasers. Opt. Commun. **269**, 166–173 (2007)
8. Atsushi, M.: Synchronization of feedback-induced chaos in semiconductor lasers by optical injection. Phys. Rev. A **65**, 65–72 (2002)
9. Mun, F.: Haoticheskie kolebanija. Perevod s anglijskogo. M.: Mir, 46 p. (1990)
10. Kalyaev, I.A.: Homogeneous neuronlike structures for optimization variational problem solving. In: 5th International Conference on Parallel Architectures and Languages Europe, PARLE 1993. Lecture Notes in Computer Science (including subseries Lecture Notes in Artificial Intelligence and Lecture Notes in Bioinformatics), LNCS, Munich, Germany, vol. 694, pp. 438–451 (1993)

Lifetime Adaptation in Genetic Programming for the Symbolic Regression

Jan Merta[✉] and Tomáš Brandejský

University of Pardubice, Studentská 95, 532 10 Pardubice, Czech Republic
jan.merta@upce.cz

Abstract. This paper focuses on the use of hybrid genetic programming for the supervised machine learning method called symbolic regression. While the basic version of GP symbolic regression optimizes both the model structure and its parameters, the hybrid version can use genetic programming to find the model structure. Consequently, local learning is used to tune model parameters. Such tuning of parameters represents the lifetime adaptation of individuals. Choice of local learning method can accelerate the evolution, but it also has its disadvantages in the form of additional costs. Strong local learning can inhibit the evolutionary search for the optimal genotype due to the hiding effect, in which the fitness of the individual only slightly depends on his inherited genes. This paper aims to compare the Lamarckian and Baldwinian approaches to the lifetime adaptation of individuals and their influence on the rate of evolution in the search for function, which fits the given input-output data.

Keywords: Genetic programming · Symbolic regression ·
Hybrid evolutionary methods · Local learning · Lamarckian evolution ·
Baldwin effect

1 Introduction

Symbolic regression with genetic programming is a supervised machine learning method which searches for a mathematical model for a given dataset. The basic version of the method optimizes both the model structure and its parameters at the same time by evolution. There are also hybrid versions that try to find the best parameter settings for the given model structure using a variety of local optimization methods [1–4]. This extension represents a lifetime adaptation of an individual in the population and can, under the right circumstances, speed up the evolution of the correct mathematical model. However, it also has its disadvantages in the form of additional costs. There are two basic approaches to the lifetime adaptation of an individual. In Lamarckian evolution, offspring inherit learned behavior through reproduction directly from parent genes. The Baldwin effect is purely Darwinian, and the knowledge learned during life, affects the individual's fitness. This paper wants to compare Lamarckian and Baldwinian approaches to hybrid symbolic regression in search of a polynomial function that describes the given input-output data [5].

The first part of the paper focuses on the use of genetic programming for symbolic regression purposes. The second part describes different approaches to hybrid

© Springer Nature Switzerland AG 2019
R. Silhavy et al. (Eds.): CoMeSySo 2019, AISC 1047, pp. 339–346, 2019.
https://doi.org/10.1007/978-3-030-31362-3_33

evolutionary learning. The third part deals with our use of hybrid methods for symbolic regression problem and describes performed experiments. Finally, we added an evaluation of our experiments and a summary of the achieved results and acquired knowledge.

2 Background

2.1 Genetic Programming

Symbolic regression with genetic programming is a supervised machine learning method which searches for a mathematical model for a given dataset. The basic version of the method optimizes both the model structure and its parameters at the same time by evolution. There are also hybrid versions that try to find the best parameter settings for the given model structure using a variety of local optimization methods [1–4]. This extension represents a lifetime adaptation of an individual in the population and can, under the right circumstances, speed up the evolution of the correct mathematical model. However, it also has its disadvantages in the form of additional costs. There are two basic approaches to the lifetime adaptation of an individual. In Lamarckian evolution, offspring inherit learned behavior through reproduction directly from parent genes. The Baldwin effect is purely Darwinian, and the knowledge learned during life, affects the individual's fitness. This paper wants to compare Lamarckian and Baldwinian approaches to hybrid symbolic regression in search of a polynomial function that describes the given input-output data [5].

Genetic programming has a similar mechanism to genetic algorithms. First, an initial population of randomly generated syntactic trees is created. Then the individuals are evaluated using the fitness function. The fitness function must be evaluated for all inputs and corresponding outputs. Subsequently, individuals go through the selection and the reproduction process according to their fitness values. Genetic operators include a variety of crossover methods, such as a subtree crossover (random subtrees exchange between individuals), and various kinds of mutations, such as subtree mutation (randomly selects the tree vertex and replaces its subtree with a new randomly generated subtree). Thanks to varying chromosome lengths, trees can grow quickly and generate overcomplicated solutions. This phenomenon is called bloat [6, 7].

2.2 Symbolic Regression

Symbolic regression with genetic programming is a supervised machine learning method for finding a mathematical model that matches a given dataset by evolution. The syntactic tree corresponds to the mathematical equation. The set of non-terminals consists of mathematical operations (+, −, *, /, …) and functions (sin, cos, …), while the set of terminals contains constants (1, 2, 3.05, π, …) and variables (x, y, distance, …). Fitness is often defined as the difference (sum of squared residuals) between outputs of the model and the desired outputs [7].

The hybrid versions of GP symbolic regression try, for a given structure, to optimize its constants with local learning methods. One of the techniques is an adaptive program called STROGANOFF [1, 2], which uses multiple regression analysis for constant tuning. In article [3] the gradient ascent was tested and in paper [4] authors tried simulated annealing.

More on genetic programming and symbolic regression can be found in [6] and [7]. The next part focuses on hybrid evolutionary techniques and two different approaches to local learning.

2.3 Hybrid Evolutionary Methods

The hybrid genetic algorithm combines the global search of an evolutionary approach at the population level with local search at the individual level. Local learning represents a lifetime adaptation of individuals in the population and moves them towards the optimum [5, 8–10]. Hybrid algorithms often distinguish between genotype and phenotype. Learning affects the phenotype of an individual and his final fitness [5, 9]. Lifetime learning smooths the fitness landscape and simplifies evolution [10]. The two main forms of hybrid genetic search include Lamarckian evolution and Baldwin effect [5].

Local learning can accelerate evolution under certain conditions [9, 10]. It can also improve variation in the population [11], which also can improve the rate of evolution. It has not only the benefits but also the additional costs [10]. Learning is expensive, it costs CPU time and sometimes when it is too strong, it can reduce the rate of evolution.

2.4 Lamarckian Evolution

Lamarckian evolution is based on learned knowledge during life, which is written back into the genes of an individual, and modified genes are inherited by offspring during reproduction [5]. It is not biologically accurate, but is suitable for evolutionary calculations [10]. Whitley in the paper [5] argues that this approach distorts the population and is not compatible with Holland's Schema theorem. Lamarckism needs inverse mapping from the phenotype (and the environment) back to the genotype [10, 12]. This is sometimes challenging and this technique is therefore particularly suited to problems where the genotype and phenotype have the same structure and meaning. These limitations are overcome by the Baldwin effect.

2.5 Baldwin Effect

Hinton and Nowlan used the Baldwin approach to local learning in their paper in 1987 [12]. They used a hybrid genetic algorithm for a complex (needle in a haystack) problem of finding the right connections in the neural network. Their chromosome had 20 genes for each possible connection, and the genetic alphabet contained values 1 (for the presence of the connection), 0 (for the absence of the connection) and "?" (for undefined condition). Defined states have remained unchanged, undefined states could be tuned during local learning.

The Baldwin effect has similar results to Lamarckian evolution, but has different mechanisms [10]. The Baldwin effect is purely Darwinian, and does not require inverse genotype mapping because modified genes from the phenotype are not inherited by offspring and knowledge learned during life only affects the individual's fitness. It does not change the evolution mechanism, so the Holland's Schema theorem is still valid [5, 10, 12].

Baldwinian adaptation affects the behavior of individuals indirectly. Good learners and individuals who are closer to the optimum have greater fitness, and therefore have more offspring on average. The Baldwin effect gives the chance for good (but not great) genes to resist accidental exclusion during the selection and remain in the population [13].

The Baldwin effect is most important in more complex domains (needle in a haystack problems), where it is difficult to search for a solution by the evolutionary method only [5]. It can also have negative effects. Excessively strong local learning can slow the rate of evolution. The fitness of an individual depends minimally on the inherited genes. This reduces the role of the genotype in evolutionary selection and the evolutionary character of the method disappears. This phenomenon is called a hiding effect [9].

2.6   Stochastic Hill-Climbing Search Algorithm

The Stochastic Hill-Climbing Search Algorithm at each step generates a limited set of random neighbor states, evaluates their objective functions and chooses the neighbor state with the lowest value (moving in the direction of the steepest descent) regardless of the next steps. It is a greedy local algorithm that tends to be stuck in a local optimum [14].

Hill-Climbing, due to these features, is not a good technique for global optimization where genetic algorithms are better. It is better suited for local optimization [14]. By combining the features of these global and local search methods, we can get a robust algorithm that works much better than just the individual methods alone [15].

3   Methodology

The aim of the experiment was to compare the basic version of GP symbolic regression without adaptation, to hybrid versions with lifetime learning based on the Lamarckian and Baldwinian approaches. For the purpose of the experiment, we created a dataset of inputs and outputs corresponding to the selected polynomial function (1) with integer coefficients.

$$y = 2x^3 + 3x^2 + 9x \tag{1}$$

For our experiments, we used a classic implementation of the generational genetic programming. We have generated the initial population by a grow method with a maximum depth of tree set to 2. We chose Tournament selection as the selection mechanism. The individual's fitness was the sum of the differences (sum of squared residuals) between the function outputs and the desired outputs. We have implemented and tested the following genetic operators:

- subtree crossover,
- subtree mutation.

In the hybrid version, a discrete stochastic hill-climbing search algorithm was chosen for local learning of integer coefficients to test both basic approaches to hybrid evolution search. Fitness in the Lamarckian approach is the fitness of a phenotype with modified constants, while fitness in the Baldwinian approach is composed of the fitness of the phenotype with modified coefficients and the fitness of the genotype reduced to 10 percent of the original value.

We used the MersenneTwister random number generator (the outer and inner algorithms had their own random number generator). The population size for the outer genetic programming algorithm was set to 20, elitism to 1. The set of terminals consisted of constants 1.0, 2.0, 3.0, variable x, and the set of non-terminals contained mathematical operations of multiplication (*) and addition (+).

The Hill-Climber Search Algorithm for generating the set of neighbor states has decreased or increased the one random polynomial coefficient by 1. In experiments, we tried two different sets of parameters for the stochastic hill-climbing algorithm. The termination condition of the experiment was minimization of fitness value (sums of deviation quadrates) to 0.0.

4 Results

The performance of the individual configurations was measured as the number of generations needed to find the correct mathematical model. In every experiment we ran the GP symbolic regression 10,000 times.

First we set the number of neighbors of the Stochastic Hill-Climbing Search Algorithm to 4 and the number of steps to 3, and we compared the hybrid approaches (same values for Lamarckian and Baldwinian approach) with classic symbolic regression. The Table 1 shows the results of the first experiment:

Table 1. Results of the first experiment

Regression type	Median successful generation	Average successful generation
No adaptation	287	465
Lamarckian	186	299
Baldwinian	274	426

In the second experiment, we set the number of steps to 8 and 8 (same values for Lamarckian and Baldwinian approach). The Table 2 shows the results of the second experiment:

Table 2. Results of the second experiment

Regression type	Median successful generation	Average successful generation
No adaptation	297	472
Lamarckian	182	298
Baldwinian	264	415

We can also compare the number of successful solutions under given generation limits (from the second experiment). The results are showed in the Table 3.

Table 3. Successful solutions under the given generation limits

Limit	No adaptation	Lamarckian adaptation	Baldwinian adaptation
5	5	6	9
25	223	325	194
50	875	1204	756
250	4457	6020	4819
500	6719	8141	7152
2500	9916	9992	9949

5 Discussion

The first experiment showed that lifetime adaptation can accelerate the evolution of the mathematical model. For our experiment setting, the symbolic regression of the third degree polynomial was faster with Lamarckian evolution. The Baldwin effect also reduced the number of generations needed to find the right model, but the difference was much smaller. It is possible that the problem was not complex enough to demonstrate the effects of the Baldwinian approach.

The second experiment with stronger local learning confirmed dominance of the Lamarckian evolution over the Baldwinian approach. Both hybrid methods improved their results, but the computational time grew enormously. Although evolution has been accelerated, the computational time has increased. These costs are also influenced by the implementation of the algorithm because genetic programming was implemented in parallel, while local learning was not.

6 Conclusion

In this article, we have firstly done an overview of existing research in the field of hybrid evolutionary methods. Then we tried the possibilities of Lamarckian and Baldwinian adaptation to find a mathematical model via symbolic regression with genetic programming. Our experiments showed that lifetime adaptation can accelerate the evolution, but also added the additional costs. For a more accurate comparison of individual lifetime adaptation approaches, multiple experiments are needed.

In the future, we would like to try to find models of different kinds of functions. More complex mathematical functions could better show the behavior of hybrid methods. We would also like to test the behavior of multiple algorithms for local learning and use the findings from the experiments to solve other interesting problems.

Acknowledgments. The work has been supported by the Funds of University of Pardubice (by project "SGS 2019" No: SGS_2019_021), Czech Republic. This support is very gratefully acknowledged.

References

1. Iba, H., Sato, T., de Garis, H.: Recombination guidance for numerical genetic programming. In: Proceedings of 1995 IEEE International Conference on Evolutionary Computation, p. 97. IEEE (1995). https://doi.org/10.1109/icec.1995.489292. http://ieeexplore.ieee.org/document/489292/. ISBN 0-7803-2759-4
2. Nikolaev, N.Y., Iba, H.: Adaptive Learning of Polynomial Networks: Genetic Programming, Backpropagation and Bayesian Methods. Springer, New York (2006). ISBN 978-0-387-31239-2
3. Schoenauer, M., Lamy, B., Jouve, F.: Identification of mechanical behaviour by genetic programming part II: energy formulation. Technical report, Ecole Polytechnique, 91128 Palaiseau, France (1995)
4. Sharman, K.C., Esparcia-Alcazar, A.I., Li, Y.: Evolving signal processing algorithms by genetic programming. In: Zalzala, A.M.S. (ed.) First International Conference on Genetic Algorithms in Engineering Systems: Innovations and Applications, GALESIA, Sheffield, UK, 12–14 September 1995, vol. 414, pp. 473–480. IEE (1995). http://www.iti.upv.es/~anna/papers/galesi95.ps. ISBN 0-85296-650-4
5. Whitley, D., Gordon, S., Mathias, K.: Lamarckian evolution, the Baldwin effect and function optimization. In: Davidor, Y., Schwefel, H.P., Manner, R. (eds.) Parallel Problem Solving from Nature - PPSN III, pp. 6–15. Springer, Berlin (1994)
6. Koza, J.R.: Genetic Programming: On the Programming of Computers by Means of Natural Selection. Bradford Book, Cambridge (1992). ISBN 0-262-11170-5
7. Poli, R., Langdon, W.B., McPhee, N.F.: A Field Guide to Genetic Programming. Lulu Press, Burton (2008). ISBN 978-1-4092-0073-4
8. Le, N., Brabazon, A., O'Neill, M.: How the "baldwin effect" can guide evolution in dynamic environments. In: Fagan, D., Martín-Vide, C., O'Neill, M., Vega-Rodríguez, M. (eds.) Theory and Practice of Natural Computing. Lecture Notes in Computer Science, pp. 164–175. Springer, Cham (2018). https://doi.org/10.1007/978-3-030-04070-3_13. ISBN 978-3-030-04069-7
9. Red'ko, V.G., Mosalov, O.P., Prokhorov, D.V.: A model of evolution and learning. Neural Networks **18**(5–6), 738–745 (2005). https://doi.org/10.1016/j.neunet.2005.06.005. https://linkinghub.elsevier.com/retrieve/pii/S0893608005001358. ISSN 08936080
10. Turney, P.D.: Myths and legends of the Baldwin effect. arXiv preprint cs/0212036 (2002)
11. Anderson, R.W.: Learning and evolution: a quantitative genetics approach. J. Theor. Biol. **175**(1), 89–101 (1995). https://doi.org/10.1006/jtbi.1995.0123. http://linkinghub.elsevier.com/retrieve/pii/S0022519385701233. ISSN 00225193
12. Hinton, G.E., Nowlan, S.J.: How learning can guide evolution. Complex Syst. **1**, 495–502 (1987)

13. French, R., Messinger, A.: Genes, phenes and the Baldwin effect: learning and evolution in a simulated population. In: Brooks, R., Maes, P. (eds.) Artificial Life IV. MIT Press, Cambridge (1994)
14. Russel, S.J., Norvig, P.: Artificial Intelligence: A Modern Approach, 3rd edn. Pearson Education, Harlow (2014). ISBN 978-1-29202-420-2
15. Ackley, D.H.: Stochastic iterated genetic hill-climbing. Doctoral dissertation, Carnegie-Mellon University, Pittsburgh (1987)

Forecasting of Migration Processes by Integrating Probabilistic Methods and Simulation

V. V. Bystrov[1(✉)], M. G. Shishaev[1], S. N. Malygina[1,2], and D. N. Khaliullina[1]

[1] Institute for Informatics and Mathematical Modelling of Technological Processes of the Kola Science Center Russian Academy of Sciences, Apatity, Murmansk Region, Russia
{bystrov, shishaev, malygina, khaliullina}@iimm.ru
[2] Apatity Branch of Murmansk Arctic State University, Apatity, Murmansk Region, Russia

Abstract. The paper is devoted to the development of a methodology for forecasting migration processes. It is proposed to integrate probabilistic methods of study of migration flows with simulations, each of which has its advantages and disadvantages. The joint use of such approaches will allow, on the one hand, to ensure the accuracy of modeling in accordance with the quality of the source data, and on the other – to analyze the studied system at the required level of abstraction. The results of approbation of the proposed combination of analytical entropy and simulation models of migration are given on the example of studying the migration processes of the Arctic countries of Eurasia.

Keywords: Forecasting · Probabilistic methods · Simulation · Analytical entropy models · Migration processes · Arctic regions

1 Introduction

The modern world is characterized by a distinctive feature - the constant movement of people with different goals, between regions, countries, continents. For the movement of people, there is a tendency to increase the number of people involved in interregional migration. It should be noted that migration processes have a significant impact on demographic indicators (population size, age structure, mortality and fertility), thereby also changing the amount of labor resources of national economic systems. In turn, such changes lead to fluctuations in economic indicators and living conditions of the population. The national labor market essentially depends on migration flows. Under the influence of migration there is a transformation of socio-economic spatial relations. The speed of such changes varies and is not always consistent with the adaptive capacity of migrants and indigenous peoples, which can lead to increased social tensions.

For example, for the countries of Scandinavia, issues related to the increase in immigration flows from the countries of the Middle East, North Africa and Asia, leading to a variety of negative consequences in society, have recently become urgent.

R. Silhavy et al. (Eds.): CoMeSySo 2019, AISC 1047, pp. 347–359, 2019.
https://doi.org/10.1007/978-3-030-31362-3_34

Thus, according to the research of the Arctic University "UArctic" [1], important problems of migration policy in the Nordic countries are associated with the following phenomena:

- slow assimilation of visitors in conditions of strong cultural differences;
- difficulties in integrating immigrants into local communities;
- the emergence of problems in the formation of a balanced regional labor market.

At the same time, the Arctic regions of Russia have their own problems in migration, namely: the outflow of young people from the Northern regions; the growth of "personnel hunger" of industry enterprises; the successful provision of personnel logistics for infrastructure projects and others.

Forecast modeling is being used to develop mechanisms for the rational management of migration flows between different regions. Various models serve as the main means of predicting the movement of human resources in different scenario conditions. Despite the large number of studies in the field of migration modeling, this problem is still relevant today. This is due to the fact that the development of tools for forecasting migration processes is a non-trivial task [2–4]. The lack of objective information on the composition and relationships between the components of migration is a basic reason for the complexity of migration modeling.

2 Background

At present, analytical models based on differential equations and other mathematical tools are used to study migration processes. Let's take a look at the most interesting relatively modern research work in this area.

The paper by Schmertmann [5] proposes a model of population change in conditions when the stable fertility is below the level of reproduction and there is a constant migration flow. The results of theoretical analysis show that the population in such a country eventually becomes constant. The model simulates the process of gradual "demographic forgetting", which is that the size and structure of the original population does not affect its size and structure in the long term, since each next generation of their descendants is smaller than the previous one, and asymptotically tends to zero. It was shown that under these conditions, the migration flow leads to unusual age structures, which depend on the distribution of the ages of arriving immigrants. Numerical calculations using statistical data show that in the limit the ratio between migrants and indigenous population will be minimized.

In the work of Feichtinger, Prskawetz, Veliov [6] the problems of optimal management of demographic policy related to the distribution of migration by age groups are considered. For this purpose, the authors developed a macro-model, which is based on a combination of the Lotka model of population dynamics and the Solow economic model. The controls in the model are migration values and savings rates, and the maximization of unit consumption is accepted as a criterion of usefulness. The problem is solved by the principle of the maximum Pontryagina, through which qualitative ideas about the rational values of migration and the rate of accumulation, depending on age, are obtained. Results are illustrated by numerical experiments on Austrian statistics.

The work of Simon, Skritek, Velova [7] provides the results of simulations of demographic policy. The model developed is the task of optimal management of the age density of the population with restrictions. The behavior of the control object is described by a first-order differential equation in partial derivatives of Kendrick – von Foerster on an infinite period of time.

In the work of Genies, Volpert, Auger [8] the integro-differential reaction-diffusion equation is studied in the dynamics of populations, including the consumption of resources. It proves that the equilibrium in the model can lose stability and this leads to the emergence of stationary spatial structures and at the same time can be observed once-personal types of running waves.

Analytical models, as a rule, are based on observational data on the most significant parameters, presented in the form of time series, and provide a sufficiently high accuracy. At the same time, they have a number of drawbacks. First, with the complexity of the structure of the simulated system complexity of the analytical model increases dramatically. Secondly, the problem with the availability of source data (for example, for modeling socio-economic systems) most often only macro-level observational data are available. This forces the model to scale up, which inevitably reduces the accuracy of the simulation and limits the ability to use the models at the lower levels of decision-making.

An alternative approach to migration research is simulation and, in particular, agent-based modeling. This approach initially describes the structure of the system as a whole, and then identifies the parameters of the model using statistical, expert and other methods. In comparison with analytical approaches, agent-based models make it possible to reproduce the structure of the system with the required detail without significant increase in labor costs.

In this paper, it is proposed to integrate agent-based and analytical modeling to create computer models of migration. According to the authors, this will allow, on the one hand, to ensure the accuracy of modeling in accordance with the quality of the source data, and on the other - to analyze the study system at the required level of abstraction.

3   Approaches and Methods

The paper presents some results of the research project "Development of methods for identification of dynamic models with random parameters and their application to forecasting migration in Eurasia". This project is carried out within the framework of scientific collaboration of researchers - the first group led by RAS Academician Yu. S. Popkov from the Institute of Systems Analysis of the Russian Academy of Sciences, the second group led by RAS Professor M. G. Shishaev from the Institute for Informatics and mathematical modeling of the KSC RAS.

To solve the problem of studying migration processes in the framework of this project, it is proposed to use two basic approaches: mathematical demo-economics and simulation. Consider each of these approaches in more detail.

Many mathematical models of migration reflect the relationship of demographic and migration processes with economic ones. The approach to the study of such demo-economic systems was proposed by the academician Yu. S. Popkov. Modeling of demo-economic systems aims to detect some of their states that arise as a result of the interaction of the processes of spatial-temporal evolution of the population and the economy in terms of macro-indicators. When studying these processes, it is assumed that the population is economically motivated, i.e. the decisions of individuals are mainly determined by economic indices. In aggregate form, the structure of the demo-economic model is represented by two subsystems – "Population", "Economy" and the auxiliary subsystem "Interaction", which models the direct and inverse connections between the population and the economy. Subsystems of the demo-economic model are included in a closed loop, and therefore the designation of forward and backward linkages in it depends on the priority given to the subsystem "Population" in relation to the subsystem "Economy".

The mathematical demo-economics uses one of the types of probabilistic models, namely entropy models. Yu. S. Popkov proposes to consider the task of modeling migration processes as a task of maximizing entropy function, which is based on the demographic and economic indicators that affect migration. In this case, the mathematical description of migration is a randomized model, the parameters of which are random variables characterized by the corresponding probability density distribution. In the opinion of Popkov Yu.S., the principle of maximizing the information entropy on sets defined by observing the "entry" and "exit" of a randomized model guarantees the best solutions with maximum uncertainty.

Currently, simulation is a flexible and powerful mean for creating computer models in various fields of knowledge. Simulation has proved to be an effective tool for solving practical problems related to the analysis and forecasting of various scenarios of systems development. The project used three methods of simulation:

- Agent-based modeling. It was used to encapsulate the characteristic properties and actions of the domain entity in separate software agents. Each agent has its own behavior and decision-making algorithm.
- System dynamics. It allows you to present the process as a network of interacting levels and flows. This method was used to describe the dynamics of entities that are characterized by the accumulation of certain resources. For example, to describe the process of formation of goods flow between countries.
- Discrete-event modeling. It was used in the form of state diagrams to describe the behavior of some agents. In fact, it is an algorithm for making decisions by the agent depending on its current state and input information. An example is the definition of a person's decision-making algorithm to migrate to another country.

As part of this work, it is proposed to integrate analytical entropy and simulation models for the study of migration processes between countries. At the current stage of the study, such integration is presented so far in the form of using the structure and parameters of some entropy models as part of the developed simulation models.

4   Analytical Entropy and Simulation Models of Migration

Within the framework of the project "Development of methods for identification of dynamic models with random parameters and their application to forecasting migration in Eurasia", a combination of original analytical and simulation models for fore-casting inter-regional migration flows was proposed. Consider each type of developed models in more detail. It is worth noting that this paper focuses more on the complex of simulation models of migration, as they are directly within the area of responsibility of the authors of the paper.

Analytical Entropy Models of Migration

The results of studies of socio-economic macro-systems using the demo-economic approach of the academician Yu.S. Popkov were taken as a mathematical model of interregional migration. Below are some basic designs showing the basic principles of the applied concept of mathematical modeling.

Consider the following system of spatial and temporal distribution of immigration flows. Let there be N_s regions - sources of emigration. At the moment n there are $M_1[n], \ldots, M_{N_s}$ - potential economically motivated immigrants. This population is going to migrate to the $\mathcal{K}_1, \ldots, \mathcal{K}_N$ run-off regions, where at the time of n lives $K_1[n], \ldots, K_N[n]$ people. Here n - whole numbers denoting discrete values of calendar time.

Consider the mathematical description of this model. The model of the local-stationary state is based on the stochastic hypothesis: migrants from the source region i choose the run-off region k randomly, independently of each other with a priori probabilities $v_{ik}[n], i = \overline{1, N_s}; k = \overline{1, N}$, which depend on the ratio of the economic potentials of the run-off regions and the regions-sources. We introduce the following notations:

- $x_{ik}[n]$ - flow of migrants from the i-th source region to the k-th run-off region;
- $E_1[n], \ldots, E_N[n]$ - distribution of numbers of immigrants by run-off region.

You can build a mathematical model of immigration, the stationary state of which is described by the following entropy model:

$$H_i(X[n]) = -\sum_{k=1}^{N} x_{ik}[n] \ln \frac{x_{ik}[n]}{ev_{ik}[n]} \Rightarrow \max, \tag{1}$$

$$\sum_{k=1}^{N} x_{ik}[n] = M_i[n], i = \overline{1, N_s}.$$

This task of maximizing entropy has an analytical solution to this problem:

$$x_{ik}^*[n] = M_i[n] \frac{v_{ik}[n]}{\sum_{j=1}^{N} v_{ik}[n]}, \qquad i = \overline{1, N_s}; \quad k = \overline{1, N}. \tag{2}$$

It follows that the following number of immigrants is formed in the run-off regions:

$$E_k^*[n] = \sum_{i=1}^{N_s} x_{ik}^*[n], k = \overline{1, N}. \tag{3}$$

This section provides only a fragment of the analytical model of interregional migration in terms of maximizing the entropy function. In the future, this model becomes more complicated due to the inclusion of balance equations in its composition, interpreting the processes of forming indicators of economic development of regions and their impact on the attractiveness of the region for migration. The entropy randomized machine learning algorithm is used to adjust the interaction between the "Demography" and "Economics" subsystems.

Simulation Models of Migration

A complex of simulation models has been develop based on the analytical migration model. Structurally, the complex reproduces the basic principles of mathematical demo-economics [9] and is a set of model agents. Anylogic was used to implement simulation models, allowing you to combine system dynamics, discrete-event and agent-based modeling.

In the developed complex of simulation models, the macro region (e.g. the Eurasian continent) divided into separate territorial and administrative zones (e.g. countries). Each zone is a model agent and includes the "Population" (collection people) and "Economics" (collection sectors) subsystems, which are set in the programming implementation in Anylogic as a collection of agents of the appropriate type (Fig. 1).

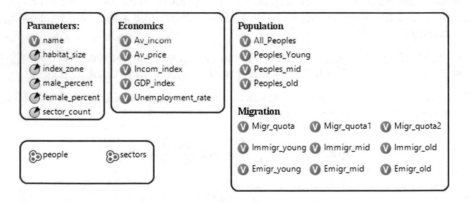

Fig. 1. Structure of agent "Zone"

Thus, simulation models contain three types of agents: Zone - region (within the study, country); Person – an individual (directly involved in migration processes); Sector – sector of the economy (reflects the results of a separate economic industry).

The filling of the zone by people is in accordance with the percentage division of women and men, as well as the probability distribution of types of women. These parameters can be set in the model configuration. In the complex of simulation models, depending on the fertile characteristics, there are two types of female persons - Western and Eastern. The Western type of women characterized by the birth of an average of 1–2 children, and the Eastern – 4–6.

The parameters of the agent "Zone" (Fig. 1) are the name of the zone (name), the initial value of the population (habitat_size), the percentage distribution of men and women of different types (male_percent, female_percent), the number of sectors of the economy (sector_count).

Indicators describing the state of the subsystems calculated for each zone in the simulation process (Fig. 1). For a subsystem "Population" such indicators are the total number of the population (All_Peoples), the number of children (People_Young), economically active population (People_mid) and number of elderly (People_old). For characteristic of migration processes the quota migration volumes (Migr_quota, Migr_quota1, Migr_quota2), emigratory (Emigr_young, Emigr_mid, Emigr_old) and immigration streams (Immigr_young, Immigr_mid, Immigr_old) various age groups are calculated.

The average per capita income (Av_incom), the average equilibrium price (Av_price), index per capita income (Incom_index), index per capita GDP (GDP_index), unemployment rate (Unemployment_rate) are calculated for the subsystem "Economy".

For agent type Person (Fig. 2) the following characteristics are defined: age (age), sex (male_female), number of children (birth), probability of death (death), type of migration (for economic reasons or for family reunification, migration), mobility (probability of migration mobility) and current area of residence (zone).

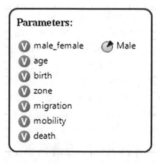

Fig. 2. The parameters of the agent "Person"

The model for the Person type defines three main actions: birth, death, migration, each of which is programming implemented as a function or method of the appropriate class – type of agent. In the simulation, the birth of a person is interpreted as the creation of a new instance of an agent of type Person and the function people.add (Person o) is called. In this case, regardless of the gender of the created instance, a connection with the zone of its birth, as well as with the parent agents by recording the corresponding values in certain parameters of the instance automatically is established. In order to preserve the genetic affiliation of women to a certain type of fertility at the birth of a girl, her type automatically becomes the type of fertility of her mother. When an event "Death" occurs, a particular instance of the Person agent is called the people. remove(o) method, which results in the removal of all connections with other agents and the release of memory.

The age update for all agents of the Person type occurs automatically at each model step equal to one calendar year. The Age parameter of the Person agent allows you to adjust the processes associated with its life activity, in particular:

- to monitor the occurrence of the event "Death" of a person, as this parameter is fixed and lies in the range from 0 to 85 years;
- to change the age category of the agent of the Person type (for example, when this parameter is reached >17 years, the agent is transferred to the category of the economically active population);
- to manage the reproductive process for agents representing women in the model. At the moment, the fertile age for any type of women is set in the form of an interval from 15 to 45 years.

When the event "Birth of a child" appears in an agent of type Person female, the parameter birth responsible for storing information about the number of children is increased by one. The probability of the birth of an agent of type Person male and female is set in the configuration of the complex simulation models (by default, they are equal to 50%). The parameter birth allows you to adjust the number of children in a particular woman during imitation according to the type of her fertility. The relationship between the agent-child and the agents of his parents is take into account when organization of migration processes. It is liquidated when the agent reaches the age of 21.

Two types of migration are implemented in the complex of simulation models: labor migration and migration motivated by the desire to reunite with family (family migration). Traditionally, the first type of migration is subject to a mobile economically active population. The second type of migration does not depend on the economic indicators of the region, but is determined depending on the presence of family ties.

There are some constraints to consider when modeling the migration process. These are imposed on so-called desirable and undesirable migrations. Desirable immigration for each age group depends on the change in the host population at a certain age range. Undesirable immigration in its age profile is associated with the preservation of the number of certain age groups. Families and even family groups realize quite a large part of the immigration flow. They are always present children of younger ages and relatives of middle and older ages. Therefore, the number of these age groups are trying not to increase.

Hence, there is the following constraint on the number of immigrants to the zone n:

$$\sum_{m=1}^{N} \text{migr}(n, m, a, t) \leq \text{Migr\_quota}(n, a, t), \qquad (4)$$

With: Migr_quota(n, a, t) is the quota migration volume for zone n in a certain age range a, migr(n, m, a, t) is the migration flow in the age range a from zone m to zone n.

There is also a constraint on the number of emigrating from the zone (to leave the zone can not be more than it is home to the mobile population):

$$\sum_{i=1}^{N} \mathrm{migr}(i, m, a, t) \leq \mu K(m, a, t) \tag{5}$$

With: $\mathrm{migr}(i, m, a, t)$ - migration flow in the age range a from zone m to zone i, μ - mobility coefficient ($\mu < 1$), $K(m, a, t)$ – total population of the age group a in zone n.

Migration flows are the result of numerous individual decisions made by individuals under the influence of various factors. Migration decisions are made by individuals from age groups whose age is higher than a certain value a_{migr}. Children of younger age groups do not make migration decisions, but participate in family migration. That is, the migration decision of an individual in the age groups from 0 to a_{migr} is determined by his belonging to the family of an individual from the age group older than a_{migr}.

The migration solution is characterized by two components – the mobility (probability of migration mobility) of an individual from a certain age group and the direction of migration, i.e. the probability that an indie species from zone n will go to zone m. For labour migration, the choice of destination is determined by the per capita income of the entry zone, for family – by the presence of relatives in the entry zone.

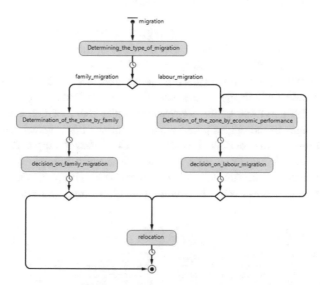

Fig. 3. Statechart of migration

The decision to migrate by an individual agent of type Person is programmatically implemented as a statechart (Fig. 3). The first step is to determine the type of migration (labour or family) depending on the age group (Determining_the_type_of_migration), the second step defines the direction of migration for each type of migration (Determination_of_the_zone_by_family, Definition_of_the_zone_by_economic_performance), after that, the decision on migration is made (decision_on_family_migration, decision_on_labour_migration), namely, the restrictions on entry into the selected zone are checked, if entry is possible, then the event "relocation" occurs, simulating the move from one zone to another.

Interaction within a separate zone of such subsystems as "Population" and "Economy" has been previously published and discussed in more detail in the paper [10].

5 Results

In order to check the efficiency of the proposed analytical entropy and simulation models, they were tested on the example of solving the problem of forecasting migration flows of the Arctic countries of Eurasia. Such countries include Norway, Denmark, Sweden, Finland, Iceland and Russia. The study conducted a number of computational experiments to determine the number of immigrants and emigrants between the designated countries. The set of models was set up on the basis of retrospective data on demographic, social and economic indicators presented in official statistics of the European Union and the Russian Federation [11, 12]. Statistical data in the time interval from 2000 to 2013 were used to configure the models, in the interval from 2014 to 2017 to verify the models. Forecast values were computed in the range from 2018 to 2026.

The results of forecasting two types of migration for each of the Arctic countries are presented in the Figs. 4, 5, 6, 7 and 8. On the graphs, the dotted line shows the time series taken from the official statistics, the solid line shows the data obtained in the results of computational experiments. At the same time, it is worth noting that according to the applied concept of entropy-robust modeling, the result of a series of computational experiments is the ensemble of all permissible under the specified conditions of trajectories of changes in the output parameters of models. For a visual representation of the simulation results was conducted by the procedure of averaging and selection of the median of all possible values of a parameter in a specific point in time. The step of presenting the results of the simulation is one year. As can be seen from the graphs, the most active migration flows occur between Norway, Sweden and Finland.

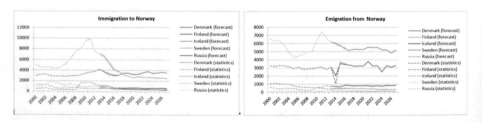

Fig. 4. Forecast and statistic date of migration in Norway, person

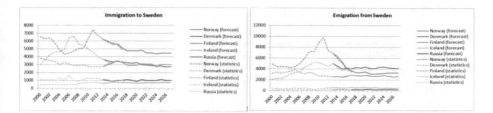

Fig. 5. Forecast and statistic date of migration in Sweden, person

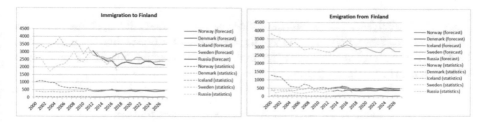

Fig. 6. Forecast and statistic date of migration in Finland, person

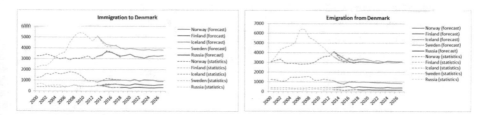

Fig. 7. Forecast and statistic date of migration in Denmark, person

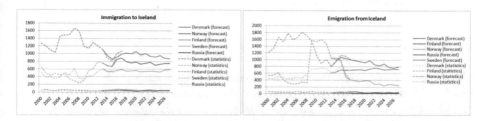

Fig. 8. Forecast and statistic date of migration in Iceland, person

To assess the adequacy of the models, the simulation results were compared with the statistics for the Arctic countries for 4 years (2014–2017). For each country, a relative error was calculated of the immigration forecast from a specific country and on the basis of this, on average, over the year. The average relative error in immigration by country is in the range of 3.8% (Sweden) to 6.3% (Iceland). The results are presented in Table 1.

Table 1. Average relative error of immigration forecast, %

	2014	2015	2016	2017	Average for the country
Immigration to Norway	4%	4%	6%	8%	5,50%
Immigration to Sweden	6%	3%	3%	3%	3,75%
Immigration to Finland	3%	7%	6%	7%	5,756%
Immigration to Denmark	4%	8%	4%	1%	4,25%
Immigration to Iceland	5%	5%	8%	7%	6,25%
Average value (total)					5,11%

A similar analysis and assessment of the accuracy of the modeling results was made for emigration flows between the Arctic countries of Eurasia and is presented in Table 2. The average relative error for emigration by country lies in the range from 4.3% (Denmark) to 10.3% (Norway).

Table 2. Average relative error of emigration forecast, %

	2014	2015	2016	2017	Average for the country
Emigration from Norway	22%	5%	6%	8%	10,25%
Emigration from Sweden	5%	3%	5%	5%	4,50%
Emigration from Finland	4%	7%	9%	13%	8,25%
Emigration from Denmark	4%	6%	4%	3%	4,25%
Emigration from Iceland	4%	14%	7%	12%	9,25%
Average value (total)					7,30%

As can be seen from the evaluation results presented in Tables 1 and 2, on average, the accuracy of the proposed migration models is within 10%, which is a good enough result for forecasting socio-economic processes.

6   Conclusion

The research considered in the paper is aimed at the development of the methodology for predictive modeling of migration processes. The integration of demo-economic entropic and simulation models is one of the ways to improve the quality of methodological support for forecasting. Such a combination allows the flexibility to change the scale of consideration of the socio-economic system under study while maintaining the level of accuracy of the forecast results determined by the mathematical apparatus used and the quality of the source data. The article describes the results of testing this approach on the example of the study of migration of Arctic countries of Eurasia. Further development of the work is the improvement of the integration mechanisms of analytical and simulation models in the direction of automating the processing of collective expert knowledge and their application to set up models.

The lack of solutions proposed by the authors is that the methods used to study migration processes are computationally very expensive. The computational algorithm for computations of all possible ensembles of system trajectories used in analytical entropy models is very demanding on computing power. The same high demands on the performance of computing systems and available RAM are made by the developed simulation models, which is due to the large number of instances of human-type agents. However, according to the authors, these problems with computing power are coming to naught due to the rapid development of distributed computing based on various hardware platforms and information and communication technologies (cloud computing, supercomputers, computational clusters, GPU-based computing, and more).

Acknowledgements. The study was financially supported by RFBR project № 16-29-12878 (ofi_m) "Development of methods for identification of dynamic models with random parameters and their application to forecasting migration in Eurasia".

References

1. Official site The University of the Arctic. https://research.uarctic.org/. Accessed 15 May 2019
2. Denisenko, M.B.: Mathematical models of population migration. In: The Book Modern Demography. Under the editorship of A.I. Kvasha, V.A. Iontsev. MSU, Moscow (1995)
3. Lukina, A.A.: Analysis and mathematical modeling of international labour migration. In: Lukina, A.A., Prasolov, A.V. (eds.) Management Consulting, vol. 10, no. 82, pp. 146–156 (2015)
4. Yudina, T.N.: Predictive models of migration to Russia: approaches and analysis. In: Yudina, T.N. (ed.) In MIGRATION 6, Tbilisi, Ivane Javakhishvili Tbilisi State University Migration Research Center, pp. 25–43 (2013)
5. Schmertmann, C.P.: Stationary populations with below-replacement fertility. Demogr. Res. **26**, 319–330 (2012). Article 14. https://doi.org/10.4054/demres.2012.26.14
6. Feichtinger, G., Prskawetz, A., Veliov, V.M.: Age-structured optimal control in population economics. Theor. Popul. Biol. **65**(4), 373–387 (2004)
7. Simon, C., Skritek, B., Veliov, V.M.: Optimal immigration age-patterns in populations of Fixed size. J. Math. Anal. Appl. **405**, 71–89 (2013)
8. Genieys, S., Volpert, V., Auger, P.: Pattern and waves for a model in population dynamics with nonlocal consumption of resources. Math. Model. Natur. Phen. **1**(1), 65–82 (2006)
9. Popkov, Y.S.: Mathematical demoeconomic: Macrosystem approach. In: Popkov, Y.S. (ed.) LENAND, Moscow (2013)
10. Bystrov, V.V., et al.: Development of the forecasting component of the decision support system for the regulation of inter-regional migration processes. Comput. Stat. Methods Intell. Syst. **859**, 60–71 (2019)
11. Database – Eurostat. http://ec.europa.eu/eurostat/data/database. Accessed 16 Apr 2019
12. Federal State Statistics Service Russian Federation. http://www.gks.ru/wps/wcm/connect/rosstat_main/rosstat/ru/statistics/accounts/. Accessed 16 Apr 2019

Social Networking Services as a Tool for State and Regional Governance (On the Example of the Arctic Region)

Igor O. Datyev$^{(\boxtimes)}$, Andrey M. Fedorov, Andrey L. Shchur,
and Vitaliy V. Bystrov

Institute for Informatics and Mathematical Modeling - Subdivision of the Federal
Research Centre "Kola Science Centre of the Russian Academy of Sciences",
184209 Apatity, Russia
{datyev,fedorov,shchur,bystrov}@iimm.ru

Abstract. The paper explores the problem of using social networking services in state and regional governance. The specific features of social networking services are presented, as well as the experience of organizational and legal support of interaction through social networking services of citizens and government officials in various countries. The current and potential technical and ethical difficulties of the full implementation of the e-participation concept are highlighted. The use of online social networks is analyzed on the example of one of the Arctic region of the Russian Federation. Conclusions are drawn about the current state of using social networking services for solving problems of state and regional government, in particular, about the prevalence of using social networking services only to inform the public. There is a need for further development of information technology to control the real working activity of government officials, taking into account their rank and position.

Keywords: Digital economy · E-Government · E-participation · Arctic region government · Officials · Social networking service

1 Introduction

The aspiration of the modern world for the digital economy implies the widespread introduction of information technologies in production, political, government, administrative and other processes. There are many definitions of digital economy and digitalization. Some of them were examined in detail in the previous research [1]. Analysis of the digital economy development programs implementation in USA, India, China and European Union shows difference in selected paths and the reached outcome, confirming that there is no universal program for this process [2]. To measure the level of digitalization of various countries and regions, digitalization indices are being developed [3]. Such indices often carry specific indicators that evaluate the introduction of information technologies in state and regional government processes, as well as the level of digital literacy of the population. Many of the digitalization index components are closely related to the use of specialized Internet resources, in particular, Social Networking Services (SNS).

© Springer Nature Switzerland AG 2019
R. Silhavy et al. (Eds.): CoMeSySo 2019, AISC 1047, pp. 360–370, 2019.
https://doi.org/10.1007/978-3-030-31362-3_35

Today, many projects are being implemented in the world, aimed to reduce the time and material resources expenditures of the state and state service users through the introduction of information technologies. The largest of these projects is e-government, and one of its key elements is e-participation. E-participation is the term referring to "ICT-supported participation in processes involved in government and governance" [4]. In turn, the important components of e-participation include e-inform, e-consult and e-decisions [5].

The obvious effect of digitalization is the ubiquitous use of SNS by the public. Some researchers equate such services to mass media, highlighting the only difference: the Internet access dependency of SNS [6]. For many active users these social services become the only source of news information. It is difficult to overestimate the potential role of such SNS in the political life on all levels, be it a separate state or even the whole world [7]. Some modern political events that have been repeatedly occurring in the world during the last decade, have become possible explicitly because of the SNS help, starting with the events of the Arab Spring [8] and ending with the recent United States Presidential election [9].

The potential of SNS is also very high in terms of state and regional governance. It should be noted that the active use of SNS is particularly relevant in remote Arctic regions (where Internet access is available, of course) due to harsh natural conditions, leading to adaptation difficulties and problems with creating a comfortable living in such environment. As a result, the problems of providing personnel needs and population of northern regions, medical services for the population and biomedical studies of the "northern" people specifics and their health, development and accessibility of housing and transport infrastructure are particularly acute.

The main objective of this study is to analyze the current experience of using open (public) SNS profiles of government officials to solve problems of regional management, using a selected Arctic region of the Russian Federation as an example. The main research questions are as follows: do SNSs have specific features that need to be taken into account in state and regional governance? How one organizes the interaction between the government and the population by using the SNS? What legal regulation is used/necessary for the implementation of this interaction? What technical and ethical difficulties arise during the legal regulations monitoring process?

2   The Specifics of SNS in State and Regional Governance

SNSs are used everywhere nowadays but they have a number of specific features that users (both officials and citizens) rarely think about. For example, social connections in SNS are quite stable, but at the same time impersonal and fleeting [10]. These social connections are virtual, but are shaped by reality and, at the same time, influence the reality in return.

Society itself becomes a society of real virtuality [11] – the reality of communication and information.

The following SNS functions can be distinguished [12]: self-expression; communication and socialization ("cyber socialization"); entertainment; education; informational.

The papers [13–16] state that the profiles of SNS users fairly correctly reflect their real identity.

Despite the discussions about the possibility or impossibility of treating various SNSs as mass media [17], SNSs, in fact, perform all the functions corresponding to the mass media – informative, cultural and regulatory [18]. Moreover, the regulatory function has a wide impact on mass audience, starting from contact establishing and ending with society control in the way of forming group and individual public consciousness, public opinion and the creation of social stereotypes.

It should be noted that the content of traditional mass media has largely become determined and supplanted by the news from SNS.

J. Nesbitt noted that "In an environment where the computer has expanded the power of an individual, citizens can more effectively monitor the actions of their governments than governments can monitor the actions of citizens... Computers, satellite telephony, telefaxes endow a person with power and not oppress him, how it was feared before." [19]. Already in 2000, Naisbitt and Aburdene predicted that with the development of information and communication technologies the modern society would transit from representative democracy to participating democracy, where the most important decisions in the life of citizens would be made interactively with the involvement of the population.

The authors [20] note that politics inevitably migrate to the origin of any mass audience. The presence of political actors in the SNS is a unique opportunity to combine the promotion of a politician and the party and the main feature of the political entity image – full accessibility.

As a result, the SNS hide the ability to manipulate and control public consciousness (i.e. perform as a form of social control) and therefore become one of the tools for the political and state governance [21].

At the same time, SNS are a vault of user content and information about the structure of user connections. When using various technologies for analyzing data from SNS and comparing them with other data sources, hidden patterns about both the account holder and third parties can be revealed.

One should also not forget about the ethical side of the issue: even with all the features of online social networks, they are a reflection of the real world people and the connections between these people, so there human norms, laws and rules should act accordingly. That is why the legal and technical support for monitoring the application of SNS in solving the tasks of state and regional government needs to be developed.

3   Organizational Experience of the Governmental Services in SNS

Social Networking Services is one of the technologies used in the E-Government of transformative 2030 Agenda for Sustainable Development. The UK government uses social networks to collaborate within its system and interact with the public. Within the government, a dedicated Civil Pages website is predominantly used. At their workplace, civil servants can create profiles, update their status, and exchange messages with each other only through Civil Pages and LinkedIn and are not allowed (except for the authorized personnel) to use any other SNS – but they can still do so off work. Appointed commissioners speak on behalf of the organization and interact with the public through various websites, including Facebook, Twitter, LinkedIn, etc. [22].

Almost 91% of local executive and administrative personnel in US are registered on Facebook, 71% on Twitter, 50% on Youtube, 22% on Skype, 17% on Google Docs. It has been revealed that local authorities use SNS to inform the public, involve citizens in public affairs and as an inner tool for the joint work of public servants. The experience in creating a digital registry should be noted. The digital registry is designed to confirm the official status of public profiles in SNS, mobile applications, etc., for several purposes, including prevention of cybercrime [23]. In United States and Canada, special supervising departments have been formed to monitor the activity of civil servants in the virtual space. They have the authority to request not only links to profiles, but also passwords to them.

In 2012, Twitter was the most popular website for the police in the Netherlands for the daily publication of thousands of reports about detained criminals, traffic situation, possible crimes, etc., that millions of users read [24].

The monitoring of the official SNS pages of the federal executive branch in Russia has revealed [25] that "among 78 federal executive agencies, 30 are not represented in any of the online social networks". At the same time, not all federal agencies respond to user comments and conduct surveys. The republican government agencies and other state organizations of the Republic of Belarus use six social networks: Facebook, Twitter, VKontakte, Youtube, Instagram and Odnoklassniki. Links to social networks are present on 28 official sites out of 38, which is comparable to the Russian experience [26].

Many countries are trying to overcome the challenges of using SNS to increase the public administration transparency and involve citizens in governmental decision-making. The analysis shows that SNS is an important channel of communication between government and citizens. At the same time, there are specialized websites designed exclusively for conducting general business within the state apparatus.

However, when solving the tasks of state and regional governance, there are problems of blurring the boundaries between professional and personal goals in using SNS by government officials, legalizing feedback from citizens, documenting this activity, etc.

4   The Experience of Legal State Regulation Use of SNS by the Officials

In many countries the legal regulation of SNS use by the citizens engaged in governmental service is carried out through the Social Media Policy.

For example, Swedish set of regulations highlight several important issues, such as [27]: control over the activity in SNS, study of citizens' opinions about the organization, the procedure for documenting this activity, the period of time required to respond to the user, consistency with other Internet resources.

The Office of Government Ethics in the United States has adopted a standard of conduct for personal use of SNS, involving issues of rational use of working time, job position in a personal profile, job search, maintaining an official profile, etc. [28].

In the UK, the SNS Guide for Civil Servants has been adopted to encourage the proper use of SNS in accordance with the code of ethics governing the interaction of public servants with citizens during working and non-working hours, as well as regulating the discussion of decision-making [29].

The European Commission has adopted SNS Recommendations for employees of the European Union. It emphasizes the blurring of boundaries between personal and official use of profiles in social networks. This document lists the positions that allow to speak on behalf of the European Commission officially, separating it from the rest of the employees, who can only express their personal opinion.

In Belarus, e-participation in the conduct of governmental affairs through SNS is not a direct subject of legal regulation [26]. However, according to the law "On Mass Media", SNS, as a non-state information resource (website), is mass media. Therefore, a government agency is entitled to carry out e-inform via SNS in the manner provided for mass media. In Russia in 2015, the author [25] notes the absence of any legal acts stipulating the creation or use of official SNS profiles by the federal agencies. However, it should be noted that the Federal Law N 8-FZ dated 09.02.2009, partly regulates the organization of access to information about the activities of government agencies and local governments, posted on the Internet. However, neither in Russia nor in Belarus the procedure for conducting e-consult and e-decision is defined.

In 2016, in Russia, Federal Law No. 224-FZ dated June 30, 2016, had become one of the most discussed among the legislation experts. According to it, all state and municipal employees are required to annually provide data about the information they have published on the Internet. Such information includes only publicly available data, which allows to identify a specific user, as well as links to personal pages (profiles) in the SNS. The said legal act does not contain requirements for the disclosure of personal correspondence, even if it was strictly work-related. It is worth noting that a similar report should also be submitted by citizens applying for filling vacant posts. In France and a number of European countries, the declaration of accounts is a mandatory requirement at the signing of the government employee job contract.

It is worth noting the use of SNS control in relation to not only state and municipal employees, but also individual citizens. In particular, one of the mandatory requirements for those who want to get a US visa can be the provision of passwords from all available personal profiles on the Web [30].

The German authorities in April 2017 enacted a legal provision imposing heavy fines of up to 50 million against SNS owners who allowed the placement of inaccurate news information. In Russia, there is also a similar legislative initiative. In addition, Russia is planning to create an open registry of unreliable news, which will also include their authors.

Sometimes new rules are directly aimed at countering corruption, disclosing interpersonal relationships (which can lead to a conflict of interest), ensuring effective counteraction against hate speech and the placement of inaccurate information, as well as activating and structuring the online activity of officials.

5   Technical and Ethical Difficulties in Monitoring Legislation Compliance

One of the biggest scandals of 2018 was centered around Cambridge Analytica, owner of Facebook social network Mark Zuckerberg and the results of the Donald Trump presidency election.

Despite the fact that many countries have already developed norms for using SNS and even anchored some of them at the legislative level, regularly occurring scandals around the use of SNS by the officials strongly demonstrate the imperfection of the legal regulation, and perhaps the impossibility of exercising comprehensive control over the implementation of legal norms.

For example, in Russia for the sake of executing Federal Law No. 224-FZ dated June 30, 2016 (detailed in the previous section), the Ministry of Labor has formed special guidelines and a report form on the conducted Internet activity. Despite this, even with the hypothetical presence of special supervisory department, a number of problems arise [31]. Taking into account the fact that all information is provided by state and municipal servants independently, it will be difficult to practically control the degree of its completeness and authenticity. Even if hidden profiles are revealed, how can one prove their belonging to a specific official within the legal terms? In addition, one must consider the possibility of creating fake profiles by any Web user.

The next problem is more technical. According to the analytics of the Russian Association of Electronic Communications, about 1.5 million officials fall under the scope of Federal Law No. 224-FZ, 70% of whom use social networks [32]. The reliability verification of such a large amount of information requires the development of an automated intelligent information technology, and its creation may require a large amount of time and resources.

In addition, when developing such technology, a number of new ethical problems may arise. If an SNS is used by an official not only for working purposes or for posting official information, but also for personal communication and representational purposes, then excessive attention of the controlling structures and the public will be on the verge of interfering with private life. For example, many regional mass media periodically analyze personal profiles of officials, correlating photographs of various tangible property, lifestyles (for example, travel) with income declarations.

Thus, in addition to the problem of verifying the completeness and accuracy of information, arises the problem of maintaining the balance between the sphere of professional competence of an official and his private life. It is obvious that this problem, in addition to technical complexity, is more ethical in nature.

Another serious problem, more related to the level of e-decision and the profiles of citizens themselves, lies in ensuring the authenticity of statements and "votes", blocking bots, fake accounts and detecting multiple accounts belonging to one person. This problem occurs largely due to the virtual nature of the SNS. It should be noted that the task of establishing the correspondence of events and the "population" of the real and virtual worlds is itself a separate complex research topic.

6   Study of SNS Use in Arctic Region Governance

As an experiment, open data was gathered from the public profiles of various level officials in one of the Arctic regions of Russia. The study involved profiles in SNS VKontakte (VK), Facebook (FB), Instagram (IG) and Twitter. The selected officials included the acting governor and his assistants, ex-governor, city mayors, heads of government ministries, etc.

Overall, 34 public servants were selected, based on their representation on the official government website of the Arctic region. The study involved the survey of open publications made by these officials on their personal SNS pages in a period of up to 6 months.

Table 1. Distribution and activity of SNS accounts

	Vkontakte, %	Facebook, %	Instagram, %	Twitter, %
Used actively	20%	23%	9%	3%
Not used	23%	20%	–	3%
Not found	57%	57%	91%	94%

Table 1 represents the intensity of SNS use by the government officials. The obtained data is divided into three categories, reflecting the intensity of use. It should be noted that even in the most popular Russian SNS – Vkontakte and Facebook – less than 50% of the studied group have profiles, and only half of them actively maintain their pages, while the other have blank profiles or updated them more than one year ago.

Table 2. Account distribution by the number of work and non-work related publications per SNS page

Posts quantity		75–100%	50–75%	25–50%	1–25%	No posts
SNS Profiles (%), Work related posts	VK	25%	25%	–	13%	38%
	FB	17%	33%	33%	17%	–
	IG	–	33%	–	67%	–
SNS Profiles (%), Non-work related posts	VK	50%	–	25%	13%	13%
	FB	17%	50%	17%	–	17%
	IG	67%	–	33%	–	–

Messages posted on the pages of all surveyed profiles were divided into two categories: messages related to the work activity and messages not related to the work activity, i.e. personal. All profiles have been divided into groups according to the corresponding number of work-related posts, published on their pages. From the data obtained it can be seen that 25% of the studied accounts in VK (and 17% in FB) have from 75 to 100% of the messages of a "working" nature, i.e. devoted to work activities.

Reports of the studied profiles were further divided into different topics. At the current stage of the study, four thematic sectors of direct and essential importance for the Russian Arctic were selected: personnel security ("Personnel"), transport infrastructure ("Transport"), medical topics ("Healthcare"), housing, utilities and communal services ("Housing").

Table 3. Publications distribution based on the Arctic-related topics importance

	Distribution of work-related posts				
	Personnel	Transport	Healthcare	Housing	Other
VK	17%	12%	9%	33%	29%
FB	9%	8%	2%	33%	48%
IG	14%	9%	11%	22%	44%

Basing on the data obtained (Table 1), it can be concluded that, at present, the level of SNS use by the Russian regional officials is below average. Only half of the target group is represented in Vkontakte and Facebook and is practically absent in Instagram and Twitter. Only one fifth of the analyzed profiles were active. As one of the reasons for such low activity among the public servants one can name additional SNS use regulations, imposed on the officials as public figures representing the state. It is easier to have no SNS profile or simply a blank one, rather than follow the strict content rules and regularly report on all made posts. However, the situation is gradually changing with the younger personnel taking governing posts and bringing along the way the modern culture of active online presence.

One of the peculiar research topics is the use of SNS personal profiles for the publication of work-related posts. The result presented in Table 2 shows that more than 50% of activity of the studied SNS profiles are devoted to work-related issues. Therefore, the publicity nature of the jobs that government officials have, leaves a noticeable imprint on their online activities, even when it comes to the personal SNS space. Those of them, who actively use SNS accounts, show a strong connection between their personal and work activities. The reasons for such close interweaving might be the deep involvement of an individual in the professional sphere, but at the same time it can also be an opportunistic desire to use any PR opportunities to demonstrate a "closer to the people" stance. To differentiate between these two polar manifestations, more research is needed.

Among the studied work-related publications, the most interesting are those that are directly related to the important issues of regional development of the Arctic territories. The results of these statistics are shown in Table 3: the content of more than half of work-related publications is marked with selected thematic headings.

Thus, it can be said that the government officials, through the active use of personal SNS, are trying to popularize the effectiveness of their territory governing activity among the citizens (other SNS users). This type of actions is now at an average level of intensity and mostly falls under e-inform category. But the current trends, tied with the digitalization course announced in the world, will make it possible to achieve the better indicators in the issues discussed.

7 Conclusion

VKontakte, Facebook, Instagram and others SNSs, unspecialized for solving problems of state and regional management, are initially an unofficial resource. So when they are used for informing, discussing and involving population in decision-making, it generally acquires the same shade of informality, frivolity and non-commitment. In order to fully utilize SNS in state and regional governance, the above difficulties must be overcome by not only developing existing organizational and legal norms, but also by developing information technologies to solve a number of technical and ethical problems associated with controlling the use of SNS. Of course, today the degree of elaboration of these issues differs significantly from one country to another: some of them already have an extensive legal framework, but in others, the main regulatory and legal acts are only at the early development stage. In most cases, the legal level regulates the use of SNS by officials, both for personal and professional purposes, with reference to ethical rules of conduct.

The full potential of using SNS for solving problems of state and regional governance has not yet been revealed: most countries use SNS only at the e-inform level, which is caused by a number of heterogeneous problems caused by the specifics of SNS. Since Russia has no formal requirements for the placement of content by the officials in the SNS, excessive public attention has already led to the closure of information in personal profiles, or to a reduction in the number of such profiles. However, some positive trends should also be noted in the use of SNS – with Arctic zone of Russian Federation being an example of a region where most pressing infrastructure problems are being openly discussed, including personnel, medicine, housing and transport topics. In particular, for most of the Arctic regions of Russia a very important issue for the stable development of the economy is the staffing of industrial enterprises, including mining and chemical industries. For rational management of personnel policy, it is necessary to solve a number of urgent tasks, one of which is the forecasting of personnel needs. One of the actual ways of improving the quality of forecasts of personnel needs of the regional economy is the use of SNS as a tool for obtaining new data from informal sources of information.

Further development of information technology would allow not only to monitor the implementation of the laws of conduct in the SNS, but also to control the real working activity of government individuals, taking into account their rank and position.

Acknowledgements. The study was partially supported by RFBR project № 19-07-01193 A "Methods and means of information support of personnel security management of the regional mining and chemical cluster".

References

1. Fedorov, A.M., Datyev, I.O., Shchur, A.L., Oleynik, A.G.: Online social networks analysis for digitalization evaluation. In: Silhavy, R. (ed.) Software Engineering Methods in Intelligent Algorithms, CSOC 2019. Advances in Intelligent Systems and Computing, vol. 984. Springer, Cham (2019). https://doi.org/10.1007/978-3-030-19807-7_38
2. Revenko, L., Revenko, N.: International practice for implementing the development programs of the digital economy examples of the USA, India, China and EU. Int. Process. **15**(4), 20–39 (2017). https://doi.org/10.17994/it.2017.15.4.51.2. (in Russian)
3. Panshin, B.: Digital economy: features and development trends. Sci. Innovations. **157**, 17–20 (2016). (in Russian)
4. Organization for Economic Co-operation and Development (OECD): Promise and Problems of E-democracy: Challenges of Online Citizen Engagement. OECD Publications, Paris (2003)
5. Sharma, G.: E-government, e-participation and challenging issues: a case study. Int. J. Comput. Internet Manag. **22**(1), 23–35 (2014)
6. Bitkov, L.A.: Social networks: between mass communication and interpersonal communication. Chelyabinsk State Univ. Bull. **28**(282), 36–38 (2012). (in Russian)
7. Dini, A.A., Sæbø, Ø.: The current state of social media research for e-participation in developing countries: a literature review. In: 49th Hawaii International Conference on System Sciences (HICSS), pp. 2698–2707. IEEE, January 2016
8. Naumov, A.O.: Traditional and new media as the actors of "color revolutions". Discourse-Pi **3–4**(32–33), 79–87 (2018). (in Russian)
9. Effect of Cambridge Analytica's Facebook ads on the 2016 US Presidential Election. https://towardsdatascience.com/effect-of-cambridge-analyticas-facebook-ads-on-the-2016-us-presidential-election-dacb5462155d. Accessed 12 May 2019
10. Grabowicz, P.A., Ramasco, J.J., Moro, E., Pujol, J.M., Eguiluz, V.M.: Social features of online networks: the strength of intermediary ties in online social media. PLoS ONE **7**(1), e29358 (2012). https://doi.org/10.1371/journal.pone.0029358
11. Castells, M.: The Rise of the Network Society, The Information Age: Economy, Society and Culture, vol. I. Blackwell, Cambridge; Oxford (1996, second edition 2010)
12. Selezenev, R.S., Skripak, E.I.: Social networks as a phenomenon of the information society and the specific social relations within them. Herald Kemerovo State Univ. **2**(54), 125–131 (2013). (in Russian)
13. Back, M.D., Stopfer, J.M., Vazire, S., Gaddis, S., Schmukle, S.C., Egloff, B., Gosling, S.D.: Facebook profiles reflect actual personality not self-idealization. Psychol. Sci. **21**, 372–374 (2010)
14. Gosling, S.D., Augustine, A.A., Vazire, S., Holtzman, N., Gaddis, S.: Manifestations of personality in online social networks: self-reported Facebook-related behaviors and observable profile information. Cyberpsychol. Behav. Soc. Network. **14**(9), 483–488 (2011)
15. Kluemper, D.H., Rosen, P.A., Mossholder, K.W.: Social networking websites, personality ratings, and the organizational context: more than meets the eye? J. Appl. Soc. Psychol. **42**(5), 1143–1172 (2012)
16. Kim, J., Roselyn Lee, J.E.: The Facebook paths to happiness: effects of the number of Facebook friends and self-presentation on subjective well-being. Cyberpsychol. Behav. Soc. Netw. **14**(6), 359–364 (2011)
17. Braslavets, S.A.: Social networks as media: problem statement. VGU Bull. Ser. Philol. Journalism **1**, 125–132 (2009). (in Russian)

18. Bitkov, L.A.: Social networks: between mass communication and interpersonal communication. Chelyabinsk State Univ. Bull. **28**(282). Philol. Art Criticism. **70**, 36–38 (2012). (in Russian)
19. Naisbitt, J., Aburdene, P.: What awaits us in the 90s. Megatrends. Year 2000. Republic, Moscow (1992)
20. Efimova, I.N., Makoveychuk, A.V.: Social networks as a new mechanism for forming the image of the subjects of political activity. In: Proceedings of the Altai State University, vol. 4-1, no. 76, pp. 245–248 (2012). (in Russian)
21. Kozyreva, A.A.: Why are social networks a tool of political power? Herald Kemerovo State Univ. **2–2**(62), 56–59 (2015). (in Russian)
22. Rooksby, J., Sommerville, I.: The management and use of social network sites in a government department. Comput. Supported Coop. Work **21**, 397–415 (2012)
23. U.S. Digital Registry. https://usdigitalregistry.digitalgov.gov. Accessed 28 May 2019
24. Meijer, A.J., Torenvlied, R.: Social media and the new organization of government communications: an empirical analysis of Twitter usage by the Dutch Police. Am. Rev. Public Adm. **46**(2), 143–161 (2016)
25. Dmitrieva, N.: Communications in social networks: results of the monitoring of openness of federal bodies of the executive power. Public Adm. Issues **2**, 123–146 (2015). (in Russian)
26. Parfenchik, A.A.: The use of social networks in public administration. State Municipal Adm. Issues **2**, 186–200 (2017). (in Russian)
27. Klang, M., Nolin, J.: Disciplining social media: an analysis of social media policies in 26 swedish municipalities. First Monday **16**(8) (2011)
28. United States Office of Government Ethics "The Standards of Conduct as Applied to Personal Social Media Use" (2015). https://www.oge.gov/web/oge.nsf/Resources/LA-15-03: +The+Standards+of+Conduct+as+Applied+to+Personal+Social+Media+Use. Assessed 04 Feb 2017
29. Cabinet Office "Social Media Guidance for Civil Servants: October 2014" (2014). https:// www.gov.uk/government/publications/social-media-guidance-for-civil-servants/social-media-guidance-for-civil-servants. (дата обращения: 04.02.2017)
30. Baraniuk, Ch.: US Border Authority Seeks Travelers' Social Media Details. BBC (2016). http://www.bbc.com/news/technology-36650857 free. Accessed 04 Feb 2017. (in Russian)
31. Gubanov, A.V., Zotov, V.V.: Social networks as a new tool of state and municipal management in the Russian Federation. Communicology **5**(4), 83–92 (2017). https://doi.org/ 10.21453/2311-3065-2017-5-4-83-92. (in Russian)
32. Zykov, V.: Officials Will Hand Over Their Accounts. Izvestia (2016). http://izvestia.ru/news/ 646799. Accessed 04 Feb 2017. (in Russian)

Discrete Synthesis of a Stochastic Regulator for Adaptation of a Discrete Economic Object on a Target Manifold

Svetlana Kolesnikova[(✉)]

Institute of Computational Systems and Programming St. Petersburg State University of Aerospace Instrumentation, St. Petersburg, Russia
skolesnikova@yandex.ru

Abstract. An algorithm for analytical design of a new control law is presented, which would provide stabilization of the initial stochastic economic system (inventory theory) in the neighborhood of the target manifold. The results of comparison of numerical modeling attempts of the given design control system and alternative (based on the method of local-optimal output control) are presented for a stochastic model of a single-product inventory location. The new control system possesses the following advantages: ease and clarity of the control structure synthesis technique; achievement of the vicinity of the prescribed set of states; provision of a minimal variance of the response variable.

Keywords: Stochastic economic object · Stochastic inventory theory · Robust regulator · Minimal variance of response variable · Target manifold · Multiple controls

1 Introduction

The problem complexity to synthesize control over stochastic objects is well known, since the equations, which are quite close in their form, behave absolutely differently (according to J.S. Hadamard) [1–6].

An analytical design of control over a stochastic object (sophisticated object) based on specially determined macrovariables, setting the invariants (functioning laws) of the would-be control system, ensures not only a logical and straightforward user-oriented structure of the synthesized control system only, but also an acquisition by the control object of the properties predesigned by the user [4–7].

The purpose of this paper is to address and discuss applicability conditions in many respects [5], and an attractive method for the analytical synthesis for a deterministic control system for a stochastic object [8, 9]. This object was designed earlier and successfully implemented on some engineering objects under deterministic conditions.

The most popular and recently reviewed inventory models were deterministic in nature (see review, e.g., [10, 11]). It was assumed that demand is known as constant on the infinite/finite horizon. However, such assumptions are not very suitable in real logistic systems. The demand in reality is a random variable, which distribution may not be known.

© Springer Nature Switzerland AG 2019
R. Silhavy et al. (Eds.): CoMeSySo 2019, AISC 1047, pp. 371–378, 2019.
https://doi.org/10.1007/978-3-030-31362-3_36

This paper discusses a new approach to solve the problem of optimal supply policy of the simplest problems to manage supply to a single-product warehouse with the random demand for products.

The quality of the obtained control is also compared with a solution of a similar formulation based on the method of local-optimal output control (hereinafter referred to as LOC) (see, e.g., [12]). The principle difference between the proposed approach to stochastic control of this object from the existing ones is normality abandonment of probability distribution of the product demand process and classical quadratic control of functional quality.

Here we will continue to consider the application of a new synthesis algorithm for robust control of a stochastic discrete object previously described in [9].

The second and third sections describe the problem statement and the algorithm to synthesize a robust regulator based on the stochastic method for analytical design of aggregated regulators for a n-order object [8, 9], respectively ($n \geq 1$); the fourth section presents technique applications of stochastic control synthesis for a stochastic, discrete simple problem of inventory control; the fifth section provides the results of comparative modeling of two control systems designed based on the methods of ADAR(S) (Analytical Design of Aggregated Regulator for a Stochastic discrete object) and local-optimal output control [12].

2 Formulation of the Problem

Let the sequences of independent, equally distributed non-correlated random variables be prescribed $\{\xi_i[t]\}_{t \geq 0}, i = \overline{1, m_1}, m_1 \leq n; \mathbf{E}\{\xi_i[t]\} = 0, \mathbf{D}\{\xi_i[t]\} = \sigma_i^2, t \geq 0$, and the stochastic control vector $\mathbf{X}[t] = (X_1[t], \ldots, X_n[t])$ have the following description:

$$\mathbf{X}[t+1] = \mathbf{F}[t] + \mathbf{u}[t] + \xi[t+1] + c\xi[t], \tag{1}$$

where

$\mathbf{F}[t] := \mathbf{F}(\mathbf{X}[t]) \in R^n, \mathbf{u} \in R^m, m \leq n, \xi \in R^l, l \leq n, 0 < c < 1, t \in \{0, 1, \ldots\}$ are the non-linear function, control, noise, constant coefficient, and discrete time variable, respectively.

It is required to determine the law of control $\mathbf{u}$, ensuring the fulfillment of the following properties of the control system:

(1) $\lim\limits_{t \to \infty} \mathbf{E}\{\boldsymbol{\psi}[t]\} = 0$, where $\boldsymbol{\psi}[t] = \boldsymbol{\psi}(\mathbf{X}[t]) \in R^m$ is a certain prescribed vector function expressing the target system properties;

(2) $\mathbf{E}\{\psi_j[t]\} = 0, \mathbf{E}\{\psi_j[t+1] + \lambda_j \psi_j[t]\} = 0, j = \overline{1, m}, t \to \infty;$

(3) the minimum dispersion of the values is $\boldsymbol{\psi}[t+1] + \lambda \boldsymbol{\psi}[t], t \to \infty;$

$$\mathbf{D}\{\psi_j[t+1] + \lambda_j \psi_j[t]\} \to min, j = \overline{1, m}, t \to \infty;$$

(4) the minimum average quality functional value is as follows:

$$\Phi = \sum_{t=0}^{\infty} \sum_{j=1}^{m} \left(\alpha_j^2 \left(\psi_j[t] \right)^2 + \left(\Delta\psi_j[t] \right)^2 \right).$$

(2)

Coefficients $\alpha_j, \lambda_j, j = \overline{1, m}$ are the parameters of the synthesized control system; they have a conceptual meaning and are interrelated by a relationship determined in [5].

3   Problem Solution of Synthesizing a Robust Regulator for a Discrete Model

Let us single out the major statements of construction of a discrete, stochastic system of control over a non-linear object from among the ADAR-strategies minimizing the dispersion and present them as the following algorithm (see Fig. 1) [5, 8, 9].

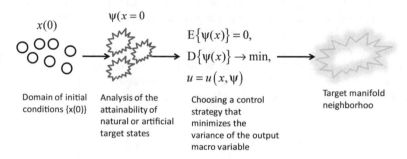

| Domain of initial conditions {x(0)} | Analysis of the attainability of natural or artificial target states | Choosing a control strategy that minimizes the variance of the output macro variable | Target manifold neighborhoo |

Fig. 1. Schematic illustration of the ADAR(S) ideology: macrovariable $\psi(t) = \psi(x(t))$ is a function of state and target requirements on the control system; $\psi(t) = 0$ is the invariant of the target system

1. Formation of the control structure in accordance with the classical method of aggregated regulator design in an assumption of a complete certainty in the description (the law of noise variation is known). Let us denote the variables in this stage as $\tilde{\mathbf{u}}^A[t]$, $t \in \{0, 1, \ldots\}$.
2. Determination of a conventional mathematical expectation $\hat{\mathbf{u}}[t] = \mathbf{E}\{\tilde{\mathbf{u}}^A[t] \,|\, \xi^t\}$, where $\xi^t = (\xi[t], \xi[t-1]\ldots\xi[0])$, $\xi[t] = (\xi_1[t], \ldots, \xi_n[t])$.

The realizability of this operation is conditioned by the peculiarities of the deterministic ADAR-synthesis (of the first state) and the initial hypotheses in description (1), with the pairwise independence of the random variables $\xi_i[t], \xi_j[t], X_i[t], i \neq j$ taken into consideration; n is the dimensionality of object (4).

Remark. Under the condition of independent noise $\{\xi_i[t]\}_{t \geq 0}$ control $\mathbf{u}[t]$ would be determined by the population of quantities $\{\mathbf{X}[s]\}_{s=\overline{0,t}}$, $\{\mathbf{u}[s]\}_{s=\overline{0,t-1}}$, or $\mathbf{u}[t] = \mathbf{U}(\mathbf{X}[t], \mathbf{X}[t-1], \ldots, \mathbf{X}[0]; \mathbf{u}[t-1], \mathbf{u}[t-2], \ldots, \mathbf{u}[0])$, $t \in \{0, 1, \ldots\}$.
Here, variable $\hat{\mathbf{u}}[t]$ depends on a random variable $\xi[t]$.

3. Elimination of variable $\xi[t]$ from expression $\mathbf{E}\{\tilde{\mathbf{u}}^A[t]\,|\xi^t\}$ based on the decomposition of the initial Eq. (1) and equations $\psi_j[t+1] + \lambda_j\psi_j[t] = 0, j = \overline{1, m}$ (left-hand parts of the Euler – Lagrange equations for functional (2)).

Thus the synthesis of a stochastic regulator with the properties listed above (1)–(3) has been completed.

Relying on the above-presented algorithm for designing a stochastic regulator minimizing the dispersion of the output variable, let us discuss the solution of two applied problems.

4 Technique Applications of Stochastic Control Synthesis for a Stochastic, Discrete Problem of Inventory Control

Let us look at the simplest model of a single-product inventory location given by the following: $x(t+1) = x(t) + u(t) - s(t)$, where $x(t)$ is the storage on hand (at the inventory location) at time t, $u(t)$ is the scope of delivery at time t, $s(t)$ is the current demand, and $x(0)$ is the initial quantity of goods.

We use the following demand model: $s(t+1) = as(t) + b + c\xi(t) + \xi(t+1)$, where $\xi(t)$ is the normal stochastic process with a zero mean and a dispersion equal to σ, and a and b are the object's parameters. Thus the description of the inventory model would become a system of stochastic difference equations

$$
\begin{aligned}
x(t+1) &= x(t) + u(t) - s(t), \\
s(t+1) &= as(t) + b + c\xi(t) + \xi(t+1).
\end{aligned}
\tag{3}
$$

The requirements for control would be included in the following form: $U_{\min} \leq u(t) \leq U_{\max}$. Quantities $x(0), s(0), U_{\min}, U_{\max}$ have been set (model initialization). Let us formulate the natural rule of control:

$$
u(t) = \begin{cases}
0, & \text{if } u^{As}(t) < U_{\min}, \\
u^{As}(t), & \text{if } U_{\min} \leq u^{As}(t) \leq U_{\max}, \\
U_{\max}, & \text{if } u^{As} > U_{\max}.
\end{cases}
\tag{4}
$$

The upper index in u^{As} indicates an extension of the classical method as the stochastic ADAR (ADAR(S)).

A solution of the problem of inventory control based of the ADAR-S algorithm.

Step 1. Formation of the structure of the synthesized control system for object (3). Let us formulate the control objective according to ADAR as follows:

$$
\psi(t) = x(t) - x^* \to 0, t \to \infty
\tag{5}
$$

Denote the control achieved in this step as $\tilde{u}$ (which indicates the presence of control uncertainties in this step).

A problem is posed (Φ, ψ), which accompanies the synthesis of control

$$\Phi = \sum_{t=0}^{\infty} \left((\alpha)^2 (\psi(t))^2 + (\Delta\psi(t))^2 \right) \to \min, \psi = 0$$

Control $\tilde{u}$, in accordance with the classical deterministic ADAR (for the fixed variables $\{\xi(t)\}_{t \geq 0}$), is found from the following equation:

$$\psi(t+1) + \omega\psi(t) = 0, 0 < \omega < 1, t \in \{0, 1, \ldots\} \tag{6}$$

On the solutions of (6), in turn, the unconstrained minimum of functional Φ is achieved.

Let us substitute the equations of object (3) into (6), considering the form of (4), to obtain the structure of control given by

$$x(t) + \tilde{u}(t) - s(t+1) - x^* + \omega\psi(t) = 0,$$
$$\tilde{u}(t) = -\omega\psi(t) + x^* + s(t+1) - x(t).$$

Substitute the available relations from (3) into the last of the above expressions to obtain

$$\tilde{u}(t) = -(\omega+1)\psi(t) + as(t) + b + c\xi(t) + \xi(t+1).$$

Step 2. Derivation of a stochastic regulator for object (3). Control for the object under the conditions of unknown (restricted) noise will be sought for in the form of a conditional mathematical expectation (CME) $u(t) = \mathbf{E}\{\tilde{u}(t) \,|\, \xi^t\}$, $t \in \{0, 1, \ldots\}$, where ξ^t denotes a fixed random vector up to the moment of time t $\xi^t = (\xi(t), \xi(t-1) \ldots \xi(0))$:

$$u(t) = \mathbf{E}\{\tilde{u}(t) \,|\, \xi(t)\} = \mathbf{E}\{-(\omega+1)\psi(t) + as(t) + b + c\xi(t) + \xi(t+1) \,|\, \xi^t\}.$$

Upon taking the CME,

$$u(t) = -(\omega+1)\psi(t) + as(t) + b + c\xi(t). \tag{7}$$

Step 3. Now eliminate noise from Eq. (7) based on decomposition of the initial description (3) upon substituting control (7) into (3)

$$u^{As}(t) = -(\omega+c+1)\psi(t) + as(t) + b - c\omega\psi(t-1), \tag{8}$$

The final description of the stochastic system of control represents a family of Eqs. (3)–(5), and (8) (see Fig. 2).

The simulation was performed for the following parameter values: $a = 0, 9; b = 1, 2; c = 0, 2; \omega = 0, 6; x^* = 180;$ $U_{\min} = 180; U_{\max} = 200; s(0) = 5; x(0) = 160$ in uniform-noise assumption, $\sigma^2 = 3$.

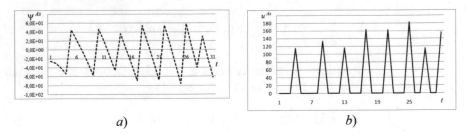

Fig. 2. Trajectories of the variables, $\psi^{As}(t) : \psi(t)$, $u^{As}(t)$, (a) and (b), respectively.

5 Comparative Simulation Results

Based on the solution of the control problem (1) with the control target (2), we obtain a system of stochastic control given by (9), which behavior is illustrated by the trajectories given in Fig. 3(a) (d), which suggests some preference to use a stochastic regulator ADAR(S) over the stochastic LOC one under the conditions, where noise is Gaussian process.

Example. Let us take a stochastic control object

$$x(t+1) = x(t) + u(t) - s(t),$$
$$s(t+1) = as(t) + b + \zeta(t), \tag{9}$$
$$J = (x(t+1) - x^*)^2 C + u(t)^2 D \rightarrow \min,$$

where constant parameters $a = 0,9$, $b = 1,2$ are the proportionality coefficients; $C = 1, D = 0$ are the regulator parameters; J is quality criterion; $\zeta(t) \sim N(0,5)$.

The structure of the regulator formed in accordance with the method of local-optimal control [12] in Gaussian-noise assumption has following form here

$$u^L(t) = -x(t) + s(t) + x^*(t).$$

According to algorithms of ADAR(S) and LOC, we have behavior of target variables $\psi(t)$ is illustrated by the trajectories given in Fig. 3.

Remark. In case of any other noise, the use of a LOC-regulator will be incorrect. At the same time, the energy cost of ADAR(S) control in the conditions of normal (uniform) noise is approximately 15% (10%) less than that spent on LOC control. We also have the following gain in the variances of the variables in stocks and control:

$$\mathbf{D}(x^{As}) = 0,35 \cdot \mathbf{D}(x^L), \mathbf{D}(u^{As}) = 0,86 \cdot \mathbf{D}(u^L).$$

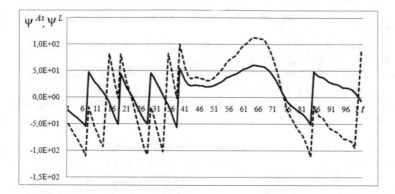

Fig. 3. Trajectories of the target macrovariables $\psi(t)$ on the base ADAR(S) - $\psi^{As}(t)$ (solid line) and LOC - $\psi^{L}(t)$ (broken line), respectively.

The simulation was performed for the same data as in Fig. 2 in normal-noise assumption, $\sigma^2 = 4$.

6   Conclusion

In support of the technology for synthesizing control over stochastic objects of different applications discussed in this study, let us present a number of arguments, as follows:

- according to N.N. Moiseev, 'there are no optimal solutions in the nature', so it makes sense to address the flow of any process 'as a certain extremal, for which there always exists its own functional which this extremal minimizes';
- control models, induced by numerous versions 'target – control – consequence', offer a possibility of evaluating the degree of significance of the quantities (parameters, variables) describing the object;
- control models, designed analytically, have the advantages of speed and efficiency, since their program implementation, conditioned by the known target of control set by the company administration and developers, is quite straightforward and clear.

The results of numerical simulation have been presented, which validate the consistency and performance of the regulator constructed in this study.

References

1. Astroem, K.J.: Introduction to Stochastic Control Theory. Academic Press, New York and London (1970)
2. Astroem, K.J., Wittenmark, B.: Adaptive Control. Dover Publications, Нью-Йорк (2008)
3. Emelyanov, S.V., Korovin, S.K.: Control of Complex and Uncertain Systems: New Types of Feedback. Springer, Dordrecht (2012)

4. Krasovskiy, A.A.: Mathematical and Applied Theory. Selected Works. Nauka, Moscow (2002)
5. Kolesnikov, A.A.: Synergetics and problems of control theory: collected articles edited by A. A. Kolesnikov. FISMATLIT, Moscow (2004)
6. Tyukin, I.Yu., Terekhov, V.A.: Adaptaciya v nelinejnyh dinamicheskih sistemah [Adaptation in nonlinear dynamical systems], 384 p. SPb.: LKI (2008). (in Russian)
7. Arcak, M., Teel, A., Kokotovic, P.: Robust nested saturation redesign for systems with input unmodeled dynamics. In: Proceedings of the 2000 American Control Conference, vol. 1, no. 6, pp. 150–154 (2000)
8. Kolesnikova, S.I.: A multiple-control system for nonlinear discrete object under uncertainty. Optim. Methods Softw. (2018). https://doi.org/10.1080/10556788.2018.1472258
9. Kolesnikova, S.: Stochastic discrete nonlinear control system for minimum dispersion of the output variable. In: Cybernetics and Automation Control Theory Methods in Intelligent Algorithms, CSOC 2019, AISC 986, pp. 1–7. Springer, Cham (2019). https://doi.org/10.1007/978-3-030-19813-8_33. Accessed 27 June 2019
10. Zipkin, P.H.: Foundations of Inventory Management. Burr Ridge, Irwin (2000)
11. Simchi-Levi, D., Chen, X., Bramel, J.: Stochastic inventory models. In: The Logic of Logistics. Springer Series in Operations Research and Financial Engineering. Springer, New York (2014). https://doi.org/10.1007/978-1-4614-9149-1_9
12. Smagin, V.I., Paraev, Yu.I.: Synthesis of servo control systems by quadratic criteria. Publishing House Tom. University, Tomsk (1996)

Integrated Model of Bandwidth Optimization and Energy Efficiency in Internet-of-Things

Bhagyashree Ambore[✉] and L. Suresh

Department of Computer Science and Engineering, CiTech, Bengaluru, India
ambore.bhagyashree@gmail.com, suriakls@gmail.com

Abstract. The Internet-of-Things (IoT) is an inevitable technology that is going to be transform the automation process by massively connected communicating devices that has limited resources. However, in order to carry out fault tolerant communication in IoT, it is essential to ensure that each IoT nodes should be able to perform seamless communication in a defined level of restricted bandwidth along with better energy retention capability. Review of existing solution shows that it offers solution to highly narrowed scale of problem. Therefore, the proposed system offers a novel framework which uses scheduling approaches in order to optimize the bandwidth utilization as well as retain maximum amount of residual energy among the communicating IoT nodes. The simulation of the model shows that proposed scheme offers better bandwidth management and efficient energy conservation in contrast to existing system.

Keywords: Internet-of-Things · Bandwidth management · Energy efficiency · Scheduling · Communication

1 Introduction

Internet-of-Things (IoT) consists of massive scale of networked devices (also called as IoT nodes/devices) which are meant for assisting in device-to-device communication [1]. However, there are challenges involved in it with various respect. The primary problem is bandwidth management system and second problem is energy efficiency [2]. With the inclusion of connection among heterogeneous IoT nodes (that are majorly wireless in its operation) with different physical and networking demands, managing an effective speed of data delivery is obviously a challenging task in IoT [3, 4]. Hence, non-effective bandwidth management will also result in violation of service level agreement as well as quality of services relayed in IoT [5]. The second issue of energy efficiency is connected with the usage of the resource constrained IoT nodes/ device which demands more and more energy supply [6]. Owing to independent deployment of the IoT nodes in a wireless environment it is obviously anticipated that there will be exponential depletion of energy factor. Irrespective of various research-based approaches towards energy efficiency in IoT, the existing schemes are found to claim energy efficiency considered very much narrowed networking and device parameters [7–10]. Such approaches are not able to cater up the energy demands when near real-time IoT applications are considered. Apart from this, bandwidth and energy efficiency problem has never been studied jointly in any of the existing system. Therefore, the proposed

© Springer Nature Switzerland AG 2019
R. Silhavy et al. (Eds.): CoMeSySo 2019, AISC 1047, pp. 379–388, 2019.
https://doi.org/10.1007/978-3-030-31362-3_37

system introduces a novel framework that uses a scheduling-based approach in order to perform communication among IoT devices leading to IoT gateway node. Section 2 discusses about the existing research work followed by problem identification in Sect. 3. Section 4 discusses about proposed methodology followed by elaborated discussion of algorithm implementation in Sect. 5. Comparative analysis of accomplished result is discussed under Sect. 6 followed by conclusion in Sect. 7.

2 Related Work

At present, there has been various research works towards improving the research section of IoT [11]. The research approaches towards bandwidth in IoT are as follows-The work carried out by Ji et al. [12] have discussed various traits of visual IoT application and its demand over constructing smart city. A unique scheme of sharing content was presented by Lee [13] where multiple terminals are selected for delivering data. Bandwidth management over IoT can be also carried out using network virtualization as seen in the work of Cao et al. [14]. Hou et al. [15] have introduces another bandwidth management scheme using edge devices along with trajectory based formulation to control bandwidth usage. Kua et al. [16] and Miao [17] have used a queue management over the gateway system considering small application usage. Power controlling can be carried out using narrowband IoT system also. The work of Xu and Darwazeh [18] has used multiplexing-based approach in presence of noise to prove effective bandwidth utilization. Approach using buffer occupancy for selection of the streamed content bitrate for an effective control of bandwidth was presented by Zhao et al. [19]. Han et al. [20] have used temporal-based approach for managing channel capacity between heterogeneous IoT nodes. The research approaches towards energy efficiency in IoT are as follows-In existing system, there is an evolution of energy harvesting scheme as the prime mechanism for resisting higher energy drainage. Studies in such direction were carried out by Huang et al. [21] where software defined radio was used along with game theory for energy efficiency in IoT. Approach using energy harvesting was also carried out by Hou et al. [22], Akan et al. [23], Garlatova et al.[24], etc. Subjective-based solution for energy efficiency problems were presented by Javed et al. [25] where machine learning is used. Routing-based approach was also reported to conserve energy in IoT as seen in work of Long et al. [26]. Moghaddam and Leon-Garcia [27] have developed an architecture using fog computing for energy conservation in IoT. Association of clustering with energy has been discussed by Ju et al. [28]. Studies towards energy was also carried out by Lin et al. [29] and Pan et al.[30]. Hence, there are various schemes in existing times that has contributed towards solving bandwidth and energy problems in IoT eco system. The next section briefs of the problems.

3 Problem Description

Bandwidth plays a critical role while performing communication in IoT environment and it has its direct effect on the energy factor too. The problems with the existing research approaches are as follows: (i) bandwidth factor is completely controlled by traffic load, hence if the traffic of IoT devices (or even a smart part of) gets compromised due to security reason or it gets affected for other non-security reasons (e.g. congestion, bottleneck condition etc) then bandwidth is adversely affected, (ii) there are very few studies focusing on dynamic bandwidth system that results in downtime in cloud networks thereby affecting the IoT device performance, (iii) poor bandwidth allocation will results in degradation of quality of service performance and such degradation will affect the network in every way.

4 Proposed Methodology

The current work adds to incorporate more supportability of quality of service to our prior model [31, 32]. Adopting analytical research methodology, the proposed system offers solution for IoT nodes against the problem of energy depletion and bandwidth optimization. The architecture adopted for implementation is as shown.

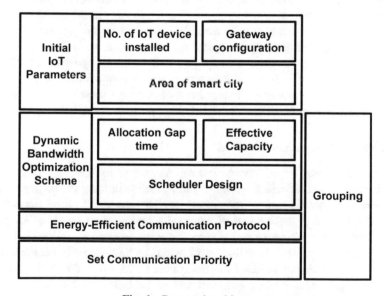

Fig. 1. Proposed architecture

Figure 1 shows the architecture where it begins with initial parameterization of IoT nodes followed by dynamic bandwidth optimization scheme. Using new parameters, the system ensure effective bandwidth utilization over dynamic traffic condition using a novel scheduler design. Apart from this, an energy efficient communication protocol is developed along with priority based data transmission scheme over IoT. The next section offers illustrative information about the system implementation.

5 System Design Implementation

The target of the proposed system implementation design is to evolve up with a simplified dynamic optimization of bandwidth considering higher degree of energy efficiency for the resource constraint IoT devices. The system design is constructed on the basis of network parameters that directly impacts channel capacity in IoT. This section briefs of all the aspects of system design.

5.1   Implementation Strategy

The primary formulation of the proposed system design is to consider that a particular amount of bandwidth is assigned for each IoT devices. This is carried out on the basis of the condition of ongoing state of traffic. The *primary* implementation strategy is carried out on the basis of association between the data packet scheduling and bandwidth allocation. The logic is – every IoT device performs the task of data forwarding on the basis of specific communication time allotment towards the IoT gateway and this operation is carried out by exchanging specific forms of internal messages. The *secondary* implementation strategy is to formulate the exploration process of the neighboring IoT devices followed by grouping the communicating IoT nodes. The *third* implementation strategy is to adopt an active-passive mode of operation of IoT devices as a part of scheduling in order to balance energy consumption as well as optimize bandwidth usage. The complete implementation strategy targets to deliver reduced delay and better energy conservation with reduction in computational complexity too.

5.2   Essential Parameters Involved

The design will consist of IoT parameters e.g. number of IoT device that are installed and performing communication within a smart city, positioning of the gateway node, and dimensional specification of the smart city. The second block of the design discusses about the novel mechanism of bandwidth optimization process. For this purpose, the study will consider adopting certain new parameters e.g. allocation gap time and effective capacity. The allocation gap time is the time duration between the allocations of the dynamic bandwidth from the system to the IoT nodes. This parameter is used to investigate what exactly is the impact of the time over bandwidth allocation system on IoT nodes. Effective capacity is the amount of bandwidth that could always offer successful data transmission irrespective of any traffic situation. The proposed system will also introduce a novel scheduler design that will assists in scheduling the bandwidth dynamically according to various dynamic network attributes in IoT. A grouping based communication protocol will be designed to ensure energy efficiency among the nodes. The proposed system also prioritizes the communication process so that no urgent information is hold back due to the implication of proposed implementation plan. Inclusion of this system parameters are the novelty factors of proposed modeling where the underlying principle is to mechanism the communication system using scheduling approach for dynamically using the bandwidth appropriately. Moreover inclusion of these parameters doesn't offer system complexity.

5.3   Execution Flow

The complete flow of the execution of the proposed system is as follows:

- **Deployment of Parameters:** The new parameter plays a crucial role in the design process viz. number of IoT devices, positioning of IoT gateway node, dimensional specification of smart city, allocation gap time, and effective capacity. The proposed system constructs an internal message frame which consists of (i) meta-data, (ii) data, and (iii) priority flag (Fig. 2).

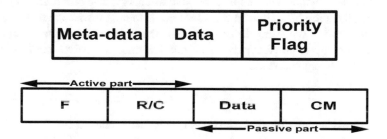

Fig. 2. Constructs of internal message

- **Initiating Communication:** The communication is carried out by grouping all the IoT nodes where the IoT nodes communicate finally with IoT gateway using multi-hop communication. In this operation, each IoT nodes constructs a buffer that consists of list of all the communication vector of neighboring IoT devices that is used for constructing multi-path leading to IoT gateway. This process ensures reliable delivery of data to IoT gateway node (Fig. 3).

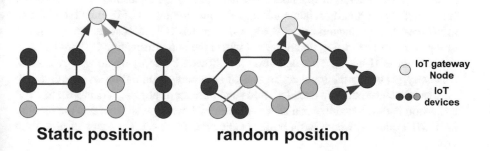

Fig. 3. Initiating communication system

- **Scheduling Packets:** The complete scheduling operation is carried out by revising the frame splitted into two parts (i) active part and (ii) passive part. The active part of the message consists of internal messages for request (R) and conformity (C) for forwarding the message from one IoT node to another. The passive part of the message consists of original data (pkt) to be forwarded and confirmation message (CM). In order to support a better updating process, the active part of the message also consists of a distinct field of update frequency (F). The proposed system is also uses a priority flag in the internal messages which is basically set by the IoT devices while forwarding. Depending upon the priority flag, the proposed scheduler re-arranges the queue. The scheduler is designed in such a way that passive duration is good enough to forward the message as well as its conformation.
- **Energy Management:** The IoT device switch off its reception operation when it is in passive state. After the saturation point of the passive mode, the IoT device regains back its active state and thereby save its energy. The proposed system offers dynamic allocation of the transmittance energy to the IoT devices to make necessary usage of power. According to this process, the transmitting node computes the necessary energy required for data transmission once it receives conformity message of C. It is computed as,

$$E_{nec} = \Delta E \cdot E_{co}$$

In the above expression, E_{nec} represents necessary energy for forwarding data packet from IoT device to another IoT device (single-hop)/ IoT gateway (multi-hop). The second variable ΔE empirically represents maximum energy required per received energy. The third variable E_{co} represents cut-off energy computed by multiplying minimum required strength of signal with network constant.

- **Algorithm Implementation:** The algorithm takes the input of n (number of IoT nodes), p (IoT node), q (gateway node) which after processing yields E_{tx} (transmittance energy), *buffer* (Allocated buffer). The initial step of algorithm is to compute Euclidean distance ed for IoT node and gateway node (Line-2). It then checks for the entire possible communication vector d which is found less than communication range cr (Line-3). It then constructs a matrix θ (Line-4) that bears all the hop information of IoT node. The next part of the algorithm is to construct a function $f(x)$ as a routing scheme taking the input arguments of θ, effective distance xy) (Line-6). The outcome of this algorithm is basically a cost factor. Finally, a group is constructed using function $\psi(x)$ where the hop information is shared among the IoT nodes (Line-7). This operation is followed by applying an energy control function $f_1(x)$ taking the input arguments of communication radius cr, packet length p_{len}, and distance between each IoT node and number of hops n_{hop} (Line-8). Finally, a function is developed that can access the internal message and extracts the identity id of IoT nodes on which basis buffer is allocated (Line-9). The steps of algorithm are:

Algorithm for Energy Efficient Bandwidth Management
Input: n (number of IoT nodes), p(IoT node), q(gateway node)
Output: E_{tx} (transmittance energy), *buffer* (Allocated buffer)
Start
1. **For** i=1: n
2. $d=ed(p, q)$
3.   **If** $d<cr$
4.   construct θ
5.   **End**
6.   derive *cost* → $f(\theta,$ xy, i)
7.   Apply $\psi(x)$
8. $[E_{tx}\ E_{rx}]$ → $f_1($cr, p_{len}, d(i, n_{hop}))
9.   *alloc* buffer → dataPriority(*id*)
10. **End**
End

6   Results Discussion

The analysis of the proposed study is carried out over 100–1000 IoT nodes configured randomly over 1000×1100 m^2 deployment area. For extreme analysis, the study uses allocation gap time ranging between 1–10 s as the test environment while the proportion of active period of operation is kept between 0.05–0.5 as standard statistical value. The analysis outcome is compared with the REL protocol [33] assessed over residual energy and delay.

The outcome shows that proposed system offers better energy control (Fig. 4(a)) and better bandwidth management (Fig. 4(b)) in contrast to existing system. It is because existing system involves too many steps for obtaining energy efficiency while proposed system offers a novel approach of allocation of buffer on the basis of current traffic condition. Hence, irrespective of any size of data packet, proposed system can offer dynamic allocation of packet for a given channel capacity. Apart from this, the prioritization of packet further assists in supporting time critical application with increased allocation gap time.

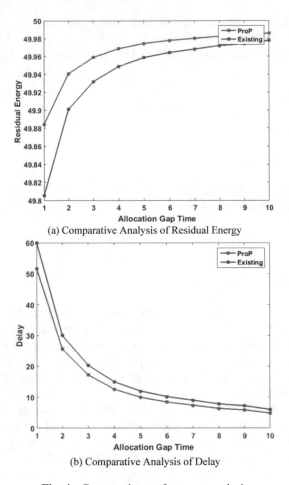

(a) Comparative Analysis of Residual Energy

(b) Comparative Analysis of Delay

Fig. 4. Comparative performance analysis

7 Conclusion

IoT nodes are characterized by restricted resource availability as well as limited computational capability. At present, existing IoT nodes are programmed to captured the data and forward directly to a gateway system via certain communication channel. However, in large scale deployment, the routing system has to be multi-hop system. Unfortunately, maintaining an effective bandwidth utilization is going to be big problem in multi-hop network of massive number of connected devices while good chances are for faster energy drainage. Therefore, this problem has been addressed by proposed model that implements a specific form of scheduling mechanism for jointly addressing both the problem. The contribution of proposed system are (i) it uses a temporal factor of communication system where internal messaging policy has been formulated for better transparency about their capability to forward data, (ii) energy is significantly conserved as IoT nodes are now scheduled to be operated over both active and passive mode without any issues, (iii) an effective clustering scheme has been introduced which is not only easy but also dynamic in its approach, and (iv) energy and bandwidth allocation scheme has been introduced which offers the capability to process any size of data irrespective of traffic situation.

References

1. Lea, P.: Internet of Things for Architects: Architecting IoT Solutions by Implementing Sensors, Communication Infrastructure, Edge Computing, Analytics, and Security. Packt Publishing Ltd., Birmingham (2018)
2. Alioto, M.: Enabling the Internet of Things: From Integrated Circuits to Integrated Systems. Springer, Berlin (2017)
3. Gravina, R., Palau, C.E., Manso, M., Liotta, A., Fortino, G.: Integration, Interconnection, and Interoperability of IoT Systems. Springer, Berlin (2017)
4. Vermesan, O., Bacquet, J.: Next Generation Internet of Things: Distributed Intelligence at the Edge and Human Machine-to-Machine Cooperation. River Publishers, Gistrup (2019)
5. Tomar, P., Kaur, G.: Examining Cloud Computing Technologies Through the Internet of Things. IGI Global, Hershey (2017)
6. Siu, C.: IoT and Low-Power Wireless: Circuits, Architectures, and Techniques. CRC Press, Boca Raton (2018)
7. Akkaya, K., Guvenc, I., Aygun, R., Pala, N., Kadri, A.: IoT-based occupancy monitoring techniques for energy-efficient smart buildings. In: 2015 IEEE Wireless Communications and Networking Conference Workshops (WCNCW), New Orleans, LA, pp. 58–63 (2015)
8. Jalali, F., Khodadustan, S., Gray, C., Hinton, K., Suits, F.: Greening IoT with fog: a survey. In: 2017 IEEE International Conference on Edge Computing (EDGE), Honolulu, HI, pp. 25–31 (2017)
9. Das, K., Das, S., Darji, R.K., Mishra, A.: Survey of energy-efficient techniques for the cloud-integrated sensor network. J. Sens. **2018**, 1–17 (2018)
10. Kaur, N., Sood, S.K.: An energy-efficient architecture for the Internet of Things (IoT). IEEE Syst. J. **11**(2), 796–805 (2017)
11. Ambore, B., Suresh, L.: Assessing trends of existing research contribution towards Internet-of-Things. Int. J. Adv. Comput. Sci. Appl. **9** (2018). http://doi.org/10.14569/IJACSA.2018.090924
12. Ji, W., Xu, J., Qiao, H., Zhou, M., Liang, B.: Visual IoT: enabling Internet of Things visualization in smart cities. IEEE Netw. **33**(2), 102–110 (2019)
13. Lee, B.M.: Multi-point media content sharing scheme in Internet of Things networks. IEEE Access **6**, 71360–71367 (2018)
14. Cao, H., Yang, L., Zhu, H.: Novel node-ranking approach and multiple topology attributes-based embedding algorithm for single-domain virtual network embedding. IEEE IoT J. **5**(1), 108–120 (2017)
15. Hou, W., Li, W., Guo, L., Sun, Y., Cai, X.: Recycling edge devices in sustainable Internet of Things networks. IEEE IoT J. **4**(5), 1696–1706 (2017)
16. Kua, J., Nguyen, S.H., Armitage, G., Branch, P.: Using active queue management to assist IoT application flows in home broadband networks. IEEE IoT J. **4**(5), 1399–1407 (2017)
17. Miao, Y., Li, W., Tian, D., Hossain, M.S., Alhamid, M.F.: Narrowband Internet of Things: simulation and modeling. IEEE IoT J. **5**(4), 2304–2314 (2018)

18. Xu, T., Darwazeh, I.: Non-orthogonal narrowband Internet of Things: a design for saving bandwidth and doubling the number of connected devices. IEEE IoT J. **5**(3), 2120–2129 (2018)
19. Zhao, P., Yu, W., Yang, X., Meng, D., Wang, L.: Buffer data-driven adaptation of mobile video streaming over heterogeneous wireless networks. IEEE IoT J. **5**(5), 3430–3441 (2018)
20. Han, Y., Yan, C., Wang, B., Liu, K.J.R.: Enabling heterogeneous connectivity in Internet of Things: a time-reversal approach. IEEE IoT J. **3**(6), 1036–1047 (2016)
21. Huang, X., Yu, R., Kang, J., Xia, Z., Zhang, Y.: Software defined networking for energy harvesting Internet of Things. IEEE IoT J. **5**(3), 1389–1399 (2018)
22. Hou, Z., Chen, H., Li, Y., Vucetic, B.: Incentive mechanism design for wireless energy harvesting-based Internet of Things. IEEE IoT J. **5**(4), 2620–2632 (2017)
23. Akan, O.B., Cetinkaya, O., Koca, C., Ozger, M.: Internet of hybrid energy harvesting things. IEEE IoT J. **5**(2), 736–746 (2017)
24. Gorlatova, M., Sarik, J., Grebla, G., Cong, M., Kymissis, I., Zussman, G.: Movers and shakers: kinetic energy harvesting for the Internet of Things. In: ACM SIGMETRICS Performance Evaluation Review, vol. 42, no. 1, pp. 407–419. ACM (2014)
25. Javed, A., Larijani, H., Wixted, A.: Improving energy consumption of a commercial building with IoT and machine learning. IT Prof. **20**(5), 30–38 (2018)
26. Long, N.B., Tran-Dang, H., Kim, D.-S.: Energy-aware real-time routing for large-scale industrial Internet of Things. IEEE IoT J. **5**(3), 2190–2199 (2018)
27. Moghaddam, M.H.Y., Leon-Garcia, A.: A fog-based internet of energy architecture for transactive energy management systems. IEEE IoT J. **5**(2), 1055–1069 (2018)
28. Ju, Q., Zhang, Y.: Clustered data collection for Internet of batteryless things. IEEE IoT J. **4**(6), 2275–2285 (2017)
29. Lin, C.-C., Deng, D.-J., Liu, W.-Y., Chen, L.: Peak load shifting in the Internet of energy with energy trading among end-users. IEEE Access **5**, 1967–1976 (2017)
30. Pan, J., Jain, R., Paul, S., Vu, T., Saifullah, A., Sha, M.: An Internet of Things framework for smart energy in buildings: designs, prototype, and experiments. IEEE IoT J. **2**(6), 527–537 (2015)
31. Ambore, B., Suresh, L.: Game-Decision Model for Isolating Intruder and Bridging Tradeoff Between Energy and Security (2019). http://doi.org/10.1007/978-3-030-19813-8_2
32. Ambore, B., Suresh, L.: Novel model for boosting security strength and energy efficiency in Internet-of-Things using multi-staged game. Int. J. Electr. Comput. Eng. (IJECE) **9**, 4326–4335 (2019). Scopus Indexed and UGC approved Journal
33. Machado, K., Rosario, D., Cerqueira, E.: A routing protocol based on energy and link quality for Internet of Things applications. Open-Access Sens. **13**, 1942–1964 (2013)

Modeling of Secure Communication in Internet-of-Things for Resisting Potential Intrusion

Nasreen Fathima[1(✉)], Reshma Banu[2], and G. F. Ali Ahammed[3]

[1] Department of Computer Science Engineering, ATME College of Engineering,
Mysuru, India
nasreen16fathima@gmail.com
[2] Department of Information Science and Engineering, GSSSIETW,
Mysuru, India
reshma127banu@gmail.com
[3] Department of Computer Science Engineering, VTU Post Graduate Centre,
Mysuru, India
aliahammed78@gmail.com

Abstract. Smart appliances running under Internet-of-Things (IoT) offers excellent interconnected network and gives a true encapsulation of machine-to-machine communication system. However, due to various inevitable situation as well as novelty in the IoT implementation concept, it is shrouded by various loopholes in its security system. Review of existing approaches too shows that there are less modeling and exhaustive framework to deal with this problem. Therefore, the proposed study introduces a framework that offers secure communication system between IoT nodes and internet host (i.e. gateway) by harnessing finite field encryption system of public key cryptography. The proposed system offers a unique key generation system as well as novel digital signature generation securing the entire communication. The simulation outcomes show that proposed system offer better performance than existing system.

Keywords: Internet-of-Things · Security · Secure routing · Gateway security · Encryption · Wireless sensor network

1 Introduction

The Internet-of-Things (IoT) is considered as big boon for pervasive computing as it offers higher degree of connection with increased vulnerability [1, 2]. The prime cause of security threats in IoT ecosystem are vulnerabilities of software, attacks in computer system, interception of data [3]. There are various protocols and approaches towards securing communication in IoT [4–6]. However, the biggest challenges lies in internet host which is connected with internet and is a pivotal point of intrusion from various cyber security threats. Another biggest problem associated with the IoT is the communication system between the IoT devices and the gateway nodes or internet host [7–10]. Such communication system is actually very much dynamic where there is no much robust existing model to address such problem. Therefore, the proposed system

© Springer Nature Switzerland AG 2019
R. Silhavy et al. (Eds.): CoMeSySo 2019, AISC 1047, pp. 389–398, 2019.
https://doi.org/10.1007/978-3-030-31362-3_38

offers a discussion of a security model that uses public key encryption and a novel digital signature to offer communication security in IoT. The organization of the paper is as follow: Sect. 2 discusses about the existing research work followed by problem identification in Sect. 3. Section 4 discusses about proposed methodology followed by elaborated discussion of system implementation in Sect. 5. Comparative analysis of accomplished result is discussed under Sect. 6 followed by conclusion in Sect. 7.

2 Related Work

At present, there are various research work carried out toward securing communication system in IoT. Existing system has a report of approach towards securing communication system over Software Defined Network in IoT as seen in work of Liu et al. [11] where the authors have presented a *secure data flow* model using *undirected graph*. The work of Ban et al. [12] emphasizes on encryption over the application data for reducing complexity of conventional Datagram Transport Layer Security (DTLS) protocol in IoT. Ambrosin et al. [13] have investigated the strength of attribute-based encryption over smart devices. Study towards exploring attack strategies in IoT environment has been carried out by Choi and Kwak [14]. Association of security and data analytics has been closely studied by Kumarage et al. [15]. Security upgradation using trusted third party is presented in the work of Tsai et al. [16] where the authors have also used key exchange protocol with multiple keys. Xu et al. [17] have presented a unique relay strategy to circumvent eavesdropping attack over physical layer in IoT. The authors have used multi-hope communication along with a strategy to increase rate of secrecy. A unique approach of investigation towards IoT security has been initiated by Schurgot et al. [18] where both encryptions based and data manipulation based approach has been studied. The work of Premnath and Hass [19] has emphasized on potential of using smaller key sizes for controlling the computational complexity associated with IoT nodes by analyzing the cost of cryptanalysis. Adoption of elliptical curve cryptography has been discussed in work of Yalçin et al. [20] where the study has applied synthesis-based approach for assessing the security scheme Study towards addressing eavesdropping problem in IoT has been carried out by Zhang et al. [21] using both probability theory and transformation techniques. Existing system has also witnessed design of unique adversary model mapping with real-world intrusion. Study toward group key has been witnessed in the work of Porambage et al. [22] where hash function is used along with signature scheme. Literature has also witnessed the trust management utilization towards addressing specific form of attacks as seen in the work of Mendoza and Kleinschmidt [23]. Kim et al. [24] have carried out a model design that focuses on reduction of security overhead in IoT connected with a specific form of security framework using conventional cryptography. Li et al. [25] have studied the impact of packet dropping adversary in an IoT environment in presence of different challenges of wireless networks. Apart from this, there are various other security solutions towards IoT e.g. authentication based [26–28], attack-specific solution [29], Elliptical curve cryptography [30], accomplishing interoperation [31]. Hence, there is availability of various solutions towards solving security problems in IoT; however, it is still in very nascent stages of implementation as there is no robust or benchmarked solution ever reported in existing approaches towards safeguarding security problems in communication of IoT. The next section outlines the problems identified to be solved in proposed system.

3 Problem Description

It is found from existing studies that there is maximum utilization of elliptical cryptography but without addressing the problems associated with excessive generation of prime fields. Apart from this, there has been no novel digital signature scheme being implemented towards securing privacy problems in IoT environment. Possible implication of the existing security scheme over the resource consumption in presence of an iterative scheme has never been investigated. Deployment of dynamic nodes over its IoT environment is less investigated with respect to the key-based attacks when IoT nodes communicate with a gateway. Hence, the statement of the problem considered in study is "*Developing a novel security framework that can offer joint management of secret key generation and novel signature scheme to offer reduced resource consumption with non-iterative encryption scheme*".

4 Proposed Methodology

The prime motive of the proposed study is to develop a framework that offers secure communication among the IoT devices and internet host. The core idea is also to ensure that proposed system offers capability to address dynamic communication demands with full security towards key-based attacks. Figure 1 highlights the block diagram of proposed scheme.

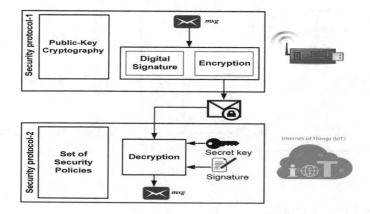

Fig. 1. Proposed schema of implementation

The proposed system implement public key encryption where key management is followed by generation of digital signature unlikely any of the existing system. The complete implementation is carried out in two phases where 1st protocol emphasizes on encryption while 2nd protocol emphasizes on decryption. As decryption is just a reverse step of encryption, hence the proposed manuscript elaborates only about the first stage of implementation. The next section illustrates about system design.

5   System Implementation

This section discusses about the process flow of the system implementation to understand each and every operation involved in the proposed modeling.

5.1   Network Setup

For a given deployment area, there are various IoT devices dispersed in random fashion. The IoT device captures the data and forward it to internet host, which is a gateway system connected to cloud services. The internet hosts are positioned in grid fashion with proper reachability of the IoT devices. If the number of IoT nodes are n and number of internet host (gateway) are m than the study considers that $n \gg m$. The IoT devices are classified into two types i.e. devices with maximum capability i.e. MaxCap nodes and devices with minimum capabilities MinCap nodes. It should be known that prior to deployment; all the IoT nodes have equal capability which depends upon their resource availability. However, in progression of processing event, rate of resource consumption varies and so are the capabilities. Therefore, in order to offer better form of routing, MinCap nodes forward data to the MaxCap nodes which then forward the data to its nearest internet host. However, if MinCap node is located near to internet host than it can directly forward to it only if there is no availability of MaxCap Node (Fig. 2).

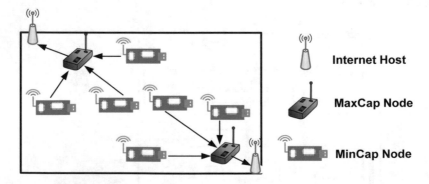

Fig. 2. Network setup

5.2   Key Management

A trusted authority (residing within the internet host) generates parameters associated with public key encryption. The proposed system uses 4 different types of keys managed within internet host. The *default key* is allocated by the internet host to the IoT nodes for mutually authenticating each other. It could be both public as well as private.

A *discrete key* is pre-assigned key prior to deployment and shared by each IoT devices with the internet host which is basically used for ciphering any information forwarded by both MaxCap/MinCap nodes to the internet host. The *sync key* is another form of key which is used for ciphering any information exchanged between the MaxCap nodes and it's connected MinCap Nodes. A *group key* is used for encrypting all the information exchanged between the MinCap nodes and MaxCap nodes. It is also used for updating MaxCap nodes if there is any change in the MinCap Nodes (Fig. 3).

^{Q} Default key-Prior Deployment ^{Q} Discrete key-During Comm   ━━◾ Sync Key   ◾━ Group Key

Fig. 3. Proposed key management

5.3 Confirming System Set-up

The first step in this process is to generate all the essential parameters of system. In this step, the internet host performs concatenation of following parameters (i) a *security attributes* viz. $\alpha = [\alpha_1, \alpha_2, \alpha_3, \alpha_4]$ where each parameters represents finite field, chosen pattern of prime fields, monogeneous group, and point generator, (ii) *security encoders* $e = [e_0, e_1, e_2, e_3]$ where each encoders represents a hash function of security attribute α_3, and (iii) *public key* P_{key}. The next step is selection of random confidential value (integer type) by the MinCap nodes for computing the partial key of MinCap nodes. This trusted authority than chooses another random number for computing the other part of the key which is later validated by the MinCap nodes followed by generation of the full key by MinCap nodes which is re-forwarded to the internet host to generate *discrete key*. Once all the secret key is generated, the internet host maintains a confidential matrix consisting of tags as well as public keys of all IoT nodes (Fig. 4).

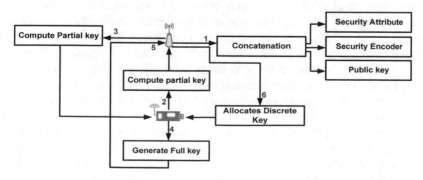

Fig. 4. Confirming system setup

5.4   Key & Signature Generation

This process begins only after the node starts communicating (i.e. after network deployment). Each beacon consists of the identity information of IoT device and its respective P_{key}. When node-X receives beacon from node-Y than it initially generates gateway secret key for long term usage and then it extracts session key for data encryption. In such case, node-X chooses a random integer number multiplied with 4th security attribute (α_4) to obtain σ_1. It is followed by computation of a temporary key using the random umber used to obtain σ_1, first encoder i.e.e_0, public key P_{key}, a prime number. The final secret key is then generated by applying 2nd encoder e_1 on σ_1 and temporary key. This process is followed by generation of digital signature which is obtained by applying 3rd encoder i.e.e_2 on σ_1, temporary key (Fig. 5).

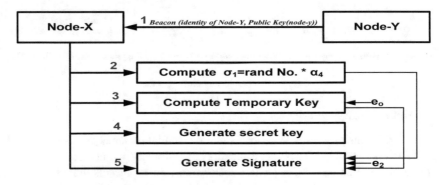

Fig. 5. Key & signature generation system

5.5   Secure Group Communication in IoT

After the deployment of the node is over, all the MaxCap nodes receive beacons (for joining) from MinCap nodes. If the mutual authentication is successful, than a group is formed by the MaxCap nodes with its neighboring MinCap and a mutual-equivalent key is shared among each other. This initiates the group communication followed by regular update of the group keys as well as other respective keys. The unique part of the group communication mechanism is that it acts uniquely for new node joining the group and old node leading the group. For old node leaving the group, the digital signature as well as assigned keys is disposed from its shared matrix maintained in trusted authority and it updates the same to the trusted authority. On the other hand, for new node joining the group, it obtains its information from the request message as well as it also access its old information maintained within the trusted authority. If the node is completely new, the complete node registration process followed by all the other key management process is implemented. However, if the new node joining is actually an old node for other group than the cluster head performs only one-way handshaking mechanism to generate full length secret key and checks the validity of the digital signature. If the digital signature is found invalid, it than re-perform the process to re-generate new digital signature and completes the final secure routing from IoT nodes to internet host.

6 Results Discussion

From the briefing of the algorithm description, it can be seen that proposed system uses public key encryption along with the novel digital signature in a very different manner to assist in secure communication in IoT environment. The scripting of the proposed logic is carried out in MATLAB considering 100–1000 sensors with 4–25 internets host spreaded over $1000 \times 1000 \ m^2$ simulation areas. The study outcome is assessed using energy consumption factor associated with update frequency and pause time. Figure 6 shows the graphical visualization of the energy consumption over increment of the update frequency. The overall inference of the outcome is that proposed system successfully reduces the rate of energy consumption irrespective of the variations of the velocity. The prime reason behind this logic is that proposed system has spontaneous usage of updating the cluster key and amends the authentication system dynamically for both new node joining and old node leaving. Therefore, higher dissipation of the energy doesn't has much adverse effect on the energy consumption and hence network lifetime too increases. This fact also offer potential immunity to the node as they have more energy needed to execute the algorithm for resisting various key-based attacks in an IoT environment.

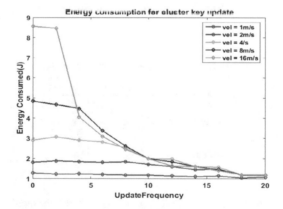

Fig. 6. Analysis of energy consumption w.r.t update frequency

The next part of the analysis is related to the energy consumption with respect to the increasing pause time. Increment of pause time indicates a situation where there is ongoing processing over the routing activity among the sensors and the internet host. It can be expected that there will be energy drainage if the pause time is increased. However, Fig. 7 shows that proposed system has significant control over the energy consumption even with the increasing pause time. The prime reason behind this fact is that complete implementation of the proposed system is carried over both MaxCap node as well as MinCap node with more emphasis on the MinCap nodes for better optimization. This result in more power conservation for MaxCap nodes while upgrading the capability of the MinCap nodes. Therefore, proposed system balances various communication demands with the proposed secured routing scheme

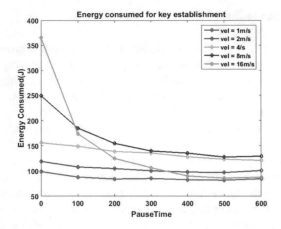

Fig. 7. Analysis of energy consumption w.r.t pause time

Apart from this, the proposed system is hypothetically compared with some of the related work e.g. Ding et al. [32], Liu et al. [33] and Liu et al. [34]. All these approaches have used elliptical curve as a medium of public key encryption as well as digital signature scheme. Table 1 highlights the qualitative findings of comparative analysis.

Table 1. Comparative performance analysis

Factors	Proposed	Existing
Key generation	Finite field	Elliptical curve
Usage of digital signature	Simplified	Complex
Iterative	No	Yes
Resource evaluation	Yes	No

A closer look into the above Table 1 highlights that proposed system is still the better alternative as compared to the existing related approaches. The prime reason behind this are (i) proposed system uses the essentials of finite field encryption with more control over the finite point generator that is found missing in the existing approaches using elliptical curve cryptography, (ii) the mechanism of using digital signature and deploying it for validation is extremely different from any of the existing system till date as proposed system constructs the signature using system attributes and security key attributes/encoders. Another distinct part of the digital signature in proposed system is that it a step of the validation doesn't require any dependency of external attributes of pre-defined prime fields or other security attributes and rather it computes everything when it requires deploying it. It will mean that there are very less number of steps for both signature generation as well as signature validation in proposed system whereas the existing system mechanism is quite iterative and is computationally complex. The overall processing time of the proposed system is found to be 0.35512 s in core – i3 processor on normal system configuration whereas normal usage of elliptical curve as well as digital signature will consume more than it (approximately 3.76221 s). Hence, the proposed system can be claimed to offer better security performance in line with communication need.

7 Conclusion

This paper presents a discussion about the solution for resisting various forms of threats over the IoT system. Basically the IoT nodes as well as the gateway nodes are the prime target of the victim. As IoT nodes are kept in confined area therefore chances of getting it compromise is slightly less but still there are chances if the devices are purely wireless with good range of communication. On the other hand, gateway node offers connectivity with many IoT nodes and hence chances of getting gateway node being exposed to many rogue IoT nodes are quite high. Therefore, the proposed system offers a unique usage of public key cryptography as well as digital signature unlike any existing encryption system in literature. The simulation outcome of the study shows that proposed system offers better security and balances its operation well with resource consumption for IoT nodes.

References

1. Dhanjani, N.: Abusing the Internet of Things: Blackouts, Freakouts, and Stakeouts. O'Reilly Media, Sebastopol (2015)
2. Minteer, A.: Analytics for the Internet of Things (IoT). Packt Publishing Ltd., Birmingham (2017)
3. Macaulay, T.: RIoT Control: Understanding and Managing Risks and the Internet of Things. Morgan Kaufmann, Burlington (2016)
4. Sadeeq, M.A.M., Zeebaree, S.R.M., Qashi, R., Ahmed, S.H., Jacksi, K.: Internet of Things security: a survey. In: 2018 International Conference on Advanced Science and Engineering (ICOASE), Duhok, pp. 162–166 (2018)
5. Deshmukh, S., Sonavane, S.S.: Security protocols for Internet of Things: a survey. In: 2017 International Conference on Nextgen Electronic Technologies: Silicon to Software (ICNETS2), Chennai, pp. 71–74 (2017)
6. Oracevic, A., Dilek, S., Ozdemir, S.: Security in Internet of Things: a survey. In: 2017 International Symposium on Networks, Computers and Communications (ISNCC), Marrakech, pp. 1–6 (2017)
7. Solanki, V.K., Díaz, V.G., Paulo Davim, J.: Handbook of IoT and Big Data. CRC Press, Boca Raton (2019)
8. McEwen, A., Cassimally, H.: Designing the Internet of Things. Wiley, Hoboken (2017)
9. Rahmani, A.M., Liljeberg, P., Preden, J.-S., Jantsch, A.: Fog Computing in the Internet of Things: Intelligence at the Edge. Springer, Berlin (2017)
10. da Costa, F., Henderson, B.: Rethinking the Internet of Things: A Scalable Approach to Connecting Everything. Apress, New York (2014)
11. Liu, Y., Kuang, Y., Xiao, Y., Xu, G.: SDN-based data transfer security for Internet of Things. IEEE IoT J. 5(1), 257–268 (2018)

12. Ban, H.J., Choi, J., Kang, N.: Fine-grained support of security services for resource constrained Internet of Things. Int. J. Distrib. Sens. Netw. **12**, 7824686 (2016)
13. Ambrosin, M., et al.: On the feasibility of attribute-based encryption on Internet of Things devices. IEEE Micro **36**(6), 25–35 (2016)
14. Choi, S., Kwak, J.: Enhanced SDIoT security framework models. Int. J. Distrib. Sens. Netw. **12**, 4807804 (2016)
15. Kumarage, H., Khalil, I., Alabdulatif, A., Tari, Z., Yi, X.: Secure data analytics for cloud-integrated Internet of Things applications. IEEE Cloud Comput. **3**(2), 46–56 (2016)
16. Tsai, K.L., Huang, Y.L., Leu, F.Y., You, I.: TTP based high-efficient multi-key exchange protocol. IEEE Access **4**, 6261–6271 (2016)
17. Xu, Q., Ren, P., Song, H., Du, Q.: Security enhancement for IoT communications exposed to eavesdroppers with uncertain locations. IEEE Access **4**, 2840–2853 (2016)
18. Schurgot, M.R., Shinberg, D.A., Greenwald, L.G.: Experiments with security and privacy in IoT networks. In: 2015 IEEE 16th International Symposium on a World of Wireless, Mobile and Multimedia Networks (2015)
19. Premnath, S.N., Haas, Z.J.: Security and privacy in the Internet-of-Things under time-and-budget-limited adversary model. IEEE Wirel. Commun. Lett. **4**(3), 277–280 (2015)
20. Yalçin, T.: Compact ECDSA engine for IoT applications. Electron. Lett. **52**(15), 1310–1312 (2016)
21. Zhang, Y., Shen, Y., Wang, H., Yong, J., Jiang, X.: On secure wireless communications for IoT under eavesdropper collusion. IEEE Trans. Autom. Sci. Eng. **13**(3), 1281–1293 (2016)
22. Porambage, P., Braeken, A., Schmitt, C., Gurtov, A., Ylianttila, M., Stiller, B.: Group key establishment for enabling secure multicast communication in wireless sensor networks deployed for IoT applications. IEEE Access **3**, 1503–1511 (2015)
23. Mendoza, C.V.L., Kleinschmidt, J.H.: Mitigating on-off attacks in the Internet of Things using a distributed trust management scheme. Int. J. Distrib. Sens. Netw. **11**, 859731 (2015)
24. Kim, S.-K., Kim, B.-G., Min, B.-J.: Reducing security overhead to enhance service delivery in Jini IoT. Int. J. Distrib. Sens. Netw. **11**, 205793 (2015)
25. Li, X., Wang, H., Dai, H.-N., Wang, Y., Zhao, Q.: An analytical study on eavesdropping attacks in wireless nets of things. Mobile Inf. Syst. **2016**, 1–10 (2016)
26. Dong, Q., Tong, J., Chen, Y.: Cloud-based RFID mutual authentication protocol without leaking location privacy to the cloud. Int. J. Distrib. Sens. Netw. **11**, 937198 (2015)
27. Hou, J.-L., Yeh, K.-H.: Novel authentication schemes for IoT based healthcare systems. Int. J. Distrib. Sens. Netw. **11**, 183659 (2015)
28. Chi, L., Hu, L., Li, H., Chu, J.: Analysis and improvement of a robust user authentication framework for ubiquitous sensor networks. Int. J. Distrib. Sens. Netw. **10**, 637684 (2014)
29. Zhang, K., Liang, X., Lu, R., Shen, X.: Sybil attacks and their defenses in the Internet of Things. IEEE IoT J. **1**(5), 372–383 (2014)
30. Piñol, O.P., Raza, S., Eriksson, J., Voigt, T.: BSD-based elliptic curve cryptography for the open Internet of Things. In: 2015 7th International Conference on New Technologies, Mobility and Security (NTMS), pp. 1–5, Paris (2015)
31. Alam, S., Chowdhury, M.M., Noll, J.: Interoperability of security-enabled Internet of Things. Wirel. Pers. Commun. **61**, 567–586 (2011)
32. Ding, S., Li, C., Li, H.: A novel efficient pairing-free CP-ABE based on elliptic curve cryptography for IoT. IEEE Access **6**, 27336–27345 (2018)
33. Liu, Z., Seo, H.: IoT-NUMS: evaluating NUMS elliptic curve cryptography for IoT platforms. IEEE Trans. Inf. Forensics Secur. **14**(3), 720–729 (2019)
34. Liu, Z., Seo, H., Großschädl, J., Kim, H.: Efficient implementation of NIST-compliant elliptic curve cryptography for 8-bit AVR-based sensor nodes. IEEE Trans. Inf. Forensics Secur. **11**(7), 1385–1397 (2016)

A Review of the Regression Models Applicable to Software Project Effort Estimation

Huynh Thai Hoc[(✉)], Vo Van Hai, and Ho Le Thi Kim Nhung

Faculty of Applied Informatics, Tomas Bata University in Zlin,
Nad Stranemi 4511, 76001 Zlin, Czech Republic
{huynh_thai,vo_van,lho}@utb.cz

Abstract. Software Project Effort Estimation - (further only SPEE), is an essential step in a software project; related to approximating Development Effort before development is completed, and is an important software development activity. Its accuracy has a significant effect on a project's success. The major intent of this paper is to review existing Software Project Effort Estimation – (further only SPEE), exhaustively by exploring Regression Models for modern SPEEs.

Keywords: Software Project Effort Estimation (SPEE) · Regression model · Stepwise regression · Regression clustering · AOM · RCMLR · WCO

1 Introduction

SPEE is an essential phase in software development [1]. It is a method for forecasting the number of resources required to complete - or maintain, project activities to deliver outputs-products or services of specified functional and non-functional characteristics [2]. SPEE cannot be expected to be 100% accurate [2, 3]. The estimation process is not so much to solve any of the problems of a software project - but rather, to minimise the risks of the project, or to reduce the risk of surprises in the project to the minimum. It provides a fairly clear view of the project; which helps the project leader to make good controlling decisions [2].

SPEEs have been researched and their value evaluated ever since the 1950s [1]. There are many methods used for estimation purposes - for instance, Expert Judgment [1, 2], Analogy [1, 2], Wideband Delphi [1, 2], Code Source-lines [1, 2], Function Point [1, 2], Object Points [1, 2], COCOMO [1, 2], or Use Case Points [1, 2, 4]. Some of these are non-algorithmic methods - estimations based on expert experience or on analytical comparisons with similar prior projects; others are estimated by means of Algorithms [1]. Table 1 below, describes the techniques used in effort estimation software.

With the SEE methods that use algorithmic types, apart from estimations based on the original formula, the researchers have studied variant methods; for example, using the regression model to optimise the accuracy of size estimation or effort estimation based on the historical data.

© Springer Nature Switzerland AG 2019
R. Silhavy et al. (Eds.): CoMeSySo 2019, AISC 1047, pp. 399–407, 2019.
https://doi.org/10.1007/978-3-030-31362-3_39

Table 1. Summary of techniques used in Software Effort Estimation methods.

Estimation method	Type	Techniques
Expert Judgment	Non-algorithmic	Estimations based on expert experiences and/or intuition [1, 5]
Analogy	Non-algorithmic	Estimations based on the actual cost of similar completed projects [1]
Price-to-win, Bottom-up and Top-up	Non-algorithmic	Estimations based on customer budgets [1]
Wideband Delphi	Non-algorithmic	Customer and technical teams are involved in the estimation process [1]
Planning Poker	Non-algorithmic	Estimations based on collaboration/consensus among team members like Wideband Delphi [1]
SLOC	Algorithmic	Estimations based on the previous data of the completed project. It cannot compare different programming language lines of code [1]
Function Point Analysis	Algorithmic	Estimations based on counting basic software components - (EI, EO, EQ, LIF, EIF) [1]
Object Point	Algorithmic	Estimations based on the number and the complexity of objects - (e.g. screens, reports, 3GL components). Estimation Steps: counting objects, the classification of objects, the weight of objects related to complexity [1]
COCOMO	Algorithmic	There are four COCOMO methods [1]: • Simple COCOMO: Estimations based on KLOC (Kilo Lines Of Codes) • Intermediate COCOMO: KLOC and EAF (Effort Adjustment Factors) • Detailed COCOMO: This integrates all Intermediate COCOMO characters on the Analysis, Design, Coding, and Testing steps • COCOMO II: This is an extension of the Intermediate COCOMO, that predicts the amount of effort, based on Person-Month (PM) in the projects
Use Case Points	Algorithmic	Estimations derived by counting UUCW, UAW, TCF, UAW, ECF

For example, to calculate the size estimation of software as compared with the original UCP equation - Eq. 1, UCP was proposed by Karner [2, 6], UUCW, UAW, TCF, UAW, and ECF have to be identified first, and only then will the UCP be estimated by means of the original formula [2, 6, 7]:

$$UCP = (TUAW + UUCW) \times \text{TCF} \times \text{ECF} \tag{1}$$

A Variant Model of UCP might be the regression equation of existing UCP and unknown regression coefficients (1). These unknown coefficients´ regression will be found based on historical data. Section 3 presents a summary of the Regression Models Applicable to SPEE.

The major intent of this paper is to discuss existing Regression Methods applied to size estimation. The rest of the paper is structured as follows: Sect. 2 discusses a Linear Regression theoretical background; Sect. 3 describes model variants; Sect. 4 includes discussions; and Sect. 5 the Conclusion.

2 Linear Regression

Regression Models are a popular method for analysing relationships between variables - one of which is the Dependent (Response) Variable, and others include Independent (Predictor) Variables [8]. In the software project effort estimation context, Effort Estimation is the dependent variable, and environmental factors are the independent variables that have an impact on the effort estimation process; for instance, the number of developers, or the complexity of the project [2].

A regression model with one independent variable is called Simple Regression, whereas multiple independent variables are referred to as a Multiple Regression Model, and Multi-variate Regression is a model of the relationship between more than one dependent variables and one or more independent variables [2] (Table 2).

Table 2. Types of regression concerning the number of variables [2].

Type of regression	Dependent variables Y	Independent variables X	Regression coefficients β	Error terms ε
Simple regression	$[1 \times 1]$	$[1 \times 1]$	$[1 \times 1]$	$[1 \times 1]$
Multiple regression	$[n \times 1]$	$[n \times q]$	$[q \times 1]$	$[n \times 1]$
Multi-variate regression	$[n \times p]$	$[n \times q]$	$[q \times p]$	$[n \times p]$

In general, in a Multiple Regression - Y represents an $[n \times p]$ matrix containing p dependent variables, X depicts an $[n \times q]$ matrix containing q independent variables, β represents a $[q \times p]$ matrix containing regression coefficients, and ε represents an $[n \times p]$ matrix containing the error(s) of the model.

The form of the regression model is depicting as a linear equation, where all variables are linear; or as a non-linear equation, where the relationship between dependent variables and some independent variables or some other variables appear non-linearly, for example:

$$Y = \beta_0 + \beta_1 X_1 + \beta_2 X_2 + \ldots + \beta_q X_q + \varepsilon \qquad (2)$$

Equation (1) is a Multiple Linear Regression model [8], where the dependent variable is represented by Y - and the set of predictor variables by $X_{i1}, X_{i2}, \ldots, X_{iq}$; and q denotes the number of independent variables. ε is assumed to be a random error that represents any discrepancy in the approximation. It accounts for the failure of the model to fit the data exactly. β_0 is an Intercept Parameter, $\beta_1, \beta_2, \ldots, \beta_q$ called Regression Coefficients. These parameters are unknown constants that need to be estimated from the data, and ε is depicted as the error residuals.

Regression models aim to illustrate the relationship between the dependent variable and the set of independent variables. This relationship may be used for either predicting the values of the response variable for a given set of independent variables; or, to evaluate the significance of the independent variables so as to investigate the effects of a policy that concerns changes in the values of the independent variables [8].

The prediction value in the regression model can be found based on the historical data. This task is equivalent to searching for the coefficients of a regression model, including its functional form and for coefficients based on sample observational data (Historical Data). In Eq. (1) - with the historical data, the multiple regression model is defined as follows:

$$y_i = \beta_0 + \beta_1 X_{i1} + \beta_2 X_{i2} + \ldots + \beta_q X_{iq} + \varepsilon_i, i = \overline{1..p} \tag{3}$$

Where y_i is the dependent variable, $X_{i1}, X_{i2}, \ldots, X_{iq}$ are independent variables, and all of them are available in the sample data. Equation (2) could be written as follows:

$$y = X\beta + \varepsilon \tag{4}$$

Applying the Least Squares Estimation [8], Vector $\boldsymbol{\beta}$ is given by:

$$\beta = \left(X^T X\right)^{-1} X^T y \tag{5}$$

3   Variants of Models

3.1   Multiple Linear Regression (MLR)

3.1.1   AOM

In this publication [4], the authors have proposed the AOM (Algorithmic Optimization Method) algorithm, in the MLR form to optimise UCP by adding two parameters a_1 and a_2 to the UCP equation:

$$UCP = (a_1 TUAW + a_2 UUCW) \times \text{TCF} \times \text{ECF} \tag{6}$$

Equation (6) is derived from the UCP original by adding two parameters - (a_1 and a_2). The authors proposed that the process include three steps. Step 1 is the Preparation Phase, a_1 and a_2 will be found in this step. First, the historical data from n previous projects are collected. Next, the factors (TUAW, UUCW, TCF, ECF, UCP) for each project are identified. The result of this step is the collection of values (x_{i1}, x_{i2}, y_i), $i = \overline{1..n}$, with:

$$x_{i1} = TUAW_i \times TCF_i \times ECF_i \tag{7}$$

$$x_{i2} = UUCW_i \times TCF_i \times ECF_i \tag{8}$$

$$y_i = UCP_i \tag{9}$$

Finally, using the Multiple Least Squares Regression Model in Sect. 1, to then calculate a_1 and a_2 by means of the equation below:

$$\begin{pmatrix} y_1 \\ \vdots \\ y_n \end{pmatrix} = \begin{pmatrix} x_{11} & x_{12} \\ \vdots & \vdots \\ x_{n1} & x_{n2} \end{pmatrix} \times \begin{pmatrix} a_1 \\ a_2 \end{pmatrix} \tag{10}$$

$$\begin{pmatrix} a_1 \\ a_2 \end{pmatrix} = \left(X^T X\right)^{-1} X^T y \tag{11}$$

As a result, with the sample data - (Dataset 1 and Dataset 2) [4], value $a_1 = 0.96737$, and value $a_2 = 0.64416$ can be found.

$$UCP = (0.96737 \times TUAW + 0.64416 \times UUCW) \times TCF \times ECF \tag{12}$$

Step 2 and Step 3 of this process will calculate the TUAW, UUCW, TCF, and ECF of the current project and apply Eq. (6) with values a_1 and a_2 step 1 to estimate UCP.

One main advantage of AOM is that it allows one to obtain a correction value (a_1, a_2). These values were obtained by using the Least Squares Regression Model. However, these values can only be obtained from the historical project data in order to refine the estimation. They need to be verified in greater detail in the dataset expansion and clustering projects.

3.1.2   The Stepwise Approach

Stepwise Regression is a kind of Multi-variate Regression [2], or Multiple Regression, and is the model which combines forward and backward selection techniques that involve an automatic process for selecting independent variables [9, 10]; this is applied to remove all unsatisfied input variables [10, 11] in the model.

Silhavy et al. [9], have applied Stepwise MLR to search for the best coefficients of regression in variants of the UCP model. The authors [9] proposed the following models: Model A contains an intercept term only; Model B is comprised of an intercept, and the linear values of pairs of distinct predictors; Model D contains an intercept, linear terms, and squared terms; model E contains an intercepts, linear terms, interactions, and squared terms; and Model F includes terms used as they appear in the UCP equation without intercepts.

To find which is the best performing model, the authors recommended the process include six steps: Step 1 and Step 2 involve collecting data and analysing the assumptions for Linear Regression. Step 3 will analyse the correlations between the Independent and Dependent variables. Step 4 and Step 5 relate to the creation of a

Multiple Regression Model and the selection of the best-performing model - by comparing the results of all 5 models. Finally, the model selected in Step 5 will be compared with the UCP model.

This research used historical data, Dataset 1 and Dataset 2 [9], as the inputs for the Stepwise Approach algorithm in order to find the best model - based on the five models above. The real project size in UCP (person-hours/20) is described by $Real_{p20}$. In this algorithm, the significance level of $p = 0.05$ was used to evaluate the SSE criterion.

As a result, the authors' conclusion was that Model D was better than UCP, and all UCP variables features are significant for SPEE. The table below shows Model D.

Table 3. Best-performing model of UCP [9].

Data	Method D
Dataset 1	$Real_{p20} \sim 1 + UUCW + ECF + UAW \times TCF + UAW^2 + UUCW^2 + TCF^2 + ECF^2$
Dataset 2	$Real_{p20} \sim 1 + TCF + ECF + UAW \times UUCW + UAW^2 + UUCW^2 + TCF^2 + ECF^2$

Silhavy et al. proposed the Regression Clustering Method based on Stepwise Multiple Linear Regression - (RCMLR) in their publication [9, 12], for Function Point Estimation. This method is based on Functional Points Analysis and is used for the forming of clusters, the clusters contain the analogy projects, and the Cook Distance [9, 12, 13] is applied to eliminate the project from the clusters. As a result, the effort estimation of MLR is better than when using Regression Clustering. RCMLR can search equivalent projects to create clusters, suitable for the Regression Model, and the prediction capability of RCMLR is outperformed by non-clustered MLR [12].

Nassif et al. [10], proposed the Fuzzy Sugeno [14] Linear Model. This is a Fuzzy Logic model, where the regression model was used to identify the optimal number of model inputs, and the authors used a Stepwise Approach to identify the optimal number of inputs in the Fuzzy Sugeno Linear Model. The Sugeno Model is a type of Fuzzy Logic whose output is illustrated as a linear model - or constant. The Sugeno Model is given below:

If x_1 is $A_1, \ldots, x_k$ is A_k is the input group, then the output group is $y = p_0 + \sum_{i=1}^{k} p_k \times x_k$ [10, 14], where k is the number of inputs in the model and p_k is the coefficients of the linear model.

In the given publication [10], apart from the Fuzzy Sugeno Linear Model, the authors recommended three models: Sugeno Constant Fuzzy Logic, Mamdani Fuzzy Logic [15], and MLR. All of them were tried and tested using datasets from the ISBSG (International Software Benchmarking Standards Group) [10, 16]. As a result, the authors' conclude combining the Multiple Linear Regression Method with the Fuzzy Concept; the Sugeno Fuzzy Model with Linear Output led to the better design of Fuzzy models.

3.2   Whale-Crow Optimisation (WCO)

Ahmad et al. [17] proposed the WCO algorithm in order to find the optimal regression coefficients of SPEE. This algorithm combines the CSA (Crow Search Algorithm) [18] with thed WOA (Whale Optimisation Algorithm) [19]. The purpose of this algorithm is to search for the best regression coefficients of the Regression model. This is a new approach intended to find suitable coefficients and/or, other methods; for example, AOM [4], used the Least Square Method [8] to find the coefficients.

In publication [17], the authors applied WCO to find the optimal coefficients of the Linear Regression model and the Kernel Logistics Regression [17] model with four datasets (Cocomo81, Desharnais, Cocomonas_2, and Cocomonsa_v1) from the Promise Software Engineering Repository [17], and the authors have validated the accuracy of the models, based on the value of MMRE [17].

As a result, the value of MMRE of the Proposed WCO + Linear and the Proposed WCO + Kernel Regression are smaller than Linear Regression, Kernel Regression, CSA + Linear Regression, CSA + Kernel Regression, PSO-FCM [17], Tabu Search + GA [17] and PSOCP [17].

4   Discussion

Sections 3.1 and 3.2 present variants of the Regression model needed to optimise the size estimation for a software project - the, MLR, and WCO approaches respectively. Both of them were based on the historical data and assessed by statistical hypotheses to find the best model for each method.

A Stepwise Approach is used in MLR to optimise the best effort estimation results, and Table 4 summarises the various forms of the Stepwise Approach.

Table 4. The Stepwise Approach

No.	Applied on method	Description	Data tests
1.	UCP	Search for the best variant of UCP using the Stepwise Approach	Dataset 1 and Dataset 2 [9]
2.	Function Point	Stepwise and Cook Distance are proposed so as to eliminate unsuitable projects from clusters	1650 data entries from ISBSG [20]
3.	Fuzzy Logic	Stepwise is proposed to identify the best inputs for the Sugeno Fuzzy Logic model	ISBSG [10, 16]

Silhavy et al. [9, 12] proposed two techniques to estimate the SEE based on the UCP method, the AOM algorithm and the best variant of UCP with a Stepwise approach. Both of them have been assessed by means of statistical hypotheses to find the best model for each method. Using the Stepwise Approach, we found the "best" model and "best" set of independent variables for the concrete dataset. When Stepwise is used, we are looking for the ability of "adaptability" to unknown datasets in the new project because the model formula may be different for each dataset; for example, Table 3 shows two different models corresponding to Dataset 1 and Dataset 2.

In the case of a lot of project data in the history, the Clustering Method is considered to group data according to each cluster. This approach is illustrated in the publication [12], and the algorithm called RCMLR. In the future, Clustering is an interesting approach and should be tested in another method - UCP for instance, to compare with the AOM or UCP Stepwise approach in the publication [9]. The Fuzzy Sugeno Linear model is the favored method, and it should be compared with AOM or RCMLR on the same historical data to evaluate accuracy.

Another way to optimise a regression model is the WCO algorithm. It was proposed by Ahmad et al. [17]. The authors evaluations were based on the comparison of the Linear Regression Method, and the Linear Regression Method combined with WCO. The statistical results showed that the proposed WCO algorithm is more accurate. This is an interesting result.

5   Conclusion

This paper illustrated an overview of existing Regression Models for SPEE (Software Project Effort Estimation). There is not any other most popular model to estimate the size of a software project with the highest degree of accuracy. The proposed Regression models are presented and evaluated based on the individual historical data for each model. Therefore, the project manager or team leader role is important in understanding the factors that directly affect the size estimation of the software project in order to be able to apply the appropriate method when evaluating software estimates.

Future work will be to evaluate the Stepwise Approach and the AOM Approach for UCP and WCO in a large Dataset so as to identify the best regression model, and the Clustering-based Regression Model should be compared to WCO or UCP.

Acknowledgment. This work was supported by the Faculty of Applied Informatics, Tomas Bata University in Zlín, under Project RO30196021025 and under Project IGA/CebiaTech/2019/002.

References

1. Muketha, G.: A review of agile software effort estimation methods. Int. J. Comput. Appl. Technol. Res. **5**(9), 612–618 (2016)
2. Trendowicz, A., Jeffery, R.: Software Project Effort Estimation. Springer, Berlin (2014)
3. Jørgensen, M., Sjøberg, D.I.K.: An effort prediction interval approach based on the empirical distribution of previous estimation accuracy. Inf. Softw. Technol. **45**, 23–136 (2003)
4. Silhavy, R., Silhavy, P., Prokopova, Z.: Algorithmic optimisation method for improving use case points estimation. PLoS One **10**(11), e0141887 (2015)
5. Fedotova, O., Teixeira, L., Alvelos, H.: Software effort estimation with multiple linear regression: review and practical application. J. Inf. Sci. Eng. **29**(5), 925–945 (2013)
6. Karner, G.: Metrics for objectory. Diploma thesis, University of Linkoping, Sweden, No. LiTHIDA Ex- 9344:21 (1993)

7. Ochodek, M., Nawrocki, J., Kwarciak, K.: Simplifying effort estimation based on Use Case Points. Inf. Softw. Technol. **53**(3), 200–213 (2011)
8. Chatterjee, S., Hadi, A.S.: Regression Analysis by Example, 5th edn. Wiley, Hoboken (2012)
9. Silhavy, R., Silhavy, P., Prokopova, Z.: Analysis and selection of a regression model for the Use Case Points method using a stepwise approach. J. Syst. Softw. **125**, 1–14 (2017)
10. Nassif, A.B., Azzeh, M., Idri, A., Abran, A.: Software development effort estimation using regression fuzzy models. Comput. Intell. Neurosci. **2019**, 17 (2019)
11. Silhavy, P., Silhavy, R., Prokopova, Z.: Evaluating subset selection methods for Use Case Points estimation. Inf. Softw. Technol. **97**, 1–9 (2018)
12. Silhavy, P., Silhavy, R., Prokopova, Z.: Stepwise regression clustering method in function points estimation. In: Silhavy, R., Silhavy, P., Prokopova, Z. (eds.) CoMeSySo 2018. Advances in Intelligent Systems and Computing, vol. 859, pp. 333–340 (2019)
13. Stevens, J.P.: Outliers and influential data points in regression analysis. Psychol. Bull. **95**(2), 334 (1984)
14. Sugeno, M., Yasukawa, T.: A fuzzy-logic-based approach to qualitative modeling. IEEE Trans. Fuzzy Syst. **1**(1), 7–31 (1993)
15. Mamdani, E.H.: Application of fuzzy logic to approximate reasoning using linguistic synthesis, vol. C-26, no. 12, pp. 1182–2206 (1977)
16. ISBSG: International Software Benchmarking Standards Group (2017). http://isbsg.org/
17. Ahmad, S.W., Bamnote, G.R.: Whale–crow optimization (WCO)-based Optimal Regression model for Software Cost Estimation. Sādhanā **44**(4), 94 (2019)
18. Askarzadeh, A.: A novel metaheuristic method for solving constrained engineering optimization problems: crow search algorithm. Comput. Struct. **169**, 1–12 (2016)
19. Mirjalili, S., Lewis, A.: The whale optimization algorithm. Adv. Eng. Softw. **95**, 51–67 (2016)
20. ISBSG: ISBSG development and enhancement repository - release 13. International Software Benchmarking Standards Group (ISBSG) (2015)

A Review of Software Effort Estimation by Using Functional Points Analysis

Vo Van Hai$^{(\boxtimes)}$, Ho Le Thi Kim Nhung, and Huynh Thai Hoc

Faculty of Applied Informatics, Tomas Bata University in Zlin,
Nad. Stranemi 4511, 76001 Zlin, Czech Republic
{vo_van, lho, huynh_thai}@utb.cz

Abstract. The estimation of the Software Development Effort, (further only SDE), value is critical for the effective management of any software industry. Function Point Analysis, (further only FPA) - is a standardised method designed to systematically measure the functional size of the software. Although this method tool has been become widely-used by many software organisations, it still faces many problems. In this paper, we shall present a systematic review of Software Effort Estimation, (further only SEE), methods based on Functional Points Analysis, (further only FPA). The article focuses on an analysis of the limitations and accuracy of the FPA method - which was proposed many years ago.

Keywords: Functional Points Analysis (FPA) ·
Software Effort Estimation (SEE) · Effort Accuracy Improvement (EAI)

1 Introduction

Software estimation is a critical issue in the development of a software system [6]. This problem has grown into a crucial question since numerous projects continue to not be completed on time - with either under or overestimation of efforts resulting in specific issues [7]. For this reason, different software cost estimation models were developed to manage the budget and schedule of software projects [4]. Accurate SEEs are essential to project managers, as well as customers [5]. It supports the generation of applications for proposals, contract negotiations, scheduling, monitoring, and control.

However, SEE is not a simple task. Many software project failures in the past are related directly to the planning and estimation steps. According to Moløkken, only approximately 30% to 40% of software projects are completed successfully [8]. The Standish Groups CHAOS reports the high failure rate of software projects, which accounted for 70% in 1994 [9], and this was reduced to 48%–56% in 2018 [10]. Several studies have been undertaken to find the reason(s) for software project failures. Galorath et al. [11], indicated that the blame for such failure(s) is the lack of Sufficient Requirements Engineering, poor project planning, sudden decisions in the early stages of the project, and especially - inaccurate estimations. Several researchers have also shown inaccurate cost estimation is due to significant project failure causes [5, 8, 12]. Therefore, it is necessary that SEE methods needed to evaluate various types of projects are improved. One such possible research direction might be to study methods based on

© Springer Nature Switzerland AG 2019
R. Silhavy et al. (Eds.): CoMeSySo 2019, AISC 1047, pp. 408–422, 2019.
https://doi.org/10.1007/978-3-030-31362-3_40

Function Point Analysis (FPA) [1]. A crucial aspect that needs to be measured in software engineering projects [13] is the functional size of the software. FPA [1] is a standardised method for establishing a software size measurement from its functional requirements, while also considering the features to be implemented in it. It has been designed to be applied without regard to the programming language or implementation technology. Albrecht proposed FPA as a means to measure the size of data-processing systems from end-users.

One more objective was the estimation of the development effort. The International Function Point Users Group, (IFPUG) [2], was created so as to keep going efforts to describe the method and it also publishes the official Function Point, (FP), Counting Practices Manual [3]. FPA is standardised by ISO/IEC 20926:2010. However, the FPA method has encountered some disputes from different researchers - in terms of advantages and limitations. Henderson et al. [18], studied FPA from a manager's and developer's perspective- based on 13 attributes with three key findings: SLOC-count is less complicated than FP; developers are better able to comprehend the benefits of FP than managers; the difference between managers and developers is in the Values Block Communication necessary to propound informed decisions. Kampstra et al. [14], and Kemerer [15], reported that the FPA method does not create consistent results when applied by different metrics. Meli [16], pointed out a mismatch between the complexities established for the Base Functional Components, (BFC), and the possible productivity estimates.

In order to contribute to this area, we present a systematic literature review of software effort estimation methods based on FPA. This paper focuses on the potential limitations of this method, as well as the proposed solutions. In this paper, we refer to a compilation of improvements, focused on increasing the accuracy of FPA. The paper is composed as follows: Sect. 2 defines the research questions and outlines the research objectives. Section 3 presents the overview results of the study of the limitations and improvements of FPA methods. The paper ends with a discussion and future work.

2 Problem Statement

In this article, we focus on a review of different effort estimation methods based on functional software size calculated by FPA. This analysis-determined method could increase the accuracy of the functional sizes found.

2.1 Research Questions and Goals

The research questions answered by this study are as follows:

RQ1: Which aspects can be used to improve FPA estimation accuracy?
RQ2: What improvements are being proposed for FPA?

2.2 Evaluation Criteria

One of the problem in SEE is how to evaluate the estimation model accuracy [11]. In order to assess models, we need the appropriate evaluation criteria to measure these models and determine model accuracy. There are several common criteria that are accepted as standard evaluation tools for effort estimation. In this section, we enumerate the frequently-used evaluation measures for SEE (see in Table 1).

Table 1. Evaluation criteria used in previous studies.

Accuracy indicator	Studies using the accuracy indicator
The Mean Squared Error $$MSE = \frac{\sum_{i=1}^{n} Actual_i - Estimated_i}{N} \quad (1)$$	[37]
The Magnitude Relative Error $$MRE_i = \frac{|Actual_i - Estimated_i|}{Actual_i} \quad (2)$$ The Mean of Magnitude Relative Error $$MMRE = \frac{1}{N}\sum_{i=1}^{N} MRE_i \quad (3)$$	[19–24, 32, 37]
The Magnitude of Error Relative $$MER_i = \frac{|Actual_i - Estimated_i|}{Estimated_i} \quad (4)$$	[37]
The Prediction Percentage within x% $$PRED(x) = \frac{1}{N}\sum_{i=1}^{N} \begin{cases} 1 \ if \ MRE_i \leq x \\ 0 \ otherwise \end{cases} \quad (5)$$	[20, 21]
The Balanced Relative Error $$BRE_i = \frac{Actual_i - Estimated_i}{Actual_i} \quad (6)$$ The Mean of the Balanced Relative Error $$MBRE = \frac{1}{n}\sum_{i=1}^{N} BRE_i \quad (7)$$	[37]
The Relative Error $$RE_i = \frac{Estimated_i - Actual_i}{Actual_i} \quad (8)$$	[20, 22, 24]

3 Background

3.1 Functional Points Analysis

In 1986, the International Function Point User Group, (IFPUG) [2], began promoting and disseminating effective software development and maintenance management through FPA. The IFPUG is currently the FPA regulatory agency, responsible for the improvement and development of the rules set out in the Counting Practices Manual, (CPM) [3], currently in version 4.3.1. Since the creation of the IFPUG, the original FPA method is known as the IFPUG's FPA, referred in this paper only as FPA. FPA is currently standardized by ISO/IEC 20926:2010, as shown in Fig. 1.

This standard specifies the set of definitions, rules, and steps for applying it. There are similar methods, also standardised by other ISO/IEC standards, derived from the original FPA, like COSMIC, FiSMA, Mark-II, and NESMA, which are set out briefly in Sect. 3.2.

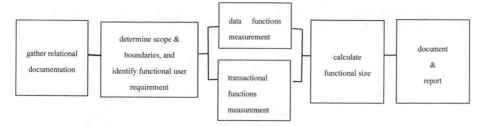

Fig. 1. A graphical overview of the function point counting process [3].

This process determines all of the standard steps. But in this section, we focus on presenting the measurement data functions and transactional functions. There are five primary types of components in the process - grouped into data functions and transactional functions. Data functions, (DF), are divided into Internal Logic Files, (ILF), and External Interface Files (EIF). The transactional functions are classified into each transactional function as an External Input, (EI), an External Output, (EO), or an External Inquiry, (EG) - as shown in Fig. 2.

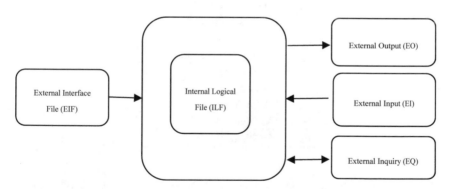

Fig. 2. IFPUG measurement components.

In order to calculate the Unadjusted Function Points, (UFP), we must determine the complexity of these components. Each of them is grouped into one of three levels of complexity - (low, average, and high), and each of these levels assigned a correspondent weight - see the following Table 2.

Table 2. Data and transactional function complexity.

		Component		
		EI	EO	EQ
Functional complexity	Low	3	4	3
	Average	4	5	4
	High	6	7	6

For the calculation of UFP, set the number of types in groups (multiplier), multiply them by complexity sizes and sum all fields in Table 2.

$$UFP = \sum_{i=1}^{n} \sum_{j-1}^{m} (multiplier_{ij} \times size_{ij}) \tag{9}$$

Where, n is the number of types, and m is the number of complexity groups.

In the next phase, we then calculate the Value Adjustment Factor (VAF). This value is based on 14 General System Characteristics, (GSCs) which rate the general functionality of the application being measured (see Table 3).

Table 3. General system characteristics (GSC).

Factor	Content
F_1	Data communications
F_2	Distributed data processing
F_3	Performance
F_4	Heavily used configuration
F_5	Transaction rate
F_6	Online data entry
F_7	End-user efficiency
F_8	On-line update
F_9	Complex processing
F_{10}	Reusability
F_{11}	Installation ease
F_{12}	Operational ease
F_{13}	Multiple sites
F_{14}	Facilitate change

Based on the stated user requirements, each factor is rated in terms of its degree of influence on a scale of 0–5. Table 4 represents the significance of influence factor ratings.

Table 4. Factor weights.

System influence	Rating
Not present or no influence	0
Incidental influence	1
Moderate influence	2
Average influence	3
Significant influence	4
Strong influence throughout	5

Then we can calculate the VAF, using the formula below:

$$VAF = 0.65 + 0.01 \times \sum_{i=1}^{14} (F_i \times rating) \tag{10}$$

The VAF thus adjusts the unadjusted functional size - (±35%), to produce the Adjusted Functional Size, (AFS). This can vary within the range of 0.65 - (when all GSCs are low), to 1.35 - (when all GSCs are high). In standard cases, this can be chosen in the smallest case. The number of AFP is calculated as follows:

$$AFP = UFP \times VAF \tag{11}$$

In the previous iteration of IFPUG, the function point was used to develop a new project but not to enhance an existing project. But, in the latest version - (4.3.1), this problem has been changed. We can now use FP for both Adjusted Development Project Function Point - (aDFP), and Adjusted Enhancement Project Functional Size (aEFP). Then use the following formula to calculate aDFP:

$$aDFP = DFP \times VAF \tag{12}$$

Where, $DFP = AFP$. In aEFP, the following formula is used to calculate function size:

$$aEFP = [(ADD + CHGA + CFP) * VAFA] + (DEL * VAFB) \tag{13}$$

Where ADD is the size of the functions being added by the enhancement project, CHGA is the size of the function being changed by the enhancement project – as they are/will be after implementation. CFP is the size of the conversion functionality, VAFA is then, the Value Adjustment Factor Application after enhancement project completion, DEL is the size of the functions being deleted, and VAFB is the value-adjustment factor before the enhancement project begins.

3.2   Method Standards

There are many FSM methods, but only five of them are standardised ISO/IEC - as shown in Table 5.

IFPUG - (The International Function Point User Group) [45]: This method is derived from the original ideal of Allan Albrecht [3]. This method will be presented with more details in the next section.

COSMIC (Common Software Measurement International Consortium) [43]: This project was commenced in November 1998, adopted in 2011 as ISO 19761, and has been proposed as a 2nd generation FSM method. The aim of COSMIC is to introduce a new rule of functional size measures that extract the best features of IFPUG, Mark II, etc. These characteristics are: (a) The size application should be measured based on user requirements; (b) The method should be academically sound, and compatible with modern ways of stating requirements - but independent of specific purposes; (c) The measurement should be consistent with ISO 14143 standards. A series of innovations is

Table 5. Base functional components.

Method	Base functional components
IFPUG	External Inputs: unique user data or control input that adds or changes data
	External Outputs: unique user data or control output that exceeds system boundaries
	External Inquiries: unique input that creates immediate output
	Internal Logical Files: an internally-maintained logical group of data
	External Interface Files: file(s) passed or shared between applications
MarkII	Input: event, user query, timed trigger
	Processing: references to retained data
	Output: report, display or response
NESMA	External Inputs: data into the application without data manipulation
	External Outputs: moves data towards the user while performing some data manipulation
	External Inquiries: moves data towards the user, and does not perform data manipulation
	Internal Logical Files: persistent data maintained by the application through the use of EIs
	External Interface Files: persistent data used by the application, but not maintained by it
COSMIC	Entry: from user to the system
	Exit: from system to user
	Write: from system to persistent data
	Read: from persistent data to system
FISMA	Interactive end-user navigation and query services
	Interactive end-user input services
	Non-interactive end-user output services
	Interface services to other application
	Interface services from other applications
	Data storage services
	Algorithmic and manipulation services

proposed: Better fit with both Real-time and Management Information Systems (MIS) environments; This is not only applied in identification systems, and the measurement of multiple software system layers, but also in different viewpoints from which the software can be observed and measured. The particular problem was that the absence of a weighting system could be estimated.

FiSMA (The Finnish Software Measurement Association) [42]: This method is service-oriented based, instead of process-oriented based, so services are defined. After the identification of each service, the size of each service was then found by applying the rules of the method. Finally, the total functional size was estimated by adding up the sizes of all services. FiSMA, accepted as an international FSM Standard in 2008, was developed by a FiSMA working group. It is a general parameterised, size measurement method which it is proposed should be applied to all software types.

MarkII [39]: This method was introduced by Symons in 1988, [38], with the primary intent to improve the original FPA method. This method proposed some suggestions that reflected the internal complexity of an application system. At this time, the Metrics Practices Committee - (MPC), of the UK Software Metrics Association - (UKSMA), is the design authority of the method [39]. It was also principally designed to measure Business Information Systems – (BIS). The Mk II FPA has been accepted as ISO/IEC 14143 and thus become an international ISO standard in 2002.

NESMA [40]: This method has the same rules as the IFPUG FPA method. ISO accepted it as an international standard in 2005, [41]. NESMA, the function points user group in the Netherlands, suggests three types of function-point counts - depending on the degree of detail possible - detailed, estimative, and indicative. The detailed function-point count is the IFPUG count. In the estimated function-point count, the steps are: (1) To determine all of the functions of all five types - ILF, EIF, EI, EO, EQ; (2) To calculate the total unadjusted function-point, assuming that every data function point is of low complexity, and every transaction point is of average complexity.

As mentioned above, the ways that one determines or classifies function-points is a difficult problem. Here is a summary of all of the above standards as relating to the Base Functional Components (BFC); see Table 5.

4 Methods Used

Many methods were proposed with the desire to improve Effort Estimation Accuracy – (EAC). According to Kitchenham et al. [36], the authors proposed a model based on three involved aspects (1) The appropriate independent variables; (2) The possible functional forms of the model; and, (3), The likely structure of the model error terms. Then, they classified these models into corresponding groups. Marcos et al. [18] have classified the proposed improvements into three categories - which are related to (1) Weights and Complexities; (2) Technological Independence Methods; and - (3), adjusted function size calculation. In the first category, the authors demonstrate that BFCs are not suitable for all cases. The second category demonstrates the fact that FPA functional measurements was not referred to any technologies; and that the functional size is generated, based on the user's view. The last category asserts that the GSC that was listed and admitted, was not a reality non-functional requirement. In answer to RQ1, we used the Marcos, et al. category. The comparison of these methods is shown in Table 6.

4.1 Weights and Complexities

Many studies showed that the FPA scored the BFC incorrectly. For example, the same data function and/or the same transactional function with different combinations of DET and RET/FTR can be categorised with the same complexity. This leads to the same number of function points for the data functions, and/or transactional functions. They also notice that - in some situations, functionalities that have very similar DET and RET/FTR can be categorised with different complexities; and thus, receive different FPA weightings.

Xia et al. [19], doubted that the Unadjusted Function Points weight values which were raised based on a study of the IBM data processing systems - (locally), could not reflect the software globally. In [20], they continually point out the existence of ambiguous classification, and the original method may not fully reflect the reality of the software complexity under the specific software application.

In [21], they proved that there is no clear boundary between two classifications in FP counting. To resolve these problems, the authors suggested the merging of three techniques - (Fuzzy Logic, Artificial Neural Networks, and Statistical Regression) in a neuro-fuzzy function point calibration model.

Ahmed et al. [22] showed that many factors could be affected by the complexity of FP weight metrics values; like methodologies to develop software, support tools, and other factors. The authors proposed that new FPA weights were measured, based on an adapted genetic algorithm. The proposed algorithm is based on a set of initial solutions - using biologically-inspired evolution mechanisms to derive new - and possibly better, solutions.

According to Hajri et al. [23], the classification of function types into simple, average, and complex, does not reflect the entire complexity necessary to develop user systems. The main improvement idea of this research is to establish a new weighting system for FP measurement using Artificial Neural Networks – (ANN), that is to say, (the back-propagation technique). In the first step, they use the original weights system as baselines in order to establish the new weights. Next, they train one of the most popular Neural Networks techniques to predict the values of the new weights. And then, they apply the new weights and the original weights in the FP model. Finally, they calculate the FP count, depending on the original and new weights.

Ya-Fanget et al. [33], reported that the IFPUG FPA is barely able to distinguish the complexity levels of the contiguous components. If this situation happens in a software program, the estimation outcome will not correspond to the practical situation. In order to resolve this disadvantage, they believed that fuzzy theory should be used to analyse the complexity of the component. They also said that the BFCs weights - which were set by IFPUG, are said to reflect the functional size of the software, but actually – today's software differs drastically from the past; so it is no longer suitable. On the same point with Ya-Fang et al. [31], said that this inconsistency in a large number of BFCs; which lies on the specified intervals' boundary areas, becomes even worse. The cause is due to the inaccurate classification of various system functionalities – which would distort its functional size. Similarly, Xia et al. [19–21]; the authors also offer conjuncture relating to fuzzy logic with artificial neural networks; such that, base upon the rules and linguistic terms defined by Fuzzy Logic, neural networks could learn from previously-estimated data.

Rao [33], showed that the Function Points Measurement Process is not accurate in some specific cases - and the number of referenced items - which determines the lower limits of the high-complexity range, can cause the same measurement accuracy difficulties - especially in systems that reference a large number of DETs. Solving these problems linked to the proposed method, the authors suggest extending from FPA to FFPA (Fuzzy Function Point Analysis), using Fuzzy Logic. Fuzzy Theory builds a formal quantitative structure capable of emulating the imprecision of human knowledge. Using FFPA, more accurately determined derived development values can be obtained. Kraji et al. [34] also pointed out that one of the weaknesses of IFPUG FPA is the limited determination of functional complexity range.

The authors built a model that helped them to determine the relationship between the standard FPA method and COSMIC-FFP and MKII FPA methods.

Additionally, Calazans et al. [24], Ferreira and Marques-Neto [26], added that data warehouses/data-marts are typical software products that have specific features - and should be taken into account in the measurement process. According to Junior et al. [31], there are two critical problems in the classification system: (a) Functions of different sizes are given the same point values; and (b) Similar functions are classified into different groups.

Prokopova et al. [37, 44], analysed the influence of selected factors on Work-Effort Estimation. These factors are independent variables – i.e. the Function Point Count Approach, business area, industrial sector, and relative size. Based on these selected factors and by experimental evidence, the authors Have proved that the selected factors have specific effects on Work Effort Estimation accuracy.

4.2 Technological Independences

The IFPUG's approach and aim is to measure the size of any software system - without involving the technology. Calazans et al. [24], argued that each technology has specific particularities - and these must be taken into account. They also argued that FPA approaches were defined many years ago; so that period' hardware and software technologies are mostly obsolete today. They showed that the functional size could be generated - based on user's views.

Xia et al. [19–21], and Harput et al. [25], also report that the Object-Oriented Paradigm has entwined itself into the software development mainstream. This has meant that traditional FPA were only able to obtain accurate estimation results with difficulty. Harput [25], suggested an FPA application approach to an object-oriented requirements model. This transforms the object-oriented model into an FPA model; where FPA experts instruct the use of heuristic rules.

4.3 Adjusted Function Size Calculation

Abdullah et al. [28, 29] argued that FPA does not cover all of the GSCs needed to calculate the Adjusted Functional Size. One of these is those aspects relating to the Information Security field. So, they suggested an extension of FPA calculation that includes security; in which security characteristics are assessed as one of the GSCs.

According to Bharadwaj [35], some of the GSCs are unimportant, and should be re-analysed and then omitted; and some others, for example: functionality, reliability, efficiency, usability, maintainability, portability - should be included. They also proposed the changing/mapping of the original factors into new factors - like GSC-12 Operation Ease to Reliability; GSC-3 Performance, GSC-8 Online Update and GSC-6 Online Data Entry to Performance; GSC-5 Transaction Rate to Scalability.

Peng et al. [29], reported that 14 GSCs have some problems: they are overlapped; the subjectivity is relatively strong in the scoring process; the weight of 14 GSCs is the same - while it should be different in different types of system: for example, real-time software, embedded software, information management software, etc.

In their opinion, the VAFs distribution should be more heterogeneous in order to better reflect some specific situations.

The authors proposed the recalculation of the Degree of Influence (DI) for each GSC and the consideration of the critical role of the GSC in the particular type of software being analysed. They also proposed the fuzzy AHP approach - which aims to define the weight of each DI assigned to a given GSC. The new DI is calculated as a result of the product between the two variables: a score of 1 to 5 - (as original FPA); and a constant, calculated by the fuzzy AHP approach, that expresses the level of importance of that GSG for the particular software.

According to Matijevic [30], the authors supposed that determining the influence of each GSC is a difficult task to be performed by those responsible for deciding each influence factor of the GSCs. It is challenging to identify the value to be assigned to a particular GSC. Furthermore, determining the wrong determination may influence the complexity of a development software project.

Calazans et al. [24], based their work on the Data Warehouse/Data Marts software analysis and determined that the calculation of FPA adjusted functional size could be improved in order to measure this type of software more accurately. They eliminated eight characteristics: data communications, heavily-used configuration, transaction rate, online data entry, online update, installation ease, multiples sites, and facilitated change), initially defined in FPA since they never influence the Data Mart development process. They also proposed the following new characteristics - as follows: Complexity, Performance, User-Support, Quality, Architecture, Interaction, Constraints, Interface, Operation, Reusability, Documentation, Reusability, Amount of Transactional System involved in the project, Source-data Structure, Development Team Knowledge-level, Source Data Update Frequency, Data-volumes, Use of Appropriate Extraction and Loading tools, and Source Systems Documentation. Apart from this, they have not only interpreted the FPA rules for that domain but also changed the way that FPA evaluate the system to be measured with new characteristics.

Table 6. Method improvement techniques

Studies	Technology used	Weights & complexities	Technological independence	Adjusted function size calculation
[19, 29, 31–33]	Fuzzy logic	✔		
[19–21, 23, 32]	Artificial neural networks	✔		
[22]	Genetic algorithms	✔		
[24–28, 30, 35]	Interpolation methods	✔	✔	✔
[19–21]	Statistical regression	✔		
[34, 37, 44]	Change affected factors	✔		

5 Discussion and Conclusion

Although many efforts are aimed, designed, oriented on improving the overall method, these are still limited.

With regard to the first type of improvement - "Weights and Complexities" – some proposals on the results show that some studies cannot solve this problem completely. Although the proposals were able to relieve the problem of allowing the same data function or transaction function with different combinations of DET and RET/FTR, and/or the same weight complexity. Entire works were devoted to proposing FPA improvements to make it more detailed and so that it can evaluate different DET and RET/FTR combinations with different values, as compared to the original method. However, such unwanted effects have not been wholly excepted.

Regarding suggestions for "technological independences"; these involve certain types of techniques in the software development process - but not for all. Thus, there should be a corresponding method for the relevant technical example. This excludes the generalisation of the method. For improvement proposals related to adjusting the functional size, the addition of many additional factors - like security, can make the estimation process more complicated and more expensive. In the case where the software only requires functional requirements, this problem becomes more serious. In addition, adding new GSCs to make the Adjusted Function Size more accurate, can also be a significant obstacle since it requires more time to evaluate. This increases costs - as well as time. Furthermore, improvements applied in a specific environment such as Data Warehouse/Data Mart have a minimal scope, and so do not represent other types.

For RQ1 answer - as we mentioned in Sect. 4, the main aspects can be considered to be improving the estimation accuracy in weights and complexities; technological independence methods; and adjusted function size calculation. Each aspect given above has provided many ideas; proposed to answer the RQ2 - as presented in Sect. 4. From this study, we have a general view of the available estimation of Function Points Analysis accuracy improvement methods.

In future works, we shall work towards suggesting a new method for improving FPA-based estimation accuracy. The model shall use back-propagation networks and consider factors that affect to estimated process.

Acknowledgment. This work was supported by the Faculty of Applied Informatics, Tomas Bata University in Zlín; under Project No.: RO30196021025, and the IGA/CebiaTech/2019/002 Project.

References

1. Albrecht, A.J.: Measuring application development productivity. In: Proceedings of IBM Application Development Symposium, Montana, CA, USA, pp. 83–92 (1979)
2. IFPUG (International Function Point Users Group). http://www.ifpug.org/

3. IFPUG: Function point counting practices manual, release 4.3.1. International Function Point Users Group, Westerville, OH, USA (2010)
4. Putnam, L.H.: A general empirical solution to the macro software sizing and estimation problem. IEEE Trans. Softw. Eng. **4**, 345–361 (1978)
5. Caper, J.: Estimating Software Cost. Mc-Graw-Hill, New York (2007)
6. Albrecht, A.J., Gaffney, J.E.: Software function, source lines of code, and development effort prediction: a software science validation. IEEE Trans. Soft. Eng. **6**, 639–647 (1983)
7. Boehm, B.W.: Software Engineering Economics. Prentice-Hall, Englewood Cliffs (1981)
8. Moløkken, K., Jørgensen, M.: A review of surveys on software effort estimation. In: International Symposium on Empirical Software Engineering, pp. 223–231. Retrieved from ACM Digital Library database (2003)
9. Moløkken, K., et al.: Project estimation in the Norwegian software industry. A summary. Simula Research Laboratory (2004)
10. The Standish Group: CHAOS Chronicles. Technical report, The Standish Group International, Inc. (2018)
11. Galorath, D.D., Evans, M.W.: Software Sizing, Estimation, and Risk Management: When Performance Is Measured Performance Improves. Auerbach, Boca Raton (2006)
12. Kemerer, C.F.: An empirical validation of software cost estimation models. Commun. ACM **30**(5), 416–429 (1987)
13. Ravichandran, T.: Organizational assimilation of complex technologies: an empirical study of component-based software development. IEEE Trans. Eng. Manag. **52**(2), 249–268 (2005)
14. Kampstra, P., Verhoef, C.: Reliability of function point counts. Department of Computer Science, VU University Amsterdam, Amsterdam, The Netherlands (2010)
15. Kemerer, C.F.: Reliability of function points measurement: a field experiment. Commun. ACM **36**(2), 85–97 (1993)
16. Meli, R.: Functional metrics: problems and possible solutions. In: Proceedings of 1st European Software Measurement Conference, Antwerp, Belgium (1998)
17. Sheetz, S.D., Henderson, D., Wallace, L.: Understanding developer and manager perceptions of function points and source lines of code. J. Syst. Softw. **82**, 1540–1549 (2012)
18. Marcos, F., Marcelo, F., Sun, V.: Improvements to the function point analysis method: a systematic literature review. IEEE Trans. Eng. Manag. **62**(4), 1–12 (2015)
19. Xia, W., Capretz, L.F., Ho, D.: Neuro-fuzzy approach to calibrate function points. In: Proceedings of the 8th WSEAS International Conference on Fuzzy Systems, pp. 116–119 (2007)
20. Xia, W., Capretz, L.F., Ho, D., Ahmed, F.: A new calibration for function point complexity weights. Inf. Softw. Technol. **50**(7–8), 670–683 (2008)
21. Xia, W., Ho, D., Captrez, L.F.: A neuro-fuzzy model for function point calibration. WSEAS Trans. Inf. Sci. Appl. **5**(1), 22–30 (2008)
22. Ahmed, F., Bouktif, S., Serhani, A., Khalil, I.: Integrating function point project information for improving the accuracy of effort estimation. In: Proceedings of the 2nd International Conference on Advanced Engineering and Computer Application in Sciences, pp. 193–198 (2008)
23. Hajri, M.A., Ghani, A.A.A., Sulaiman, M.N., Selamat, M.H.: Modification of standard function point complexity weights system. J. Syst. Softw. **74**(2), 195–206 (2005)
24. Calazans, A.T., Oliveira, K.M., Santos, R.R.: Adapting function point analysis to estimate data mart size. In: Proceedings of 10th International Symposium on Software Metrics, pp. 300–311 (2004)

25. Harput, V., Kaindl, H., Kramer, S.: Extending function point analysis to object-oriented requirements specifications. In: 11th IEEE International Software Metrics Symposium (2005)
26. Ferreira, W.G., Marques-Neto, H.T.: Estimating the size of data mart projects. In: Proceedings of 28th Annual ACM Symposium on Applied Computing, pp. 1147–1148 (2013)
27. Abdullah, N.A.S., Abdullah, R., Selamat, M.H., Jaafar, A.: Software security characteristics for function point analysis. In: Proceedings of 16th IEEE International Conference on Industrial Engineering and Engineering Management, pp. 394–397 (2009)
28. Abdullah, N.A.S., Selamat, M.H., Jaafar, A.: Extended function point analysis prototype with security costing estimation. In: Proceedings of International Symposium on Information Technology, pp. 1297–1301 (2010)
29. Peng, H., Yang, G.X., Cai, L.: Research on VAF of IFPUG method based on fuzzy analytic hierarchy process. In: Proceedings of IEEE/ACIS 11th International Conference on Computer Information Science, pp. 593–597 (2012)
30. Matijevic, T., Ognjanovic, I., Sendelj, R.: Enhancement of software projects' function point analysis based on conditional non-functional judgments. In: Proceedings of Mediterranean Conference on Embedded Computing, pp. 283–287 (2012)
31. Junior, O.S.L., Farias, P.P.M., Belchior, A.D.: Fuzzy modeling for function points analysis. Softw. Quality Control 11(2), 149–166 (2003)
32. Fu, Y.-F., Liu, X.-D., Yang, R.-N., Du, Y.-L., Li, Y.-J.: A software size estimation method based on improved FPA. In: Proceedings of 2nd World Congress on Software Engineering, pp. 228–233 (2010)
33. Rao, K.K., Raju, G.S.: Error correction in function point estimation using soft computing technique. In: Proceedings of International Conference on Advanced Computing Artificial Intelligence, pp. 194–198 (2011)
34. Kralj, T., Rozman, I., Heričkob, M., Živkovičb, A.: Improved standard FPA method: resolving problems with upper boundaries in the rating complexity process. J. Syst. Softw. 77(2), 81–90 (2005)
35. Bharadwaj, A.K., Nair, T.R.G.: Mapping general system characteristics to non-functional requirements. In: Proceedings of IEEE International Advanced Computer Conference, pp. 1821–1825 (2009)
36. Kitchenham, B., Pfleeger, S.L., McColl, B., Eagan, S.: An empirical study of maintenance and development estimation accuracy. J. Syst. Softw. 64(1), 57–77 (2002)
37. Prokopova, Z., Silhavy, P., Silhavy, R.: Influence analysis of selected factors in the function point work effort estimation. In: Intelligent Systems in Cybernetics and Automation Control Theory, CoMeSySo 2018 (2019)
38. UKSMA. MarkII Function Point Analysis Counting Practices Manual, version 1.3.1 (1998)
39. ISO/IEC 20968:2002, Software Engineering – MK II Function Point Analysis – Counting Practices Manual, International Organization for Standardization (2002)
40. NESMA, Definitions and Counting Guidelines for the Application of Function Point Analysis, v.2.0 (1997)
41. ISO/IEC 24570:2005, Software Engineering – NESMA functional size measurement method version 2.1 - Definitions and counting guidelines for the application of Function Point Analysis, International Organization for Standardization (2005)
42. ISO/IEC 29881:2010, Information technology – Systems and software engineering - FiSMA 1.1 functional size measurement method, International Organization for Standardization (2010)

43. ISO/IEC 19761:2011, Software engineering – COSMIC: a functional size measurement method
44. Silhavy, P., Silhavy, R., Prokopova, Z.: Categorical variable segmentation model for software development effort estimation. IEEE Access **7**, 9618–9626 (2019)
45. ISO/IEC 20926:2009 (IFPUG), Software and systems engineering – Software measurement – IFPUG functional size measurement method 2009

Author Index

© Springer Nature Switzerland AG 2019
R. Silhavy et al. (Eds.): CoMeSySo 2019, AISC 1047, pp. 423–424, 2019.
https://doi.org/10.1007/978-3-030-31362-3

Printed in the United States
By Bookmasters